Springer Monographs in Mathematics

For further volumes published in this series, go to
http://www.springer.com/series/3733

P.L. Sachdev · Ch. Srinivasa Rao

Large Time Asymptotics for Solutions of Nonlinear Partial Differential Equations

 Springer

P.L. Sachdev
(Deceased)

Ch. Srinivasa Rao
Department of Mathematics
Indian Institute of Technology Madras
Sardar Patel Road
Chennai-600036
India
chsrao@iitm.ac.in
drchsrao@yahoo.com

ISSN 1439-7382
ISBN 978-1-4614-2490-1   e-ISBN 978-0-387-87809-6
DOI 10.1007/978-0-387-87809-6
Springer New York Dordrecht Heidelberg London

Mathematics Subject Classification (2000): 35Kxx, 35Lxx, 35Qxx, 76xx

Printed on acid-free paper

Springer is part of Springer Science+Business Media (www.springer.com)

Preface

The present authors have been engaged in the study – both analytical and numerical – of large time asymptotic behaviour of nonlinear partial differential equations (PDEs) for many years. They had considerable interaction on this subject with Professor B. Enflo, KTH, Stockholm and Professor R. E. Grundy, University of St. Andrews, St. Andrews. Their contribution is gratefully acknowledged. The present venture was embarked upon while the first author (PLS) was Senior Scientist (Indian National Science Academy) at the Department of Mathematics, Indian Institute of Science, Bangalore, and was completed when he moved to the Department of Mathematics, University of Delhi, South Campus. He wishes to thank both the IISc and the University of Delhi for the facilities and to INSA for financial support. The second author (ChSR) is grateful to the Indian Institute of Technology Madras, Chennai for the financial support under Golden Jubilee book writing and new faculty schemes as he worked on this monograph. The authors also acknowledge the efforts of Professor K. T. Joseph, Tata Institute of Fundamental Research, Bombay and Mr. Manoj Kumar Yadav, IIT Madras, who carefully went through some parts of this book.

New Delhi, Chennai
August 2008

P. L. Sachdev
Ch. Srinivasa Rao

Contents

Chapter 1
Introduction

Nonlinear partial differential equations (PDEs) do not, in general, admit exact solutions; these solutions are even more rare when initial/boundary conditions are imposed. There are exceptional circumstances when the PDEs enjoy certain symmetries: they are invariant to a class of finite or infinitesimal transformations (Sachdev 2000). When this is the case, the PDEs are exactly reducible to ordinary differential equations (ODEs) if they are functions of two independent variables; the ODEs may occasionally be integrated in a closed form. Alternatively, one may study their qualitative properties and obtain the actual solutions numerically with reference to appropriate initial/boundary conditions. These solutions are called self-similar and belong to one of the two classes, first kind and second kind (Zel'dovich 1956, Zel'dovich and Raizer 1967, Barenblatt and Zel'dovich 1972, Sachdev 2000), and solve some degenerate problems for which 'all, or at least some, of the constant parameters in the initial and boundary conditions of the problem, having the dimensionality of independent variables, tend to zero or infinity.' These solutions describe those properties of the phenomena that do not depend on the details of the initial and boundary conditions; they do involve some nondimensional parameters which, in some integral sense, represent the memory of initial/boundary conditions. Exceptionally, there may not be any nondimensional parameter of the problem in the asymptotic solution (Barenblatt and Zel'dovich 1972). These special solutions do not describe equilibrium states; they describe intermediate stages when the process of evolution of the solution is continuing and yet the details of initial/boundary conditions have already disappeared. These solutions satisfy some singular, delta functions like, initial conditions.

An interesting denomination of this class of solutions is 'the profile at infinity' due originally to Philip (1957, 1974). These solutions give, for large time, a remarkably accurate behaviour of an entire family of solutions of initial/boundary value problems.

The notion, intermediate asymptotics, was first introduced formally by Barenblatt and Zel'dovich (1971, 1972). However, the appearance of intermediate asymptotics may be seen implicitly in the work of Kolmogorov et al. (1937); this work deals with the propagation, in a certain spatial region, of a gene whose carriers have an

P.L. Sachdev, Ch. Srinivasa Rao, *Large Time Asymptotics for Solutions of Nonlinear Partial Differential Equations*, Springer Monographs in Mathematics, DOI 10.1007/978-0-387-87809-6_1, © Springer Science+Business Media, LLC 2010

advantage in the struggle for existence. They considered the equation

$$v_t - kv_{xx} = F(v), \tag{1.1.1}$$

where the function $F(v)$, $0 \leq v \leq 1$, is sufficiently smooth and has the properties that

$$F(0) = F(1) = 0, \quad F(v) > 0 \ (0 < v < 1),$$
$$F'(0) = \alpha > 0, \quad F'(v) < \alpha \ (0 < v < 1); \tag{1.1.2}$$

here, k is a constant. Kolmogorov et al. (1937) proved that (1.1.1) has uniformly propagating waves or 'stationary solutions' of the form

$$v = \gamma(x - \lambda t + c), \tag{1.1.3}$$

where the constant c is arbitrary, satisfying the boundary conditions $\gamma(-\infty) = 0$, $\gamma(\infty) = 1$ for any speed of propagation λ such that $\lambda \geq \lambda_0 = 2\sqrt{k\alpha}$. It was further shown that, among all the solutions (1.1.3), only the solution with the minimum speed $\lambda = \lambda_0$ is stable in the following sense. The solution of the Cauchy problem satisfying the initial condition

$$v(x,0) \equiv 0, \quad x \leq a,$$
$$0 < v(x,0) < 1, \quad a < x < b,$$
$$v(x,0) \equiv 1, \quad x \geq b,$$

where a and b are arbitrary numbers and $v_x(x,0) \geq 0$, tends as $t \to \infty$, to a solution of the type (1.1.3) with $\lambda = \lambda_0$. The 'stationary' or travelling wave solution of the form (1.1.3) is possible with a continuous spectrum of speeds of propagation λ; only the solution corresponding to the extreme (minimum) value λ_0 is asymptotic to a nonstationary solution of the Cauchy problem as $t \to \infty$. The constant c in (1.1.3) remains undetermined if only travelling waves are considered. It may be found by numerically tracing the evolution of the nonstationary solution of the Cauchy problem until the travelling waveform emerges. The intermediate asymptotic character of the above solution for $\lambda = \lambda_0$ was demonstrated much later by Kanel' (1962). We may point out that the condition $F'(v) < \alpha$ is not required for the existence of a travelling wave solution of (1.1.1) with the speed of propagation $\lambda = \lambda_0$. The above investigation by Kolmogorov et al. (1937) was followed by the work of Zel'dovich and Frank-Kamenetskii (1938) who treated uniformly propagating flames in the theory of combustion. An excellent description of the Soviet contribution to this topic may be found in the review article by Barenblatt and Zel'dovich (1972).

There are two distinct types of self-similar solutions, each depending on a similarity variable, say, $\xi = rt^{-\alpha}/A$, which is a combination of the independent variables r and t. Here, A is a dimensional parameter. From a physical point of view, such solutions exist when the length and time scales cannot be constructed from the parameters of the problem so that r and t do not occur separately in the solution. The unknown functions can depend only on a dimensionless combination of r, t, and a dimensional parameter A with the dimensions $[A] = LT^{-\alpha}$, where α is real. We illustrate this matter with reference to one-dimensional equations of gas dynamics,

$$\frac{\partial \ln \rho}{\partial t} + u \frac{\partial \ln \rho}{\partial r} + \frac{\partial u}{\partial r} + (v - 1)\frac{u}{r} = 0, \tag{1.1.4}$$

$$\frac{\partial u}{\partial t} + u \frac{\partial u}{\partial r} + \frac{1}{\rho}\frac{\partial p}{\partial r} = 0, \tag{1.1.5}$$

$$\frac{\partial}{\partial t} \ln p \rho^{-\gamma} + u \frac{\partial}{\partial r} \ln p \rho^{-\gamma} = 0. \tag{1.1.6}$$

Here, p, ρ, and u are pressure, density, and particle velocity at any point r and time t. The parameter v equals $1, 2, 3$ for the plane, cylindrical, and spherical symmetries, respectively. Here, the solution would assume the form

$$p = \frac{a}{r^{k+1}t^{s+2}}P(\xi), \tag{1.1.7}$$

$$\rho = \frac{a}{r^{k+3}t^s}G(\xi), \tag{1.1.8}$$

$$u = \frac{r}{t}V(\xi), \tag{1.1.9}$$

where $\xi = rt^{-\alpha}/A$ and a is a parameter which appears in the given problem. It contains the unit of mass and has the dimensions $[a] = ML^kT^s$, where k and s are real numbers. If one substitutes the form (1.1.7)–(1.1.9) into (1.1.4)–(1.1.6), the latter reduce to a system of nonlinear ODEs and the motion is said to be self-similar. If the exponent α appearing in the similarity variable ξ and the exponents of r and t are fully determined from dimensional considerations and/or the conservation laws (1.1.4)–(1.1.6), the similarity solution is said to belong to the first kind. This class of solutions may be exemplified by the solution describing a strong explosion into a uniform medium with density ρ_0. Here we have two physical parameters, the undisturbed density, $\rho_0 \sim ML^{-3}$, and the energy of explosion, $E_0 \sim ML^2T^{-2}$, appearing in the problem. The energy of explosion is assumed to be constant and equals the energy of the moving gas. This constitutes an integral of the motion. For a strong explosion, the pressure p_0 and the sound speed c_0 in the undisturbed motion are assumed to be zero and do not enter the problem as parameters. The constants ρ_0 and E help construct the parameter A in the similarity variable: $A = (E/\rho_0)^{1/5} \sim LT^{-2/5}$. Therefore, we have the similarity variable $\xi = r/(E/\rho_0)^{1/5}t^{2/5}$. The exponent α is thus explicitly found to be $2/5$. In this case, the (reduced) system of ODEs from the system (1.1.4)–(1.1.6) is solved analytically or numerically subject to the Rankine–Hugoniot conditions at the shock and zero particle velocity at the centre of the spherical explosion (see Taylor 1950, Sedov 1946, Sachdev 2004).

In the self-similar solutions of the second kind, the substitution of the similarity form (1.1.7)–(1.1.9) into the system (1.1.4)–(1.1.6) reduces the latter to a system of nonlinear ODEs but the exponent α is not determined from dimensional considerations or conservation laws. This exponent is found from the requirement that the integral curve passes through an appropriate singular point, usually a saddle point, so that the boundary conditions, say, at the shock and the point at infinity behind the shock are simultaneously satisfied. An example of self-similar solutions of the second kind is a converging shock from an implosion. It has been discussed in some detail by Zel'dovich and Raizer (1967) and Sachdev (2004).

We have already discussed the travelling wave solutions for the equation (1.1.1) which constitute self-similar solutions of the second kind. Here, in general, the wavespeed λ is not found by solving a boundary value problem for the ODE resulting from substituting (1.1.3) into (1.1.1) and satisfying the boundary conditions $\gamma(-\infty) = 0$, $\gamma(\infty) = 1$. The wavespeed λ possesses an infinite number of possible values, the minimum of which gives a unique stable solution. Something analogous to this happens for the converging shock problem (Sachdev 2004).

The nonlinear PDEs with two (or more) independent variables may admit self-similar or travelling wave solutions which are governed by nonlinear ODEs. This comes about when the PDEs enjoy certain invariance properties subject to finite or infinitesimal transformations as we have remarked earlier (Bluman and Kumei 1989, Sachdev 2000). Using group-theoretic methods or the so-called direct method of Clarkson and Kruskal (1989), it becomes possible to exactly reduce the nonlinear PDEs with two independent variables to nonlinear ODEs and hence analyse the existence and uniqueness properties of the latter subject to appropriate boundary and/or initial conditions. It helps to understand when these ODEs possess solutions and know their behaviour for large distances. Their singular behaviour, if any, may also be discovered. This study is often motivated by a canonical equation of a certain class for which one may have access to all the relevant information. To exemplify this, one may refer to the Burgers equation which admits exact linearisation by the so-called Cole–Hopf transformation and hence yields considerable information regarding its solution. This motivates the analysis for the large class of generalised Burgers equations which do not enjoy the linearising felicity of the Cole–Hopf type transformation and must therefore be treated directly (Sachdev 1987, 2000).

When a given PDE (or a system) does not possess sufficient invariance properties and therefore cannot be reduced to an ODE (or a system of ODEs), one must turn to other methods to find analytic solutions. One intuitive approach is to mimic the solution of a simpler equation; for example, for generalised Burgers equations, one could simulate the behaviour of the solutions of the Burgers equation and obtain large time solutions which embed in them limiting forms such as inviscid and/or linear viscous ones. This becomes possible for a class of generalised Burgers equations with respect to N-wave or periodic initial conditions and involves solution of an infinite system of coupled nonlinear ODEs (see Sachdev and Joseph 1994, Sachdev et al. 1996, 2005).

A more general approach to finding asymptotic solutions, usually not leading to a closed form, is called the balancing argument. It was originally developed by Grundy and his collaborators (Grundy 1988, Grundy et al. 1994a, 1994b) and has since been exploited extensively by Sachdev and his coworkers (Sachdev et al. 1994, 1999, Sachdev and Srinivasa Rao 2000). Here, it is helpful to introduce in the given PDE a transformation such as $u = t^{\alpha}U(\eta,t)$, where $\eta = xt^{\beta}$ is the relevant similarity variable; α and β are constants found later. The given PDE in $u = u(x,t)$

is changed to one in U with η and t as the independent variables. Varied combinations of terms in the new equation are 'balanced' such that they constitute the most dominant terms as t tends to infinity. The subdominant terms are subsequently accounted for by a correction term. Several correction terms are discovered in a sequential manner which may suggest an infinite series form of the solution in descending powers of t. The large time behaviour(s) so determined prove very useful and, in conjunction with the numerical solution of the problem, provide valuable insight into the solution. A given problem may have several asymptotic behaviours depending on the parameters that appear in the equation and or the initial/boundary conditions.

To obtain a deeper analytical insight into the asymptotic nature of the special solutions, self-similar or travelling waves, one must proceed in two steps. First, one must, as pointed out earlier, carefully analyse the properties of the solutions of the (reduced) ODEs with reference to appropriate initial/boundary conditions at a finite point and/or at infinity. This is motivated by the form of solution one seeks. Correspondingly, analytic methods – initial value or shooting – may be resorted to. This analysis distinguishes different sets of parametric values for which solutions with distinct behaviour exist and enjoy the uniqueness property. The asymptotic character of these solutions may then be studied with reference to the numerical solution of the original PDEs with relevant initial/boundary conditions.

More importantly, the asymptotic nature of the exact solutions may be examined by using rigourous analytic methods suggested originally by Serrin (1967), Peletier (1970, 1971, 1972), and Oleinik (1966). These have since been extensively used and improvised upon. We briefly describe their approach. Serrin (1967) considered the asymptotic behaviour of velocity profiles for the steady Prandtl boundary layer equations. He showed that, for a power law streaming speed, $U(x) = C(x+d)^m$, where C, D, and $m \geq 0$ are constants, the velocity profile which develops downstream is asymptotically given by the similarity solution governed by the Falkner–Skan equation, a third-order nonlinear ODE. It was also found that, for a streaming speed satisfying certain bounds, the velocity profile which develops downstream is asymptotically unique, though the particular form of the resulting profile depends on the precise nature of the exterior stream. Peletier (1970) considered a simpler problem: asymptotic behaviour of temperature profiles for a class of nonlinear heat conduction problems. More specifically, he considered the conduction of heat into a semi-infinite homogeneous solid for which the coefficient of thermal conductivity depends on the temperature. The temperature at the face of the solid is held fixed whereas far away it tends to a lower value, with a complementary error function behaviour as the distance tends to ∞. It was then shown, with appropriate smoothness requirements on the solution, that whatever the initial temperature profile, the ensuing profile tends to one given by the similarity solution as time tends to infinity. These studies have made a seminal contribution to the analysis of asymptotic solutions of nonlinear partial differential equations.

The description of the asymptotic character by these special solutions – self-similar or more general obtained by the balancing argument – brings out the importance of these solutions. These solutions per se arise from some singular initial/boundary conditions; their larger significance accrues from their asymptotic nature. These particular solutions, with specific features of their own, may be exploited to demonstrate quantitatively how solutions of larger classes of initial/boundary value problems behave for large times. This information is hard to obtain by other analytic means. The best course for a scientist interested in solving a nonlinear problem, therefore, is to have access to such asymptotic analysis which, supplemented by a robust numerical scheme, can provide reliable answers for a physical problem and hence enhance its understanding.

Chapter 2 deals with the large time asymptotics for solutions of nonlinear first-order partial differential equations. We discuss the decay of discontinuous solutions of a general hyperbolic partial differential equation subject to a top hat initial condition in detail. Chapter 3 describes some constructive approaches to study the asymptotic nature of solutions of some nonlinear partial differential equations of parabolic type: Burgers equation, generalised Burgers equations, nonlinear diffusion equations, generalised Fisher's equations, and a system of nonlinear PDEs describing reaction–diffusion. The methods presented here include the balancing argument and matched asymptotic expansions. In Chapter 4, we present some problems which possess self-similar solutions as intermediate asymptotics. In the final chapter, we treat some systems of nonlinear PDEs which describe fluid flows. We consider similarity solutions of these systems and investigate in detail when they may constitute intermediate asymptotics. The topics discussed in this chapter include explosion in a power law density medium, self-similar solutions of the first and second kind, self-similar solutions for collapsing cavities, large time behaviour of solutions of compressible flow equations with damping, large time behaviour of solutions of unsteady boundary layer equations for an incompressible fluid, and asymptotic behaviour of velocity profiles in Prandtl boundary layer theory.

References

Barenblatt, G. I., Zel'dovich, Ya. B. (1971) Intermediate asymptotics in mathematical physics, *Russian Math. Surveys* 26, 45–61.

Barenblatt, G. I., Zel'dovich, Ya. B. (1972) Self-similar solutions as intermediate asymptotics, *Ann. Rev. Fluid Mech.* 4, 285–312.

Bluman, G. W., Kumei, S. (1989) *Symmetries and Differential Equations*, Springer-Verlag, New York.

Clarkson, P. A., Kruskal, M. D. (1989) New similarity reductions of the Boussinesq equation, *J. Math. Phys.* 30, 2201–2213.

Grundy, R. E. (1988) Large time solution of the Cauchy problem for the generalized Burgers equation, Preprint, University of St. Andrews, UK.

Grundy, R. E., Sachdev, P. L., Dawson, C. N. (1994a) Large time solution of an initial value problem for a generalised Burgers equation, in *Nonlinear Diffusion Phenomenon*, P. L. Sachdev and R. E. Grundy (Eds.), 68–83, Narosa, New Delhi.

Grundy, R. E., Van Duijn, C. J., Dawson, C. N. (1994b) Asymptotic profiles with finite mass in one-dimensional contaminant transport through porous media: the fast reaction case, *Quart. J. Mech. Appl. Math.* 47, 69–106.

Kanel', Ya. I. (1962) On the stability of solutions of the Cauchy problem for equations occuring in the theory of combustion, *Mat. Sb.* 59(101), 245–288.

Kolmogorov, A. N., Petrovskii, I. G., Piskunov, N. S. (1937) Investigation of the diffusion equation connected with an increasing amount of matter and its application to a biological problem, *Bull. MGU* A1(6), 1–26.

Oleinik, O. A. (1966) Stability of solutions of a system of boundary layer equations for a nonsteady flow of incompressible fluid, *Prikl. Mat. Mekh.* 30, 417–423.

Peletier, L. A. (1970) Asymptotic behaviour of temperature profiles of a class of non-linear heat conduction problems, *Quart. J. Mech. Appl. Math.* 23, 441–447.

Peletier, L. A. (1971) Asymptotic behaviour of solutions of the porous media equation, *SIAM J. Appl. Math.* 21, 542–551.

Peletier, L. A. (1972) On the asymptotic behaviour of velocity profiles in laminar boundary layers, *Arch. Rat. Mech. Anal.* 45, 110–119.

Philip, J. R. (1957) The theory of infiltration: 2. The profile of infinity, *Soil Sci.* 83, 435–448.

Philip, J. R. (1974) Recent progress in the solution of nonlinear diffusion equations, *Soil Sci.* 117, 257–264.

Sachdev, P. L. (1987) *Nonlinear Diffusive Waves*, Cambridge University Press, Cambridge, UK.

Sachdev, P. L. (2000) *Self-Similarity and Beyond–Exact Solutions of Nonlinear Problems*, Chapman & Hall/ CRC Press, New York.

Sachdev, P. L. (2004) *Shock Waves and Explosions*, Chapman & Hall/ CRC Press, New York.

Sachdev, P. L., Joseph, K. T. (1994) Exact representations of N-wave solutions of generalized Burgers equations, in *Nonlinear Diffusion Phenomenon*, P. L. Sachdev and R. E. Grundy (Eds.), 197–219, Narosa, New Delhi.

Sachdev, P. L., Srinivasa Rao, Ch. (2000) N-wave solution of modified Burgers equation, *Appl. Math. Lett.* 13, 1–6.

Sachdev, P. L., Joseph, K. T., Nair, K. R. C. (1994) Exact N-wave solutions for the nonplanar Burgers equation, *Proc. Roy. Soc. London Ser. A* 445, 501–517.

Sachdev, P. L., Joseph, K. T., Mayil Vaganan, B. (1996) Exact N-wave solutions of generalized Burgers equations, *Stud. Appl. Math.* 97, 349–367.

Sachdev, P. L., Srinivasa Rao, Ch., Enflo, B. O. (2005) Large-time asymptotics for periodic solutions of the modified Burgers equation, *Stud. Appl. Math.* 114, 307–323.

Sachdev, P. L., Srinivasa Rao, Ch., Joseph, K. T. (1999) Analytic and numerical study of N-waves governed by the nonplanar Burgers equation, *Stud. Appl. Math.* 103, 89–120.

Sedov, L. I. (1946) Propagation of intense blast waves, *Prikl. Mat. Mekh.* 10, 241–250.

Serrin, J. (1967) Asymptotic behaviour of velocity profiles in the Prandtl boundary layer theory, *Proc. Royal. Soc. London Ser. A* 299, 491–507.

Taylor, G. I. (1950) The formation of a blast wave by a very intense explosion I, *Proc. Roy. Soc. London Ser. A* 201, 159–174.

Zel'dovich, Ya. B. (1956) The motion of a gas under the action of a short term pressure shock, *Akust. Zh.* 2, 28–38, (*Sov. Phys. Acoustics* 2, 25–35).

Zel'dovich, Ya. B., Frank-Kamenetskii, D. A. (1938) Theory of uniform propagation of flames, *Doklady USSR Ac. Sci.* 19, 693–697.

Zel'dovich, Ya. B., Raizer, Yu. P. (1967) *Physics of Shock Waves and High-Temperature Hydrodynamic Phenomena*, Vol. 2, Academic Press, New York.

Chapter 2
Large Time Asymptotics for Solutions of Nonlinear First-Order Partial Differential Equations

2.1 Introduction

In this chapter, we consider the asymptotic behaviour of the solution of the (generalised) inviscid Burgers equation with damping, namely,

$$u_t + g(u)u_x + \lambda h(u) = 0, \quad \lambda > 0, \ g_u(u) > 0, \ h_u(u) > 0 \text{ for } u > 0, \qquad (2.1.1)$$

where $g(u)$ and $h(u)$ are nonnegative functions. This equation appears in several physical contexts including the Gunn effect (Murray 1970b). Equation (2.1.1) is considered subject to the initial and boundary conditions

$$u(0,t) = 0, \quad t > 0, \qquad (2.1.2)$$

$$u(x,0) = u_0(x) = \begin{cases} 0, & x < 0, \\ f(x), & 0 < x < X, \\ 0, & x > X, \end{cases} \qquad (2.1.3)$$

where

$$0 \le f(x) \le 1. \qquad (2.1.4)$$

The function $u_0(x)$ may initially be smooth or may possess a discontinuity (see Figures 2.6a–c). We closely follow the work of Murray (1970a). The term $h(u) = O(u^\alpha)$, $\alpha > 0$ for $0 < u \ll 1$ plays a crucial role in the asymptotic behaviour of the solution of (2.1.1)–(2.1.4). The initial disturbance decays (i) in a finite time and finite distance for $0 < \alpha < 1$, (ii) within an infinite time like $O(e^{-\lambda t})$ and in a finite distance for $\alpha = 1$, and (iii) within an infinite time and distance like $O(t^{-1/(\alpha-1)})$ for $1 < \alpha \le 3$, and $O(t^{-1/2})$ for $\alpha \ge 3$. The asymptotic behaviour also includes the (possible) inception and propagation of (shock) discontinuity in each case as time grows.

In a related study, Joseph and Sachdev (1994) showed how a class of first-order nonlinear PDEs

$$u_t + u^n u_x + H(x,t,u) = 0 \qquad (2.1.5)$$

P.L. Sachdev, Ch. Srinivasa Rao, *Large Time Asymptotics for Solutions of Nonlinear Partial Differential Equations*, Springer Monographs in Mathematics, DOI 10.1007/978-0-387-87809-6_2, © Springer Science+Business Media, LLC 2010

may be transformed to

$$v_\tau + v^n v_y = 0 \tag{2.1.6}$$

and hence solved subject to certain initial conditions. In conclusion, we summarise the work of Natalini and Tesei (1992) regarding the special case of (2.1.1), namely,

$$u_t + \frac{1}{m}(u^m)_x + u^p = 0 \text{ in } \mathbb{R} \times (0,\infty),$$

$$\tag{2.1.7}$$

$$u(x,0) = u_0(x), \quad x \in \mathbb{R},$$

where $m > 1$, $p > 1$, and $u_0(x)$ is a bounded nonnegative function with compact support. Here, the authors derived sharp estimates for the support of the solution of the problem (2.1.7) and the intermediate asymptotic form of its solution as $t \to \infty$. Murray (1970a) dealt with an initial boundary value problem whereas Natalini and Tesei (1992) discussed a pure initial value problem.

Section 2.2 deals with some nontrivial first-order PDEs whose asymptotic solutions may be found directly. Section 2.3 details analysis of (2.1.1)–(2.1.4), following mainly the work of Murray (1970a).

2.2 First-order nonlinear partial differential equations – Some examples

This section is largely illustrative. Here, we discuss initial value problems for some first-order nonlinear partial differential equations and show how some of them tend to special exact solutions as time becomes large. These exact solutions – similarity or product form – thus constitute intermediate asymptotics. The examples discussed here include some which actually appear in applications. We also discuss numerical solution of the equation describing spin-up and spin-down, subject to appropriate initial conditions, to demonstrate how the initial profile actually evolves with time (see Example 6).

Example 1. Consider

$$u_t + u_x = \lambda_1 u - \lambda_2 u^2, \quad x \in \mathbb{R}, t > 0, \tag{2.2.1}$$

$$u(x,0) = f(x), \quad \lambda_1 > 0, \lambda_2 > 0. \tag{2.2.2}$$

Here, $f(x) \geq 0$ and λ_1, λ_2 are constants. Equation (2.2.1) describes a population model. We solve (2.2.1)–(2.2.2) exactly, following Mickens (1988), and hence find the large time behaviour of the solution. Writing

$$u = 1/w \tag{2.2.3}$$

in (2.2.1) we obtain the linear PDE

$$w_x + w_t = \lambda_2 - \lambda_1 w. \tag{2.2.4}$$

By Lagrange's method, we have

$$\frac{dx}{1} = \frac{dt}{1} = \frac{dw}{\lambda_2 - \lambda_1 w}. \tag{2.2.5}$$

Equations (2.2.5) imply that

$$\frac{dx}{dt} = 1, \quad \frac{dw}{dt} = \lambda_2 - \lambda_1 w. \tag{2.2.6}$$

Solving (2.2.6), we get

$$x - t = c_1, \quad w e^{\lambda_1 t} = \frac{\lambda_2}{\lambda_1} e^{\lambda_1 t} + c_2. \tag{2.2.7}$$

A general solution of (2.2.4), therefore, is given by

$$w e^{\lambda_1 t} - \frac{\lambda_2}{\lambda_1} e^{\lambda_1 t} = g(x - t), \tag{2.2.8}$$

where g is an arbitrary function. Simplifying (2.2.8), we have

$$w(x,t) = g(x - t) e^{-\lambda_1 t} + \frac{\lambda_2}{\lambda_1}. \tag{2.2.9}$$

Equations (2.2.2), (2.2.3), and (2.2.9) give

$$g(x) = \frac{1}{f(x)} - \frac{\lambda_2}{\lambda_1} \tag{2.2.10}$$

and hence

$$u(x,t) = \frac{\lambda_1 f(x - t)}{\lambda_1 e^{-\lambda_1 t} + \lambda_2 f(x - t)[1 - e^{-\lambda_1 t}]}. \tag{2.2.11}$$

The solution (2.2.11), in the limit $t \to \infty$, $x - t = O(1)$, tends to a constant :

$$u \approx \frac{\lambda_1}{\lambda_2}.$$

Example 2. Consider

$$u_t + u u_x + u = 0, \quad x \in \mathbb{R}, \ t > 0, \tag{2.2.12}$$
$$u(x,0) = x, \quad x \in \mathbb{R}. \tag{2.2.13}$$

The characteristic system for (2.2.12) is

$$\frac{dx}{dt} = u, \quad \frac{du}{dt} = -u. \tag{2.2.14}$$

Suppose that a characteristic curve starts from $(\xi, 0)$. Then, we solve (2.2.14) subject to the initial condition

$$t = 0, \quad x = \xi, \quad u = \xi. \tag{2.2.15}$$

Equation (2.2.14)$_2$ and (2.2.15) imply that

$$u = \xi e^{-t}. \tag{2.2.16}$$

Solving (2.2.14)$_1$ with the help of (2.2.16), we get

$$x = 2\xi \left[1 - \frac{1}{2} e^{-t} \right]. \tag{2.2.17}$$

Eliminating ξ from (2.2.16) and (2.2.17), we arrive at the solution

$$u(x,t) = \frac{x e^{-t}}{2 - e^{-t}}, \quad x \in \mathbb{R}, \, t > 0. \tag{2.2.18}$$

This is a product solution and, in the limit $t \to \infty$, becomes

$$u(x,t) \sim \frac{x e^{-t}}{2}, \quad x \in \mathbb{R}. \tag{2.2.19}$$

Example 3. Consider

$$u_t + u u_x = u^n, \quad x \in \mathbb{R}, \, t > 0, \tag{2.2.20}$$
$$u(x,0) = u_0(x), \quad x \in \mathbb{R}. \tag{2.2.21}$$

Here, n is a positive integer. We solve this problem for $n = 1, 2$, and n general $\neq$ $1, 2$, separately. In each of these cases the solution may be obtained as a functional relation involving u, x, and t. However, these (implicit) solutions do not permit, for general $u_0(x)$, an explicit asymptotic form as $t \to \infty$.

Case (i): $n = 1$.
The characteristic equations for (2.2.20) are given by

$$\frac{dx}{dt} = u, \quad \frac{du}{dt} = u. \tag{2.2.22}$$

The initial condition (2.2.21) may be written as

$$t = 0, \quad x = \xi, \quad u = u_0(\xi). \tag{2.2.23}$$

Solving (2.2.22) and (2.2.23), we find that

$$x = u_0(\xi)[e^t - 1] + \xi, \quad u = u_0(\xi)e^t. \tag{2.2.24}$$

Eliminating ξ from (2.2.24), we have

$$u = e^t u_0 \left(x - u + u e^{-t} \right), \tag{2.2.25}$$

an implicit functional relation in x, u, and t.

Case (ii): $n = 2$.
In this case, the characteristic system for (2.2.20) is given by

$$\frac{dx}{dt} = u, \quad \frac{du}{dt} = u^2. \tag{2.2.26}$$

Equation $(2.2.26)_2$, in view of (2.2.23), gives

$$u = \frac{u_0(\xi)}{1 - tu_0(\xi)}. \tag{2.2.27}$$

Solving $(2.2.26)_1$ with the help of (2.2.27), we obtain

$$x = -\ln(1 - tu_0) + c,$$

where c is a constant of integration. Imposing the initial conditions (2.2.23), we have

$$x = \xi - \ln(1 - tu_0). \tag{2.2.28}$$

Eliminating ξ from (2.2.27) and (2.2.28), we get

$$u = \frac{u_0(x - \ln(1 + tu))}{1 - tu_0(x - \ln(1 + tu))}, \tag{2.2.29}$$

a complicated (implicit) relation involving x, t, u.

Case (iii): n a positive integer not equal to 1 or 2.
Solving the characteristic system of (2.2.20) subject to (2.2.23), we get

$$u^{1-n} = (1-n)t + u_0(\xi)^{1-n}, \tag{2.2.30}$$

$$x = \xi - \frac{u_0(\xi)^{2-n}}{2-n} + \frac{[(1-n)t + u_0(\xi)^{1-n}]^{(2-n)/(1-n)}}{2-n}. \tag{2.2.31}$$

Elimination of ξ from (2.2.30) and (2.2.31) leads to

$$u^{1-n} = (1-n)t + u_0^{1-n} \left(x + \frac{u^{2-n}}{n-2} - \frac{1}{n-2}(u^{1-n} + (n-1)t)^{(n-2)/(n-1)} \right). \tag{2.2.32}$$

We observe that the solutions given by (2.2.25), (2.2.29), and (2.2.32) are highly implicit. It does not seem possible to get large time behaviour from these expressions for arbitrary initial conditions directly.

Example 4. Here, we consider the initial value problem for the inviscid Burgers equation

$$u_t + uu_x = 0, \quad x \in \mathbb{R}, \, t > 0, \tag{2.2.33}$$

$$u(x,0) = f(x) = \begin{cases} 0, & x < 0 \\ 1, & 0 \le x \le 1 \\ 0, & x > 1. \end{cases} \tag{2.2.34}$$

The characteristics of (2.2.33) are given by

$$\frac{dx}{dt} = u, \quad \frac{du}{dt} = 0. \tag{2.2.35}$$

It is clear from (2.2.35) that the characteristics are straight lines in the (x,t) plane. Assume that

$$x = \xi, \quad u = f(\xi) \text{ at } t = 0. \tag{2.2.36}$$

Equations (2.2.35) and (2.2.36) then imply that

$$x = \begin{cases} \xi, & \xi < 0 \\ t + \xi, & 0 \le \xi \le 1 \\ \xi, & \xi > 1 \end{cases} \tag{2.2.37}$$

and

$$u = f(\xi). \tag{2.2.38}$$

Figure 2.1 gives the characteristics in the xt-plane. We observe the following.

(i)   There is a 'void' between the the lines $x = t$ $(x \le 1)$ and the t-axis. In this region no point is reached by the characteristics.

(ii)   The characteristics intersect at $x = 1$, $t = 0$ itself. Therefore, the shock is formed at $x = 1, t = 0$.

The 'void' shown in Figure 2.1 may be filled by the rarefaction wave $u = x/t$ $(0 \le u \le 1)$. Using the Rankine–Hugoniot condition at the shock, we have

$$\frac{dx}{dt} = \frac{1}{2}, \quad x = 1, \quad t = 0. \tag{2.2.39}$$

Therefore, the shock locus is given by

$$x = \frac{t+2}{2}. \tag{2.2.40}$$

The characteristic issuing from $x = 0$ intersects the shock path $x = (t+2)/2$ at $t = 2$. Therefore, for $0 \le t \le 2$, we have the following solution of (2.2.33)–(2.2.34),

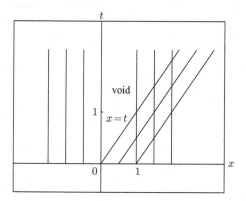

Fig. 2.1 Characteristic diagram for the problem (2.2.33) and (2.2.34).

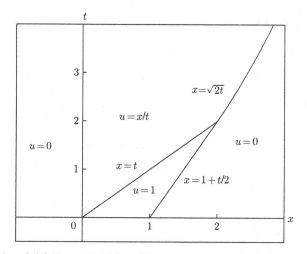

Fig. 2.2 Solution of (2.2.33) and (2.2.34) in different regions of the (x,t)-plane.

$$u = \begin{cases} 0, & x < 0 \\ \dfrac{x}{t}, & 0 \le x \le t \\ 1, & t < x < 1 + \dfrac{t}{2} \\ 0, & x > 1 + \dfrac{t}{2}. \end{cases} \tag{2.2.41}$$

For $t > 2$, the value of u on the left of the shock is x/t whereas that on the right is 0. Therefore, the shock path via the Rankine–Hugoniot condition is given by

$$\frac{dx}{dt} = \frac{x}{2t}, \quad x(2) = 2. \tag{2.2.42}$$

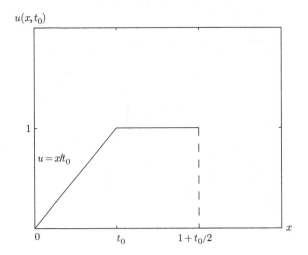

Fig. 2.3 Solution of (2.2.33) and (2.2.34) at time $t = t_0 < 2$.

Solving (2.2.42), we have

$$x = \sqrt{2t}. \tag{2.2.43}$$

Thus, for $t > 2$,

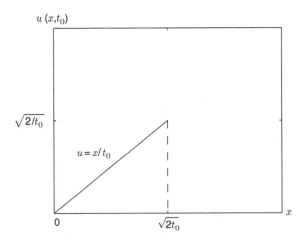

Fig. 2.4 Solution of (2.2.33) and (2.2.34) at $t = t_0 > 2$.

$$u = \begin{cases} 0, & x < 0 \\ \dfrac{x}{t}, & 0 \le x < \sqrt{2t} \\ 0, & x > \sqrt{2t}. \end{cases} \tag{2.2.44}$$

Figure 2.2 gives u for different regions of the xt-plane. Figures 2.3 and 2.4 depict u at times $t_0 < 2$ and $t_0 > 2$, respectively. Thus, it follows from (2.2.44) that, for $t > 2$,

$$\|u(\cdot,t)\|_\infty = \frac{\sqrt{2t}}{t} = O(t^{-1/2}),$$

where $\|.\|_\infty$ is the sup norm.

Example 5. We discuss here the decay of solutions of

$$u_t + u^n u_x + G(t)u + F(x)u^{n+1} = 0, \qquad (2.2.45)$$

subject to the initial condition

$$u(x,0) = u_0(x) = \begin{cases} g(x), & -\infty < a < x < b < \infty \\ 0, & \text{otherwise,} \end{cases} \qquad (2.2.46)$$

closely following Joseph and Sachdev (1994). To that end, we first transform (2.2.45) to the conservation form

$$v_\tau + v^n v_y = 0, \qquad -\infty < y < \infty. \qquad (2.2.47)$$

via the transformation

$$\tau = \tau(x,t), \qquad (2.2.48)$$
$$y = y(x,t), \qquad (2.2.49)$$
$$v(y,\tau) = f(x,t)u(x,t), \qquad (2.2.50)$$

where $f(x,t) > 0$ and

$$J = \det \begin{pmatrix} y_t & y_x \\ \tau_t & \tau_x \end{pmatrix} \neq 0.$$

Now, using the known results regarding the decay of solutions of (2.2.47), subject to the initial conditions with compact support, the decay results for (2.2.45) and (2.2.46) may be found.

Let us start with the general equation

$$u_t + u^n u_x + H(x,t,u) = 0, \qquad (2.2.51)$$

which contains (2.2.45) as a special case. A differentiation of (2.2.50) with respect to x and t gives

$$v_y y_x + v_\tau \tau_x = f u_x + f_x u, \qquad (2.2.52)$$
$$v_y y_t + v_\tau \tau_t = f u_t + f_t u. \qquad (2.2.53)$$

Solving (2.2.52) and (2.2.53) for v_τ and v_y, we obtain

$$v_\tau = -\frac{1}{J}[(y_x f_t - y_t f_x)u + y_x f u_t - y_t f u_x], \qquad (2.2.54)$$

$$v_y = -\frac{1}{J}[(\tau_t f_x - \tau_x f_t)u + \tau_t f u_x - \tau_x f u_t]. \qquad (2.2.55)$$

Using (2.2.54) and (2.2.55) in (2.2.47), we get

$$(y_x f)u_t + (y_x f_t - y_t f_x)u - y_t f u_x + f^n u^n[(\tau_t f_x - \tau_x f_t)u + (\tau_t f u_x - \tau_x f u_t)] = 0.$$
(2.2.56)

For (2.2.56) to have the same form as (2.2.51), we require that

$$y_t = 0, \quad \tau_x = 0, \quad \frac{f^n \tau_t}{y_x} = 1.$$
(2.2.57)

Combining (2.2.56) and (2.2.57), we get

$$u_t + u^n u_x + \frac{f_x}{f}u^{n+1} + \frac{f_t}{f}u = 0.$$
(2.2.58)

Furthermore, $(2.2.57)_3$ gives

$$f(x,t) = \left(\frac{dy/dx}{d\tau/dt}\right)^{1/n}.$$
(2.2.59)

Thus, equation (2.2.58) takes the form (2.2.45) with

$$G(t) = -\frac{d}{dt}\log\left(\frac{d\tau}{dt}\right)^{1/n},$$
(2.2.60)

$$F(x) = \frac{d}{dx}\log\left(\frac{dy}{dx}\right)^{1/n}.$$
(2.2.61)

On the other hand, the transformation

$$\tau = \tau(t) = \int^t \left[\exp\left(\int^s G(s_1)ds_1\right)\right]^{-n} ds,$$
(2.2.62)

$$y = y(x) = \int^x \left[\exp\left(\int^s F(s_1)ds_1\right)\right]^{n} ds,$$
(2.2.63)

$$f(x,t) = \exp\left(\int^t G(s)ds\right)\exp\left(\int^x F(y)dy\right)$$
(2.2.64)

reduces (2.2.45) to (2.2.47).

Assuming that the functions $G(t)$ and $F(x)$ in (2.2.60) and (2.2.61), respectively, are such that $\tau(+\infty) = +\infty$ and $y(\pm\infty) = \pm\infty$, we prove that the solution to (2.2.45) and (2.2.46) satisfies the estimate

$$\left|\exp\left[\int^x F(y)dy\right]u(x,t)\right| \leq \frac{c}{\exp(\int^t G(s)ds)[\int^t (\exp(\int^s G(s_1)ds_1))^{-n}ds]^{1/(n+1)}}.$$
(2.2.65)

To prove the estimate (2.2.65), we make use of the following result due to Lax (1957) and Dafermos (1985) concerning the simpler equation (2.2.47). Suppose that $v(y,\tau)$ is the solution of (2.2.47), subject to initial condition

$$v(y,0) = v_0(y) = \begin{cases} \tilde{f}(y), & -\infty < a < y < b < \infty \\ 0, & \text{otherwise.} \end{cases} \tag{2.2.66}$$

Then, $v(y,\tau)$ satisfies the estimate

$$|v(y,\tau)| \le c\tau^{-1/(n+1)}, \tag{2.2.67}$$

where c is a positive constant. Using (2.2.62)–(2.2.64) and (2.2.50) in (2.2.67), we have the required estimate (2.2.65).

Example 6. We consider an interesting first-order nonlinear partial differential equation which describes spin-up and spin-down in a cylinder of small ratio of height to diameter. It was first derived by Wedemeyer (1964) who, however, also included radial viscous effects in the equation. In a subsequent paper, Dolzhanskii et al. (1992) ignored this term and studied in some detail the resulting first-order equation

$$v_t + k(v - r)\left(v_r + \frac{v}{r}\right) = 0, \tag{2.2.68}$$

where k is a constant. They showed that the travelling waveform of the solutions of (2.2.68) describes a universal stage of spin-up; that is, they demonstrated that 'every experimental run should demonstrate one of the travelling wave solutions with specific values of the parameters arising from the initial data.' In other words, Dolzhanskii et al. (1992) showed how self-similar (or travelling wave in other coordinates) solutions may constitute large time asymptotics. Their analysis confirmed the experimental observations regarding the spin-up problem. We refer the reader to the original work of Wedemeyer (1964) for a detailed physical description of the problem.

We first find a general solution of (2.2.68). Writing $w = v/r$ in (2.2.68), we have

$$w_t - k(1 - w)(2w + rw_r) = 0. \tag{2.2.69}$$

Introducing w and r as independent variables and $t = t(w,r)$ as the dependent variable in (2.2.69), we get

$$1 - k(1 - w)(2wt_w - rt_r) = 0; \tag{2.2.70}$$

here, $w_t = 1/t_w$, $w_r = -t_r/t_w$. Equation (2.2.70) suggests the transformation

$$\xi = \ln w, \quad \rho = \ln r.$$

Therefore, we have

$$1 - k(1 - e^\xi)(2t_\xi - t_\rho) = 0. \tag{2.2.71}$$

It is possible to reduce (2.2.71) to a 'canonical' form by writing $\zeta = \xi + 2\rho$ (see Zachmanoglou and Thoe (1976), p. 137):

$$1 - 2k(1 - e^\xi)t_\xi = 0, \quad t = t(\xi,\zeta). \tag{2.2.72}$$

Equation (2.2.72) is immediately integrated to yield

$$2kt = 2kF(\zeta) - \ln(1/w - 1),\qquad(2.2.73)$$

where $F(\zeta)$ is an arbitrary function of ζ. Writing $2kF(\ln wr^2) = \ln T(wr^2)$ in (2.2.73) we may change the latter to the form

$$e^{2kt} = \frac{wT(wr^2)}{1 - w}.\qquad(2.2.74)$$

We may point out that an application of Lagrange's method to (2.2.69) quickly gives the general solution (2.2.74). Using the initial condition, $w = w_0(r)$, say, we obtain the form of the unknown function $T(x)$. The choice of the function

$$T(x) = \begin{cases} \dfrac{1-x}{x}, & x < 1 \\ 0, & x > 1 \end{cases}$$

leads to the solution, obtained earlier by Wedemeyer (1964):

$$\begin{aligned} v &= 0, \quad r \le e^{-kt} \\ v &= \frac{re^{2kt} - 1/r}{e^{2kt} - 1}, \quad r \ge e^{-kt}. \end{aligned}\qquad(2.2.75)$$

With $T(x) = x^{\alpha}$, (2.2.74) takes the form

$$t - \frac{\alpha}{k}\ln r = \frac{1}{2k}\ln\frac{w^{\alpha+1}}{1 - w}.\qquad(2.2.76)$$

We observe that the solution (2.2.76) is a travelling wave in $(t, \ln r)$ coordinates. Using local analysis (Bender and Orszag 1978), we may obtain the behaviour of the solution (2.2.76) for $-1 < \alpha < 0$ at $t = 1$:

$$w \sim r^{-2\alpha/(\alpha+1)} \quad \text{as } r \to 0,\qquad(2.2.77)$$

$$w \sim 1 - r^{2\alpha} \quad \text{as } r \to \infty.\qquad(2.2.78)$$

Dolzhanskii et al. (1992) showed that the travelling wave solution (2.2.76) of (2.2.68) represents large time asymptotic solution of the latter for a wide class of initial conditions. Figure 2.5 compares the solution (2.2.76) of (2.2.69) subject to the initial condition satisfying (2.2.77) and (2.2.78) and the experimental results for the physical problem that equation (2.2.69) simulates. It shows that in the initial stage of spin-up ($r = 0.86$), the Wedemeyer solution (short dashed curve) agrees very well with the experimental results. For smaller radii, the Wedemeyer solution fails (curve 2, $r = 0.71$) but as r decreases the experimental results (solid curves) approach one of the travelling wave solutions (long dashed curves), represented by (2.2.76). Indeed it is verified that the front of the wave propagates with a constant speed in

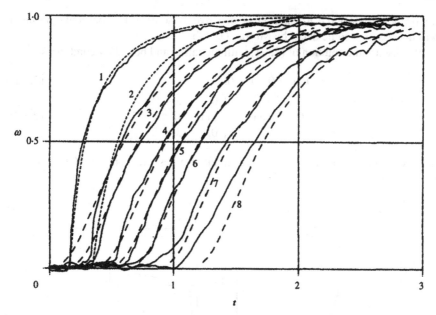

Fig. 2.5 w versus time: experimental data (—) for $r = 0.86, 0.71, 0.64, 0.5, 0.44, 0.37, 0.28, 0.21$ as given by (curves 1–8, respectively); travelling wave solution (2.2.76) (– – –); Wedemeyer solution (2.2.75) (....) (Dolzhanskii et al. 1992). Copyright © 1992 Cambridge University Press, Cambridge. Reprinted with permission. All rights reserved.)

$(t, \ln r)$ coordinates almost independent of the external parameters appearing in the problem (see Dolzhanskii et al. 1992).

In the next section, we give details of a more general initial value problem studied by Murray (1970a).

2.3 Decay estimates for solutions of nonlinear first-order partial differential equations

In this section, we study the first-order PDE

$$u_t + g(u)u_x + \lambda h(u) = 0, \tag{2.3.1}$$

where $\lambda > 0$ and $g(u)$ and $h(u)$ are nonnegative strictly monotone increasing functions for $u > 0$. Decay of solutions of (2.3.1) (for large time) subject to a finite initial disturbance are discussed following the work of Murray (1970a). We also summarise the related work of Natalini and Tesei (1992).

Equation (2.3.1) appears in many applications: stress wave propagation in a nonlinear Maxwell rod with damping, ion exchange in fixed columns, the Gunn effect in semiconductors, and so on. In a model for the Gunn effect, $g(u)$ and $h(u)$ need

not be monotonic and $h(u)$ can even be negative for some u. This particular problem
was discussed in some detail by Murray (1970b).

We discuss the solution of (2.3.1) with the initial and boundary conditions

$$u(0,t) = 0, \quad t > 0, \tag{2.3.2}$$

$$u(x,0) = u_0(x) = \begin{cases} 0, & x < 0, \\ f(x), & 0 < x < X, \\ 0, & x > X; \end{cases} \tag{2.3.3}$$

here, $0 \le f(x) \le 1$. Three types of initial profiles are considered (see Figures 2.6a–c).

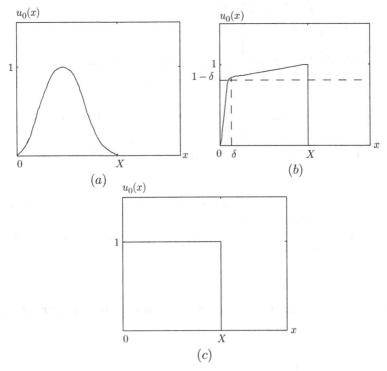

Fig. 2.6 Typical initial profiles for $u(x,t)$ (Murray 1970a. Copyright © 1970 Society for Industrial and Applied Mathematics. Reprinted with permission. All rights reserved.)

Murray (1970a) showed that the decay of the solutions of (2.3.1)–(2.3.3) for large time depends crucially on the behaviour of the function $h(u)$ as $u \to 0$. He proved that, for $h(u) = O(u^\alpha)$, $\alpha > 0, 0 < u \ll 1$, the initial disturbance decays in a finite time and finite distance when $0 < \alpha < 1$. It decays in infinite time and infinite distance like $O(t^{-1/(\alpha-1)})$ for $1 < \alpha \le 3$ and like $O(t^{-1/2})$ for $\alpha \ge 3$ as $t \to \infty$. The solution of (2.3.1)–(2.3.3) decays exponentially in an infinite time and in a finite distance for $\alpha = 1$. Following Murray (1970a), we present a detailed discussion of the above results for the problem (2.3.1)–(2.3.3). It may be pointed out that the solution of (2.3.1) for the special case $\lambda = 0$, starting from a smooth initial profile (see Figure 2.6a), becomes discontinuous after some time for $g_u(u) > 0$. However, for $\lambda \neq 0$, solutions of (2.3.1)–(2.3.3) may admit discontinuities for some cases only. For simplicity, we assume that $h(0) = 0$.

In the first instance, we derive the solution of (2.3.1) by using the method of characteristics. We then discuss the existence (or nonexistence) of discontinuities, their speed of propagation, and decay as $t \to \infty$.

The characteristic system for (2.3.1) is given by

$$\frac{dt}{d\sigma} = 1,$$

$$\frac{dx}{d\sigma} = g(u(x(\sigma), t(\sigma))), \tag{2.3.4}$$

$$\frac{du}{d\sigma} = -\lambda h(u);$$

σ is a parameter measured along the characteristics. By integration of (2.3.4) from $t = 0$ to $t = \sigma$ along the characteristics, assuming that $t = 0$ and $x = x_0$ when $\sigma = 0$, we arrive at the solution

$$t(\sigma) = \sigma,$$

$$x(\sigma) = x_0 + \int_0^\sigma g(u(x(\tau), \tau))d\tau, \tag{2.3.5}$$

$$\int_{f(x_0)}^u \frac{ds}{h(s)} = -\lambda \sigma.$$

Note that the integrals in (2.3.5) follow the characteristics. If $x = x_d(t)$ is the shock path (when it exists), then, by the Rankine–Hugoniot condition (see Courant and Friedrichs 1948),

$$\frac{dx_d}{dt} = \frac{1}{u_1 - u_2} \int_{u_2}^{u_1} g(u)du; \tag{2.3.6}$$

here, $u_1(t)$ and $u_2(t)$ are the states on the left and right of the shock, respectively. For $\lambda > 0$, it may happen that u decays sufficiently fast and shocks do not form at all. Now, we find an expression for the critical time $t = t_c$, say, when, starting from a smooth initial profile (see Figure 2.6a), the shock is formed. Let

$$H(u) = \int^u h^{-1}(s)ds.$$

By (2.3.5)$_3$,

$$\int_{f(x_0)}^u h^{-1}(s)ds = H(u) - H(f(x_0)) = -\lambda \sigma. \tag{2.3.7}$$

Monotonicity of H assures the existence of the inverse of H and hence from (2.3.7), we have

$$u(\sigma) = G(H(f(x_0)) - \lambda \sigma). \tag{2.3.8}$$

To derive (2.3.5)–(2.3.8), we have assumed that the characteristic curve initiates from $(x_0, 0)$.

We observe that differentiating (2.3.8) with respect to x partially gives an expression for u_x containing the factor $(\partial x/\partial x_0)^{-1}$, therefore, u_x becomes unbounded when $\partial x/\partial x_0 = 0$. The smallest time for which $\partial x/\partial x_0 = 0$ heralds the formation of the shock. We may point out that when g is monotonic increasing, the points on the initial profile with larger values of u move faster and therefore may lead to the multivaluedness of the profile (see (2.3.4)). This happens where $f'(x) < 0$. Using (2.3.5)$_1$ and (2.3.8) in (2.3.5)$_2$,

$$x(t) = x_0 + \int_0^t g(G(H(f(x_0))) - \lambda\tau))d\tau. \qquad (2.3.9)$$

Differentiating (2.3.9) with respect to x_0 and setting $\partial x/\partial x_0 = 0$, we have

$$1 = -\int_0^{t_c} g'(G(H(f) - \lambda\tau))G'(H(f(x_0))) - \lambda\tau)H'(f(x_0))f'(x_0)d\tau,$$

or

$$1 = \frac{1}{\lambda}h^{-1}(f(x_0))f'(x_0)\left[g(G(H(f(x_0))) - \lambda\tau))\right]_0^{t_c},$$

$$= \frac{1}{\lambda}h^{-1}(f(x_0))f'(x_0)\left[g(G(H(f(x_0))) - \lambda t_c)) - g(f(x_0))\right]. \qquad (2.3.10)$$

We now discuss the decay of solutions of the problem (2.3.1)–(2.3.3) when $h(u) = u^\alpha$, $\alpha > 0$. For large time, the decay of u is given by the behaviour of $h(u)$ for $u \ll 1$. We discuss the case $\alpha > 1$ in detail and summarise the results for $0 < \alpha \leq 1$. With $h(u) = u^\alpha$, equation (2.3.1) becomes

$$u_t + g(u)u_x + \lambda u^\alpha = 0, \quad \alpha > 1. \qquad (2.3.11)$$

The solution of the characteristic system (2.3.4) for the present case becomes

$$\begin{array}{l} x(t) = x_0 + \int_0^t g(u(x(\tau), \tau))d\tau, \\ u(t) = \left[f^{1-\alpha}(x_0) - (1-\alpha)\lambda t\right]^{1/(1-\alpha)}. \end{array} \qquad (2.3.12)$$

If we choose the initial condition corresponding to case (b) in Figure 2.6, the shock path (2.3.6) becomes

$$\frac{dx_d}{dt} = \frac{1}{u_1}\int_0^{u_1} g(u)du; \qquad (2.3.13)$$

here, $u_2 = 0$ for all $t \geq 0$ and $u_1(0) = 1$. In the following, we derive a differential equation for u_1 by making use of (2.3.12) and (2.3.13). This differential equation for u_1, in turn, helps us to discuss the decay of $u_1(t)$ for large time. Now we define

$$G(s, -t) = \left[s^{1-\alpha} - (1-\alpha)\lambda t\right]^{1/(1-\alpha)} \qquad (2.3.14)$$

(see (2.3.8)). Clearly, $G(s,0) = s$. Equations (2.3.12) and (2.3.14) imply that, along the characteristic emanating from $(x_0,0)$, we have

$$u(t) = G(f(x_0), -t);\qquad(2.3.15)$$

u exists when G is real. Because the characteristics on both sides of the shock path $x = x_d(t)$ terminate thereon, $x = x_d(t)-$ is given by $(2.3.12)_1$ and $u(x_d(t)-,t) = u_1(t)$. For the sake of convenience, we write x_d instead of x_d- in the following. Thus, equation $(2.3.12)_1$ gives

$$x_d(t) = x_0(x_d,t) + \int_0^t g(u(x(\tau), \tau))d\tau,$$

$$= x_0(x_d,t) + \int_0^t g(G(f(x_0(x_d,t)), -\tau))d\tau.\qquad(2.3.16)$$

We differentiate (2.3.16) with respect to t and obtain

$$\frac{dx_d}{dt} = \frac{dx_0(x_d,t)}{dt} + g(u_1(t)) + \int_0^t \frac{d}{dt}g(G(f(x_0), -\tau))d\tau.\qquad(2.3.17)$$

It is easy to check that

$$\frac{d}{dt}g(G(f(x_0), -\tau)) = -\frac{1}{\lambda(1-\alpha)}\frac{d}{d\tau}g(G(f(x_0), -\tau))\frac{d}{dt}f^{1-\alpha}(x_0).\qquad(2.3.18)$$

Using (2.3.18) in (2.3.17), we have

$$\frac{dx_d}{dt} = \frac{dx_0}{dt} + g(u_1(t)) - \frac{1}{\lambda(1-\alpha)}\frac{d}{dt}f^{1-\alpha}(x_0)\int_0^t \frac{d}{d\tau}g(G(f(x_0), -\tau))d\tau$$

$$= \frac{dx_0}{dt} + g(u_1(t)) - \frac{1}{\lambda(1-\alpha)}\frac{d}{dt}f^{1-\alpha}(x_0)\left[g(u_1(t)) - g(f(x_0))\right];\qquad(2.3.19)$$

remember that $x_0 = x_0(x_d(t),t)$. Using $(2.3.12)_2$ and (2.3.14) we find that

$$f(x_0(x_d,t)) = G(u_1(t),t);\qquad(2.3.20)$$

therefore,

$$x_0(x_d,t) = F(G(u_1(t),t)).\qquad(2.3.21)$$

Equations (2.3.13), (2.3.19), (2.3.20), and (2.3.21) give the following equation for $u_1(t)$.

$$\frac{1}{u_1}\int_0^{u_1} g(u)du = \frac{d}{dt}F(G(u_1,t)) + g(u_1)$$

$$- \frac{1}{\lambda(1-\alpha)}\frac{d}{dt}[G^{1-\alpha}(u_1,t)]\left\{g(u_1) - g(G(u_1,t))\right\}(2.3.22)$$

Inserting

$$G^{1-\alpha}(u_1,t) = u_1^{1-\alpha}(t) + (1-\alpha)\lambda t$$

into (2.3.22), we get an ordinary differential equation for $u_1(t)$:

$$\frac{1}{\lambda u_1^{\alpha}}\frac{du_1}{dt} = [g(u_1) - g(G(u_1,t))]^{-1}\left\{g(G(u_1,t))\right.$$

$$\left. - \frac{1}{u_1}\int_0^{u_1} g(u)du + \frac{d}{dt}F(G(u_1,t))\right\}. \tag{2.3.23}$$

Now, we discuss the solution of (2.3.1) in some detail for the top hat initial condition (see Figure 2.6c). Let t_0 be the time at which the characteristic issuing from $x_0 = 0$ intersects the shock path $x = x_d(t)$. Note that the shock forms at $t = 0$ and $x_d(0) = X$. Furthermore, when $x_0 = 0$, $f(x_0) = 1$. An integration of (2.3.13) from 0 to $t \leq t_0$ gives

$$x_d(t) = X + \int_0^t G^{-1}(1,-\tau)\int_0^{G(1,-\tau)} g(u)dud\tau, \quad t \leq t_0, \tag{2.3.24}$$

recalling that

$$u_1(t) = G(1,-t), \quad t \leq t_0 \tag{2.3.25}$$

(see (2.3.12) and (2.3.15)). Therefore, at $t = t_0$, we have

$$\int_0^{t_0} g(G(1,-\tau))d\tau = X + \int_0^{t_0} G^{-1}(1,-\tau)\int_0^{G(1,-\tau)} g(u)dud\tau, \tag{2.3.26}$$

where the LHS of (2.3.26) follows from (2.3.12) with $x_0 = 0$, $f(x_0) = 1$, and the definition of $G(s,t)$ (see (2.3.14)). Murray (1970a) has shown that a solution t_0 of (2.3.26) exists for all $\alpha \geq 2$. Now we discuss below the case $\alpha > 1$ and quote the results for $0 < \alpha < 1$.

We recall that $h(u) = u^{\alpha}$, $\alpha > 1$. We assume that $g'(0) \neq 0$; the case $g'(0) = 0$ may also be treated with some modifications. Here, again, we consider two cases: when t_0 exists and when it does not; we recall that t_0 is the time at which the characteristic curve issuing from $x_0 = 0$, $t = 0$, $u_0(0) = 1$ meets the shock which originated at $x_d(0) = X$.

(i) t_0 does not exist.

This case is possible only when $1 < \alpha < 2$ (see Murray 1970a). From (2.3.25) and (2.3.14), we have

$$u_1(t) = G(1,-t)$$

$$= [1 + (\alpha - 1)\lambda t]^{-1/(\alpha-1)}$$

$$= O(t^{-1/(\alpha-1)}), \quad t \gg 1. \tag{2.3.27}$$

Equation (2.3.13) implies that

$$\frac{dx_d}{dt} = \frac{1}{u_1} \int_0^{u_1} \left[g(0) + \frac{g'(0)}{1} u + \dots \right] du$$

$$= g(0) + \frac{g'(0)}{2} u_1 + \frac{g''(0)}{6} u_1^2 + \dots$$

$$= g(0) + O(t^{-1/(\alpha-1)}), \quad t \gg 1, \tag{2.3.28}$$

(see (2.3.27)). It follows from (2.3.27) and (2.3.28) that $u = u_1(t)$, the value of u behind the shock, tends to zero and the shock speed dx_d/dt tends to $g(0)$ as $t \to \infty$.

(ii) t_0 exists.

The initial condition in Figure 2.6b tends to that shown in Figure 2.6c as δ tends to zero. This, in turn, implies that $dF/dt \to 0$ in this limit. For $t \le t_0$, $u_1(t)$ and $x_d(t)$ may be found from equations (2.3.25) and (2.3.24), respectively. With $dF/dt = 0$, equation (2.3.23) gives

$$\frac{1}{\lambda u_1^\alpha} \frac{du_1}{dt} = [g(u_1) - g(G(u_1,t))]^{-1} \left\{ g(G(u_1,t)) - \frac{1}{u_1} \int_0^{u_1} g(u) du \right\}, \quad t \ge t_0. \tag{2.3.29}$$

To find the asymptotic form for $u_1(t)$ for $t \gg 1$, we first show that

$$G(u_1,t) = [u_1^{-(\alpha-1)} - \lambda t (\alpha - 1)]^{-1/(\alpha-1)} \tag{2.3.30}$$

decreases as t increases. We observe from (2.3.25) and (2.3.14) that

$$G(u_1,t) = 1, \quad t \le t_0. \tag{2.3.31}$$

Differentiating (2.3.30) with respect to t and simplifying, we have

$$\frac{dG(u_1,t)}{dt} = \lambda G^\alpha(u_1,t) \left[1 + \frac{1}{\lambda u_1^\alpha} \frac{du_1}{dt} \right]. \tag{2.3.32}$$

From (2.3.29) and (2.3.32), we obtain

$$\frac{dG(u_1,t)}{dt} = \lambda G^\alpha(u_1,t) [g(u_1) - g(G(u_1,t))]^{-1} \left\{ g(u_1) - \frac{1}{u_1} \int_0^{u_1} g(u) du \right\}. \tag{2.3.33}$$

Because $g_u(u) > 0$, $g(u) < g(u_1)$ for $u < u_1$, we may write the inequality

$$\frac{1}{u_1} \int_0^{u_1} g(u) du < g(u_1). \tag{2.3.34}$$

Moreover, if $u_1 < 1$ at $t = t_0$, then by (2.3.31) we have

$$g(G(u_1,t)) = g(1) > g(u_1). \tag{2.3.35}$$

Equations (2.3.33)–(2.3.35) imply that

$$\frac{dG}{dt} < 0 \quad \text{at } t = t_0;$$

that is, $G(u_1,t)$ decreases with time $t > t_0$ in a right neighbourhood of t_0. Now we show that $dG/dt < 0$ for all $t > t_0$. From (2.3.33), we infer that $dG/dt \to 0$ as $G(u_1,t) \to 0$ or $g(u_1) \to (1/u_1)\int_0^{u_1} g(u)du$. This, in turn, requires that $u_1 \to 0$. Thus, $dG(u_1,t)/dt < 0$ for $0 < u_1 < 1$ and $0 < G(u_1,t) < 1$. Because $G(u_1,t) > 0$ and $1 - \lambda t(\alpha - 1)u_1^{\alpha-1} < 1$, we arrive at the inequality

$$0 < \left[1 - \lambda t(\alpha - 1)u_1^{\alpha-1}\right] < 1 \tag{2.3.36}$$

for $u_1 > 0$. Thus, $tu_1^{\alpha-1}$ is bounded for all $t > 0$, $u_1 \geq 0$. We arrive at two subcases: (a) $tu_1^{\alpha-1} \to B$, a positive constant, as $t \to \infty$ and $u_1 \to 0$; and (b) $tu_1^{\alpha-1} \to 0$ as $u_1 \to 0$ and $t \to \infty$. From (2.3.36), we have $0 \leq B \leq 1/(\lambda(\alpha - 1))$.

Case (a): $0 < B < 1/(\lambda(\alpha - 1))$.

By (2.3.30),

$$
\begin{aligned}
G(u_1,t) &= u_1 \left[1 - \lambda t(\alpha - 1)u_1^{\alpha-1}\right]^{-1/(\alpha-1)}, \\
&\to u_1 [1 - \lambda B(\alpha - 1)]^{-1/(\alpha-1)} \quad \text{as } t \to \infty \\
&= O(u_1).
\end{aligned}
\tag{2.3.37}
$$

Using Taylor's series expansions for g in (2.3.29), we have

$$
\begin{aligned}
&\frac{1}{\lambda u_1^\alpha}\frac{du_1}{dt} \\
&\sim \left\{g(0) + u_1 g'(0) + \ldots - g(0) - u_1\left[1 - \lambda B(\alpha - 1)\right]^{-1/(\alpha-1)} g'(0) - \ldots\right\}^{-1} \\
&\quad \times \left\{g(0) + u_1(1 - \lambda B(\alpha - 1))^{-1/(\alpha-1)}g'(0) + \ldots \right. \\
&\quad \left. - \frac{1}{u_1}\left[g(0)u_1 + \frac{g'(0)u_1^2}{2} + \ldots\right]\right\} \\
&\sim \left\{u_1 g'(0)\left[1 - (1 - \lambda B(\alpha - 1))^{-1/(\alpha-1)}\right] + \ldots\right\}^{-1} \\
&\quad \times \left\{u_1 g'(0)\left[(1 - \lambda B(\alpha - 1))^{-1/(\alpha-1)} - \frac{1}{2}\right] + \ldots\right\} \\
&\to -A(\alpha, B) + O(u_1) \quad \text{as } t \to \infty, \ u_1 \to 0;
\end{aligned}
\tag{2.3.38}
$$

here,

$$A(\alpha, B) = \frac{2 - [1 - (\alpha - 1)\lambda B]^{1/(\alpha-1)}}{2 - 2[1 - (\alpha - 1)\lambda B]^{1/(\alpha-1)}}. \tag{2.3.39}$$

An integration of (2.3.38) gives

$$\frac{u_1^{1-\alpha}}{1-\alpha} \sim -\lambda t A(\alpha, B) + \text{constant}. \tag{2.3.40}$$

It follows, therefore, that

$$u_1(t) = O(t^{-1/(\alpha-1)}), \quad t \gg t_0. \tag{2.3.41}$$

Using (2.3.40), we have

$$t u_1^{\alpha-1} \rightarrow \frac{1}{\lambda(\alpha-1)A(\alpha, B)} \quad \text{as } t \rightarrow \infty. \tag{2.3.42}$$

But, by assumption (see below (2.3.36)), $t u_1^{\alpha-1} \rightarrow B$ as $t \rightarrow \infty$, $u_1 \rightarrow 0$. Therefore, it follows from (2.3.42) that

$$B = \frac{1}{\lambda(\alpha-1)A(\alpha, B)}.$$

Thus, B satisfies the equation

$$W(B) \equiv \frac{1}{\lambda(\alpha-1)A(\alpha, B)} - B = 0$$

or

$$\frac{2[1 - \lambda B(\alpha-1)] - 2[1 - \lambda B(\alpha-1)]^{1/(\alpha-1)}\left[1 - \frac{1}{2}\lambda B(\alpha-1)\right]}{\lambda(\alpha-1)\left\{2 - [1 - \lambda B(\alpha-1)]^{1/(\alpha-1)}\right\}} = 0 \tag{2.3.43}$$

(see (2.3.39)). It is easy to check that (2.3.43) has solutions $B = 0, 1/(\lambda(\alpha-1))$ for all $\alpha > 1$. Besides, equation (2.3.43) has one more solution B_i in the parametric range $2 < \alpha < 3$ with $0 < B_i < 1/(\lambda(\alpha-1))$. Furthermore, $B_i \rightarrow 1/(\lambda(\alpha-1))$ as $\alpha \rightarrow 2$.

Case (b): $B = 0$.

From (2.3.30), we have

$$G(u_1, t) = u_1 \left[1 + \left(-\frac{1}{\alpha-1}\right)(-(\alpha-1))\lambda t u_1^{\alpha-1} + \dots\right], \quad t \gg t_0$$
$$= u_1 + \lambda t u_1^{\alpha} + \dots. \tag{2.3.44}$$

On using (2.3.44) in (2.3.29), we get

$$\frac{1}{\lambda u_1^{\alpha}} \frac{du_1}{dt} = \left[g(u_1) + (\lambda t u_1^{\alpha} + \dots)g'(u_1) + \dots - g(u_1) + \frac{u_1}{2}g'(u_1) - \dots\right]$$
$$\times \left[g(u_1) - [g(u_1) + (\lambda t u_1^{\alpha} + \dots)g'(u_1) + \dots]\right]^{-1}$$

$$\sim -\frac{1}{2\lambda tu_1^{\alpha-1}}, \quad t \gg t_0.$$

This implies that

$$u_1(t) \sim \frac{\text{constant}}{t^{1/2}} \quad \text{as } t \to \infty. \tag{2.3.45}$$

It follows from (2.3.45) that $tu_1^{\alpha-1} = O(t^{(3-\alpha)/2}) \to 0$ as $t \to \infty$ only when $\alpha > 3$. Thus $B = 0$ is a possible limit only for the parametric range $\alpha > 3$. This suggests that for the parametric range $1 < \alpha \le 2$, we may choose $B = 1/(\lambda(\alpha-1))$. By choosing the value B_i for B (see below (2.3.43)) in the parametric range $2 < \alpha < 3$, a smooth variation in B with respect to α is observed. Thus,

$$B = \begin{cases} \dfrac{1}{\lambda(\alpha-1)}, & 1 < \alpha \le 2, \\ B_i, & 2 \le \alpha < 3, \\ 0, & \alpha \ge 3. \end{cases}$$

Summarising the results obtained for cases (a) and (b), we have

$$u_1(t) = \begin{cases} O(t^{-1/(\alpha-1)}), & 1 < \alpha \le 3, \\ O(t^{-1/2}), & \alpha \ge 3 \end{cases}$$

and

$$\frac{dx_d}{dt} = \begin{cases} g(0) + O(t^{-1/(\alpha-1)}), & 1 < \alpha \le 3, \\ g(0) + O(t^{-1/2}), & \alpha \ge 3. \end{cases}$$

Murray (1970a) showed that $u(x,t)$ decays in a finite time and finite distance for the parametric range $0 < \alpha < 1$. For $\alpha = 1$, it was proved that the initial disturbance decays exponentially in an infinite time but in a finite distance. Murray (1970a) also discussed the behaviour of $u_1(t)$ and $x_d(t)$ as $\lambda \to 0$ in (2.3.1); the special case $g(u) = u + a$, $h(u) = u$ was studied in considerable detail.

Natalini and Tesei (1992) studied the Cauchy problem

$$u_t + \frac{1}{m}(u^m)_x = -u^p, \quad \mathbb{R} \times (0, \infty), \tag{2.3.46}$$

$$u(x,0) = u_0(x), \quad x \in \mathbb{R}; \tag{2.3.47}$$

here $m > 1$, $p > 1$ and $u_0(x) = \sigma\chi_{[0,l]}(x)$, σ, $l > 0$ has a compact support. They derived sharp estimates for the support of the solution of the problem (2.3.46) and (2.3.47) and large time behaviour of its solution. Let supp $u(.,t) = [0, s(t)]$, $l \le s(t) < \infty$, $t \ge 0$. Regarding the support of the solution of (2.3.46) and (2.3.47), they proved the following theorem.

Theorem 2.3.1 (i) *Suppose that $1 < p < m$. Then there exists $x^* > 0$ such that*

$$l \le s(t) \le l + x^*, \quad t \ge 0. \tag{2.3.48}$$

(ii) *Suppose that $1 < m \leq p$. Then there exist constants A, $B > 0$ such that for*
      large time

$$A \ln t \leq s(t) \leq B \ln t \quad \text{for } p = m;$$
$$At^{(p-m)/(p-1)} \leq s(t) \leq Bt^{(p-m)/(p-1)} \quad \text{for } m < p < m+1; \qquad (2.3.49)$$
$$At^{1/m}(\ln t)^{-(m-1)/m} \leq s(t) \leq Bt^{1/m}(\ln t)^{-(m-1)/m} \quad \text{for } p = m+1;$$
$$At^{1/m} \leq s(t) \leq Bt^{1/m} \quad \text{for } m+1 < p.$$

In the above, $s(t)$ is the shock path starting from $x = l$ at $t = 0$. Furthermore, the constants x^*, A, B appearing in (2.3.48) and (2.3.49) may depend on l, $\|u_0\|_\infty$, m, and p. Natalini and Tesei (1992) proved the following theorem with regard to the decay of solutions of (2.3.46) and (2.3.47).

Theorem 2.3.2 (i) *Assume that $1 < p < m$. Then*

$$t\left[\frac{1}{p-1} - t\|u(.,t)\|_\infty^{p-1}\right] \to c_0$$

as $t \to \infty$ where c_0 is a positive constant.
(ii) *Assume that $1 < m \leq p$. Then*

$$t^{1/(p-1)}\|u(.,t)\|_\infty \to c_1 \quad \text{for } m \leq p < m+1;$$

$$c_2 \leq (t \ln t)^{1/m}\|u(.,t)\|_\infty \leq c_3 \quad \text{for } p = m+1;$$

$$c_2 \leq t^{1/m}\|u(.,t)\|_\infty \leq c_3 \quad \text{for } p > m+1,$$

where c_1, c_2, and c_3 are positive constants.

Here, $\|u(.,t)\|_\infty := \sup_x u(x,t)$. The constants c_i $(i = 1,2,3)$ may again depend on l, $\|u_0\|_\infty$, m, and p. Reference may also be made to the related work of Guarguaglini (1995), Reyes (2001), and Pablo and Reyes (2006).

2.4 Conclusions

In this chapter, we have discussed large time behaviour of solutions of first-order nonlinear partial differential equations subject to certain initial/boundary conditions. In Section 2.2, we have presented some examples. Examples 1 and 2 showed that special solutions such as product solutions or even constant solution may serve as asymptotics for large time. Example 3 demonstrated that an exact implicit solution may fail to provide large time behaviour of solutions. Example 4 dealt with the inviscid Burgers equation and clearly brought out the importance of self-similar solutions. In Example 5, we have discussed a transformation which reduces a more general first-order partial differential equation (2.2.45) to the conservative form (2.2.47) and hence helps us to arrive at decay properties of the solutions of more general equations (see Joseph and Sachdev 1994). In Example 6, we have presented the study of a first-order partial differential equation describing spin-up and spin-down in a cylinder of small ratio of height to diameter, following the work of Dolzhanskii et al. (1992). It was shown how travelling waves constitute large time asymptotics. In Section 2.3, we have presented the decay of solutions of a nonlinear first-order partial differential equation (2.3.1) subject to the initial profile (2.3.3) and the boundary conditions (2.3.2) for $x \geq 0$, $t \geq 0$. It was proved that the decay of the solution of (2.3.1)–(2.3.3) depends crucially on the behaviour of $h(u)$ as $u \to 0$. This section closely followed the work of Murray (1970a).

References

Bender, C. M., Orszag, S. A. (1978) *Advanced Mathematical Methods for Scientists and Engineers*, McGraw-Hill International Edition, McGraw-Hill, Singapore.

Courant, R., Friedrichs, K. O. (1948) *Supersonic Flow and Shock Waves*, Wiley Interscience, New York.

Dafermos, C. M. (1985) Regularity and large time behaviour of solutions of conservation law without convexity condition, *Proc. Roy. Soc. Edinburgh Sect. A* 99, 201–239.

Dolzhanskii, F. V., Krymov, V. A., Manin, D. Yu. (1992) Self-similar spin-up and spin-down in a cylinder of small ratio of height to diameter, *J. Fluid Mech.* 234, 473–486.

Guarguaglini, F. R. (1995) Singular solutions and asymptotic behaviour for a class of conservation laws with absorption, *Commun. Partial Differential Eq.* 20, 1395–1425.

Joseph, K. T., Sachdev, P. L. (1994) On the solution of the equation $u_t + u^n u_x + H(x,t,u) = 0$, *Quart. Appl. Math.* 52, 519–527.

Lax, P. D. (1957) Hyperbolic systems of conservation laws II, *Comm. Pure Appl. Math.* 10, 537–566.

Mickens, R. E. (1988) Exact solutions to a population model: The logistic equation with advection, *SIAM Rev.* 30, 629–633.

Murray, J. D. (1970a) Perturbation effects on the decay of discontinuous solutions of nonlinear first order wave equations, *SIAM J. Appl. Math.* 19, 273–298.

Murray, J. D. (1970b) On the Gunn effect and other physical examples of perturbed conservation equations, *J. Fluid Mech.* 44, 315–346.

Natalini, R., Tesei, A. (1992) On a class of perturbed conservation laws, *Adv. Appl. Math.* 13, 429–453.

Pablo, A. de, Reyes, G. (2006) Long time behaviour for a nonlinear first-order equation, *Nonlinear Anal.* 65, 284–300.

Reyes, G. (2001) Critical asymptotic behaviour for a perturbed conservation law, *Asymptotic Anal.* 25, 109–122.

Wedemeyer, E. H. (1964) The unsteady flow within a spinning cylinder, *J. Fluid Mech.* 20, 383–399.

Zachmanoglou, E. C., Thoe, D. W. (1976) *Introduction to Partial Differential Equations with Applications*, Dover, New York.

Chapter 3
Large Time Asymptotic Analysis of Some Nonlinear Parabolic Equations – Some Constructive Approaches

3.1 Introduction

This chapter describes some constructive approaches to the study of the asymptotic nature of solutions of some nonlinear partial differential equations of parabolic type: travelling waves, and self-similar or more general solutions obtained by the so-called balancing argument and its extensions. This is accomplished by first constructing these special solutions and hence showing their asymptotic nature analytically, numerically, or both. Sometimes it becomes possible to obtain asymptotic solutions in terms of power series in time with coefficient functions depending on the similarity variable. This approach is adopted to obtain solutions more general than self-similar or travelling waves. The analysis for such problems requires solutions of an infinite system of nonlinear ODEs. This class of solutions either 'nonlinearise' the linear solution of the given problem and/or embed in them limiting behaviour such as inviscid forms for convective–diffusive equations. An alternative approach is to use matched asymptotic expansions to obtain asymptotic solution of the initial boundary value problem. This approach was adopted by Leach and Needham (2001) for the generalised Fisher's equation. We discuss this approach in Section 3.9.

To illustrate the appoach above, we first discuss in Section 3.2 a family of travelling wave solutions of the Burgers equation which admits an exact solution of this form satisfying some definite conditions as the spatial distance tends to $\pm\infty$. It is then numerically shown how these solutions emerge from step function initial conditions as time becomes large. In Section 3.3, we take up an initial boundary value problem for the Burgers equation which arises from a minimally nonlinear form of the Fokker–Planck equation and describes vertical nonhysteretic flow in nonswelling soil. It was first studied by Philip (1969) who, quite evocatively, introduced the term 'profile at infinity' to denominate the asymptotic form of the solutions. This work was later discussed more effectively by Clothier et al. (1981). Section 3.4 describes the work of Vanaja and Sachdev (1992) relating to travelling waves in a porous medium. The asymptotic nature of these solutions is demonstrated quite rigourously and is confirmed with reference to numerical solutions. Section 3.5

P.L. Sachdev, Ch. Srinivasa Rao, *Large Time Asymptotics for Solutions of Nonlinear Partial Differential Equations*, Springer Monographs in Mathematics, DOI 10.1007/978-0-387-87809-6_3, © Springer Science+Business Media, LLC 2010

concerns the evolution of a stable profile describing cross-field diffusion in toroidal multiple plasma; here, a separable solution of an initial value problem constitutes an asymptotic solution. Section 3.6 deals with fast nonlinear diffusion and it is shown that the large time behaviour of solutions is described by an asymptotic solution in 'unusual' self-similar form (King 1993); here, the argument is intuitive but is closely related to the balancing argument which is taken up in Section 3.8. Section 3.7 illustrates how periodic solutions of the linearised form of a nonlinear PDE may be generalised to include the nonlinear effects; indeed it becomes possible to 'exactly' obtain the periodic solutions of several generalised Burgers equations in the form of infinite series, which contain that for the Burgers equation as a special case. Here, again, the asymptotic nature of the solution is verified by comparison with the numerical solution (see Sachdev et al. 2003, 2005, Srinivasa Rao and Satyanarayana 2008a). In Section 3.8, we discuss in considerable detail the balancing argument due to Grundy and his collaborators (Dawson et al. 1996, Grundy et al. 1994, Van Duijn et al. 1997); this approach delivers (approximate) asymptotic results when the given equation does not admit exact similarity or travelling wave solutions. In this approach Grundy (1988), in the manner of Bender and Orszag (1978) for nonlinear ODEs, showed how different classes of solutions may be obtained by balancing different sets of terms; the given PDE is first rewritten with time and similarity variables as the new independent variables. The generalisation of this approach due to Sachdev and his collaborators (Sachdev et al. 1999, Sachdev and Srinivasa Rao 2000, Srinivasa Rao and Satyanarayana 2008b) is also discussed in this section. Section 3.9 describes the evolution of travelling wave solutions of generalised Fisher's equations as time becomes large. We follow Leach and Needham (2001) where the method of matched asymptotic expansions is used. The final section considers periodic travelling waves for a coupled system of equations. Curiously, the analysis here is largely inspired by the asymptotic nature of the solution of the boundary value problem, obtained numerically; the latter experiments vividly bring out the independence of the asymptotic solutions (when they exist) of the details of the initial conditions.

3.2 Travelling waves as asymptotics of solutions of initial value problems – The Burgers equation

It is convenient and illustrative to consider the Burgers equation

$$u_t + uu_x = \varepsilon u_{xx}, \tag{3.2.1}$$

where $0 < \varepsilon \ll 1$ is a parameter (Sachdev 1987). First we find travelling wave solutions of (3.2.1) and then show how they arise from the solutions of step-function initial conditions as $t \to \infty$. We also show numerically how such initial conditions actually evolve from $t = 0$. Let the initial conditions be such that

$$u \to u_l \text{ as } x \to -\infty \text{ and } u \to u_r \text{ as } x \to +\infty. \tag{3.2.2}$$

The same conditions are imposed on the travelling wave solutions as $x \to \mp\infty$. We let $X = x - Ut$ and seek solutions of (3.2.1) in the form $u = \Psi(x - Ut)$, subject to the conditions (3.2.2). Thus, we have

$$-U\Psi_X + \Psi\Psi_X = \varepsilon\Psi_{XX}. \tag{3.2.3}$$

Integrating (3.2.3) and using the conditions (3.2.2), we obtain

$$\frac{1}{2}\Psi^2 - U\Psi + C = \varepsilon\Psi_X, \tag{3.2.4}$$

where

$$U = \frac{1}{2}(u_l + u_r), \quad C = \frac{1}{2}u_l u_r. \tag{3.2.5}$$

Writing (3.2.4) as

$$(\Psi - u_r)(u_l - \Psi) = -2\varepsilon\Psi_X \tag{3.2.6}$$

and integrating we have

$$\frac{X}{\varepsilon} = \frac{2}{u_l - u_r} \log \frac{u_l - \Psi}{\Psi - u_r} \tag{3.2.7}$$

or

$$\Psi = u_r + \frac{u_l - u_r}{1 + \exp\left[\dfrac{u_l - u_r}{2\varepsilon}(x - Ut)\right]}, \quad U = \frac{u_l + u_r}{2}; \tag{3.2.8}$$

the integration constant is taken to be zero. We may write (3.2.8) as

$$\begin{aligned}
\Psi &= \frac{u_l + u_r \exp\left[\dfrac{u_l - u_r}{2\varepsilon}(x - Ut)\right]}{1 + \exp\left[\dfrac{u_l - u_r}{2\varepsilon}(x - Ut)\right]} \\
&= \frac{u_l + u_r}{2} - \frac{u_l - u_r}{2} \tanh\left[\frac{u_l - u_r}{4\varepsilon}(x - Ut)\right].
\end{aligned} \tag{3.2.9}$$

We must now show how (3.2.9) emerges from the solution of the Burgers equation (3.2.1) with step function initial conditions. For this purpose we quote the existence theorem due to Hopf (1950) for (3.2.1) over $\Omega = \{-\infty < x < \infty, 0 < t < \infty\}$, subject to the initial condition

$$u(x,0) = g(x). \tag{3.2.10}$$

Suppose that $g(x)$ is integrable in every finite x-interval and that

$$\int_0^x g(s)\,ds = O\left(|x|^{1+\alpha}\right) \text{ as } |x| \to \infty \tag{3.2.11}$$

with $0 \le \alpha < 1$. Then the function

$$u(x,t) = \frac{\displaystyle\int_{-\infty}^{\infty} \frac{x-\xi}{t} \exp\left[-\frac{1}{2\varepsilon}\int_0^\xi g(s)ds\right] \cdot \exp\left[-\frac{(x-\xi)^2}{4\varepsilon t}\right] d\xi}{\displaystyle\int_{-\infty}^{\infty} \exp\left[-\frac{1}{2\varepsilon}\int_0^\xi g(s)ds\right] \cdot \exp\left[-\frac{(x-\xi)^2}{4\varepsilon t}\right] d\xi} \tag{3.2.12}$$

is a regular solution of (3.2.1) in Ω and satisfies the initial condition

$$\lim_{t\to 0}\int_0^x u(s,t)ds = \int_0^x g(s)ds \tag{3.2.13}$$

for every $x \in (-\infty,\infty)$. If, in addition, g is continuous, then

$$\lim_{t\to 0} u(x,t) = g(x). \tag{3.2.14}$$

A solution of (3.2.1) which is regular in some strip $-\infty < x < \infty, 0 < t < T$ and which satisfies (3.2.13) for each value of x necessarily coincides with (3.2.12) in that strip. If we now choose the initial function

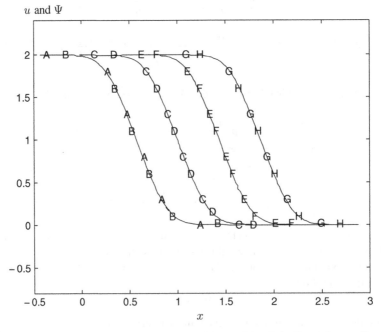

Fig. 3.1 Solution (3.2.16) of (3.2.1) subject to the initial condition (3.2.15) and the travelling wave solution (3.2.8) with $u_l = 2$, $u_r = 0$, and $\varepsilon = 0.1$ at $t = 1/2, 1, 3/2, 2$. Travelling wave solution (3.2.8) at $t = 1/2, 1, 3/2, 2$ is marked as A, C, E, and G, respectively. The solution (3.2.16) at the corresponding times is marked as B, D, F, and H. (Shih 1991. Copyright © 1991 World Scientific, River Edge, NJ. Reprinted with permission. All rights reserved.)

$$g(x) = \begin{cases} u_r, & x > 0 \\ u_l > u_r, & x < 0 \end{cases} \tag{3.2.15}$$

and substitute in (3.2.12), we obtain, after some simplification, the solution

$$u = u_r + \frac{u_l - u_r}{1 + h\exp\left[\dfrac{u_l - u_r}{2\varepsilon}(x - Ut)\right]}, \quad U = \frac{u_l + u_r}{2}, \tag{3.2.16}$$

where

$$h = \frac{\displaystyle\int_{-(x-u_rt)/\sqrt{4\varepsilon t}}^{\infty} \exp(-\xi^2)d\xi}{\displaystyle\int_{(x-u_lt)/\sqrt{4\varepsilon t}}^{\infty} \exp(-\xi^2)d\xi}. \tag{3.2.17}$$

For fixed x/t such that $u_r < x/t < u_l, h \to 1$ as $t \to \infty$ and the solution (3.2.16) approaches (3.2.8). Shih (1991) plotted (3.2.8) and (3.2.16) with $u_l = 2$ and $u_r = 0$ and varied values of $\varepsilon = 10^{-1}, 10^{-3}$. The results are shown in Figures 3.1 and 3.2. The curves A, C, E, and G denote the travelling wave Ψ given in (3.2.8) whereas B, D, F, and H refer to the solution of IVP (3.2.1) and (3.2.15) given in (3.2.16) at $t = 1/2, 1, 3/2$, and 2, respectively. The agreement between evolving and travelling

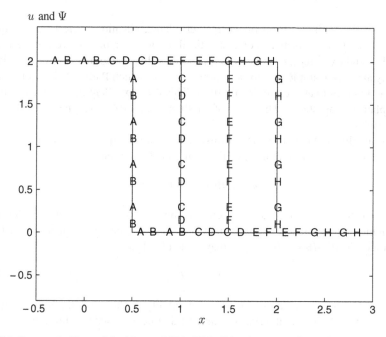

Fig. 3.2 Same as in Figure 3.1 with $\varepsilon = 0.001$. (Shih 1991. Copyright © 1991 World Scientific, River Edge, NJ. Reprinted with permission. All rights reserved.)

waves seems quite close even at an early time. We may summarise here another asymptotic result regarding the solution of the Burgers equation (3.2.1) with $\varepsilon = 1/2$, subject to the initial condition $u(x,0) = u_0(x)$ (Dix 2002). The initial condition has the form

$$u_0(x) = u_1(x) + \frac{nx}{1+x^2}, \quad x \in \mathbb{R} \tag{3.2.18}$$

when $u_1 \in L^1(\mathbb{R})$ and $n \in \mathbb{R}$, implying that $u_0(x) \sim n/x$ as $|x| \to \infty$. Two constants appear in the large time behaviour, namely, the area under the profile $\mu = \lim_{A \to \infty} \int_{-A}^{A} u_0(x)dx$ and n as in (3.2.18). Dix (2002) generalised the result of Hopf (1950) who assumed that $n = 0$. He obtained different results for $n < 1$ and $n \geq 1$. For the former case, Dix (2002) showed that

$$t^{1/2}u(\xi t^{1/2}, t) = U(\xi) + o((1+|\xi|)^{-1}) \text{ as } t \to \infty, \tag{3.2.19}$$

uniformly in ξ, where $\xi = xt^{-1/2}$ is the similarity variable. For $n \geq 1$, there are two different similarity solutions $U(\xi)$ 'that simultaneously "attract" the quantity $t^{1/2}u(\xi t^{1/2}, t)$ and each one wins in its own range of ξ.' Sharp estimates were obtained to describe asymptotic behaviour of the solution in different regions.

3.3 Profile at infinity – Initial boundary value problem for Burgers equation

A minimally nonlinear form of the Fokker–Planck equation – the Burgers equation – describes the vertical nonhysteretic flow of water in nonswelling soil. An initial boundary value problem, which describes this phenomenon, can be solved exactly and has a simple asymptotic form as $t \to \infty$, which Philip (1969) quite figuratively referred to as the profile at infinity. This is an excellent illustrative example to explain asymptoticity. The work of Clothier et al. (1981) succinctly deals with this problem.

The nonlinear Fokker–Planck diffusion–convection equation, which describes nonhysteretic infiltration in nonswelling soil, may be written as

$$\theta_t = (D\theta_z)_z - [k(\theta)]_\theta \theta_z, \tag{3.3.1}$$

where θ is the volumetric water content, t is the time, and z is the depth of the soil. Assuming that the diffusivity coefficient D is constant and the conductivity $k(\theta)$ is quadratic, $dk(\theta)/d\theta = A\theta + B$, we may write (3.3.1) as

$$\theta_t = D\theta_{zz} - (A\theta + B)\theta_z, \tag{3.3.2}$$

where A and B are constants. The physical validity of these assumptions has been discussed by Clothier et al. (1981). It may further be assumed that initially

$$\theta = \theta_n \text{ for } z > 0 \text{ at } t = 0 \tag{3.3.3}$$

and
$$B = -A\theta_n, \quad k(\theta_n) = 0, \tag{3.3.4}$$

where θ_n is the antecedent water content. By integrating the conductivity equation (see above (3.3.2)) and using (3.3.4) we obtain

$$k(\theta) = A(\theta - \theta_n)^2/2. \tag{3.3.5}$$

The boundary condition at $z = 0$ implying constant flux may be stated as follows,

$$-D\theta_z + \frac{A}{2}(\theta - \theta_n)^2 = V_0 = \frac{A}{2}(\theta_0 - \theta_n)^2 \ \text{ at } z = 0 \ \text{ for } t \geq 0. \tag{3.3.6}$$

By introducing the nondimensional variables

$$\zeta = A(\theta_0 - \theta_n)z/2D, \tau = A^2(\theta_0 - \theta_n)^2 t/4D, \tag{3.3.7}$$

where θ_0 is a notional water content such that $k(\theta_0) = V_0$ (see the boundary conditions below) and the Cole–Hopf transformation (see Sachdev 1987)

$$A\theta + B = A(\theta - \theta_n) = -2D(\ln u)_z, \tag{3.3.8}$$

we change (3.3.2) to the heat equation

$$u_\tau = u_{\zeta\zeta}. \tag{3.3.9}$$

The initial condition (3.3.3) at $\tau = 0$ via (3.3.8) becomes

$$(\ln u)_\zeta = 0 \tag{3.3.10}$$

and the boundary condition (3.3.6) at $\zeta = 0$ assumes the form

$$\frac{\partial u}{\partial \tau} = u. \tag{3.3.11}$$

The conditions (3.3.10) and (3.3.11), without loss of generality, may be written as

$$u = 1 \ \text{ at } \tau = 0 \tag{3.3.12}$$

and

$$u = \exp\tau \ \text{ at } \zeta = 0. \tag{3.3.13}$$

The solution of (3.3.9), subject to (3.3.12) and (3.3.13), may be found by the Laplace transform or otherwise (Carslaw and Jaeger 1959). The reduced water content

$$\Theta = \frac{\theta - \theta_n}{\theta_0 - \theta_n} \tag{3.3.14}$$

is related to u via (3.3.8),

$$\Theta = -\frac{1}{u}u_\zeta, \tag{3.3.15}$$

and may, therefore, be found from the solution of the heat equation and so on as

$$\Theta = \frac{f[(\zeta-2\tau)/2\tau^{1/2}] - f[(\zeta+2\tau)/2\tau^{1/2}]}{f[(\zeta-2\tau)/2\tau^{1/2}] + f[(\zeta+2\tau)/2\tau^{1/2}] - 2f[\zeta/2\tau^{1/2}] + 2\exp(\zeta^2/4\tau)}, \tag{3.3.16}$$

where

$$f[x] = (\exp x^2)\text{erfc}\,(x).$$

The derivation of (3.3.16) requires the following result,

$$\int e^{-a^2x^2 - b^2/x^2}\,dx = \frac{\sqrt{\pi}}{4a}\left[e^{2ab}\text{erf}(ax+b/x) + e^{-2ab}\text{erf}(ax-b/x)\right] + \text{constant}$$

(see Abramowitz and Stegun 1972, p. 304). Two interesting results follow from (3.3.16). First, the rise of water content at the surface $z = 0$ is easily found to be

$$\theta(0,\tau) - \theta_n = (\theta_0 - \theta_n)\text{erf}(\tau^{1/2}). \tag{3.3.17}$$

The asymptotic form as $\tau \to \infty$, 'the profile at infinity,' comes out from (3.3.16) to be simply

$$\lim_{\tau\to\infty}[\theta(\zeta,\tau) - \theta_n] = \frac{(\theta_0 - \theta_n)}{(1+\exp(\zeta-\tau))}. \tag{3.3.18}$$

The flux (see (3.3.6)), relative to the prescribed flux $V = V_0$ at $z = 0$, is

$$\frac{V}{V_0} = \tau^{-1}\int_0^\infty \frac{\theta(\zeta,\tau) - \theta_n}{\theta_0 - \theta_n}\,d\zeta \tag{3.3.19}$$

which, on using the profile at infinity, becomes

$$\frac{V}{V_0} = 1 + \frac{\ln[1+\exp(-\tau)]}{\tau}, \tag{3.3.20}$$

showing that $V \to V_0$ as $\tau \to \infty$. Equation (3.3.20) shows that, for $\tau > 2$, the profile at infinity satisfies the inequality $1 < V/V_0 < 1.0635$. The surface water content from the profile at infinity (3.3.18) is found to be

$$\lim_{\tau\to\infty}\Theta(0,\tau) = \Theta' = [1+\exp(-\tau)]^{-1} \tag{3.3.21}$$

and the exact solution (3.3.16) gives (3.3.17). If we call the latter Θ_0, we get the ratio

$$\frac{\Theta_0}{\Theta'} = \text{erf}(\tau^{1/2})[1+\exp(-\tau)]. \tag{3.3.22}$$

We conclude that $\Theta' \to \Theta_0$ as $\tau \to \infty$. It is easily seen that for all times $\tau > 2.0$, $1 < \Theta_0/\Theta' < 1.08319$. Thus the asymptotic profile at infinity is quite accurate for all times τ larger than 2.

Clothier et al. (1981) compared the full solution (3.3.16) with the profile at infinity (3.3.18) at different (nondimensional) times for $V_0 = 3.40 \times 10^{-6} \mathrm{ms}^{-1}$ and $k_s = 12 \times 10^{-6} \mathrm{ms}^{-1}$ for fine sand. The convergence of the general solution to the profile at infinity is quite remarkable. The 'wet-front' penetration was also found to be in good agreement with the field experimental data in spite of the simple assumptions regarding the soil, namely, a constant diffusivity and a quadratic conductivity–water content relationship referred to earlier.

A more general initial boundary value problem for the Burgers equation which includes the above work of Clothier et al. (1981) as a special case was mathematically treated by Joseph and Sachdev (1993) (see also Broadbridge et al. 1988, Broadbridge and Rogers 1990). Specifically, the Burgers equation

$$u_t + uu_x = \frac{\varepsilon}{2}u_{xx}, \quad x > 0, \ t > 0 \tag{3.3.23}$$

was solved subject to the initial condition

$$u(x,0) = u_0(x) \tag{3.3.24}$$

and the boundary condition

$$-\frac{\varepsilon}{2}u_x(0,t) + \frac{u^2(0,t)}{2} = \lambda(t) \tag{3.3.25}$$

where the functions $u_0(x)$ and $\lambda(t)$ are bounded; they are assumed in the form

$$u_0(x) = u_\infty + u_e(x), \tag{3.3.26}$$

$$\lambda(t) = \lambda_\infty + \lambda_e(t), \tag{3.3.27}$$

with

$$\lim_{x \to \infty} u_e(x) = 0, \quad \lim_{t \to \infty} \lambda_e(t) = 0; \tag{3.3.28}$$

u_∞ and λ_∞ in the above are constants. The functions $u_e(x)$ and $\lambda_e(t)$ are chosen such that the integrals

$$\bar{u} = \int_0^\infty u_e(x)dx, \quad \bar{\lambda} = \int_0^\infty \lambda_e(t)dt \tag{3.3.29}$$

exist. The solution $u^\varepsilon(x,t)$ of (3.3.23)–(3.3.29) is shown to be unique.

To obtain the asymptotic behaviour of the solution of (3.3.23)–(3.3.25) as $t \to \infty$, Joseph and Sachdev (1993) first solved this problem explicitly for the function

$$u^* = u - u_\infty. \tag{3.3.30}$$

Then, equation (3.3.23) becomes

$$u_t^* + \left[\left(\frac{u^*}{2} \right)^2 \right]_x + u_\infty u_x^* = \frac{\varepsilon}{2} u_{xx}^*. \tag{3.3.31}$$

Now we define the function

$$W(x,t) = - \int_x^\infty u^*(y,t)\,dy \tag{3.3.32}$$

so that (3.3.31) and (3.3.24) become

$$W_t + \frac{1}{2}\left(W_x^2 \right) + u_\infty W_x = \frac{\varepsilon}{2} W_{xx}, \quad x > 0,\ t > 0, \tag{3.3.33}$$

$$W(x,0) = - \int_x^\infty u_e(y)\,dy \tag{3.3.34}$$

(see (3.3.26) and (3.3.30)). The Cole–Hopf transformation

$$V(x,t) = \exp\left[-\frac{1}{\varepsilon} W(x,t) \right] \tag{3.3.35}$$

changes the problem (3.3.33) and (3.3.34) to

$$V_t + u_\infty V_x = \frac{\varepsilon}{2} V_{xx}, \quad x > 0,\ t > 0, \tag{3.3.36}$$

$$V(x,0) = \exp\left[\frac{1}{\varepsilon} \int_x^\infty u_e(y)\,dy \right]. \tag{3.3.37}$$

Now writing

$$Z = \exp\left(-\frac{u_\infty x}{\varepsilon} + \frac{u_\infty^2}{2\varepsilon} t \right) V \tag{3.3.38}$$

in (3.3.36) and (3.3.37), we obtain

$$Z_t = \frac{\varepsilon}{2} Z_{xx}, \quad x > 0,\ t > 0, \tag{3.3.39}$$

$$Z(x,0) = Z_0(x) = \exp\left\{ \frac{1}{\varepsilon} \left[\int_x^\infty u_e(y)\,dy - u_\infty x \right] \right\}, \tag{3.3.40}$$

$$Z(0,t) = Z_b(t). \tag{3.3.41}$$

$Z_0(x)$ and $Z_b(t)$ are now expressed in terms of the original variables. We find from (3.3.30), (3.3.32), (3.3.35), and (3.3.38) that

$$u_x = -\varepsilon \frac{(ZZ_{xx} - Z_x^2)}{Z^2}, \tag{3.3.42}$$

$$-\varepsilon u_x + u^2 = \varepsilon^2 \frac{(ZZ_{xx} - Z_x^2)}{Z^2} + \varepsilon^2 \frac{Z_x^2}{Z^2} = \varepsilon^2 \frac{Z_{xx}}{Z}. \tag{3.3.43}$$

Combining (3.3.25) and (3.3.43), we have

$$\varepsilon^2 Z_{xx}(0,t) = 2\lambda(t) Z(0,t). \tag{3.3.44}$$

We also have

$$\frac{\varepsilon}{2}Z_{xx}(0,t) = Z_t(0,t) \tag{3.3.45}$$

from (3.3.39) which, in view of (3.3.44), becomes

$$Z_t(0,t) = \frac{1}{\varepsilon}\lambda(t)Z(0,t). \tag{3.3.46}$$

On integration of (3.3.46), we have

$$Z(0,t) = \exp\left[\frac{1}{\varepsilon}\int_0^t \lambda(s)ds\right] Z(0,0). \tag{3.3.47}$$

From (3.3.29), (3.3.34), (3.3.35), and (3.3.38), we have

$$Z(0,0) = \exp\left[\frac{1}{\varepsilon}\bar{u}\right]. \tag{3.3.48}$$

We thus have the following initial boundary value problem for $Z = Z(x,t)$ (see (3.3.39), (3.3.40), (3.3.47), and (3.3.48)),

$$Z_t = \frac{\varepsilon}{2}Z_{xx}, \quad x > 0, \ t > 0, \tag{3.3.49}$$

$$Z(x,0) = Z_0(x), \tag{3.3.50}$$

$$Z(0,t) = Z_b(t), \tag{3.3.51}$$

where

$$Z_0(x) = \exp\left\{-\frac{1}{\varepsilon}\left[u_\infty x - \int_x^\infty u_e(y)dy\right]\right\}, \tag{3.3.52}$$

$$Z_b(t) = \exp\left\{\frac{1}{\varepsilon}\left[\lambda_\infty t + \int_0^t \lambda_e(s)ds + \bar{u}\right]\right\}. \tag{3.3.53}$$

We may recover $u^\varepsilon(x,t)$ from $Z(x,t)$ as follows.

$$\begin{aligned}
u^\varepsilon(x,t) &= u_\infty + u^*(x,t) = u_\infty + W_x = u_\infty - \varepsilon(\log V)_x \\
&= u_\infty - \varepsilon\left[\frac{u_\infty}{\varepsilon} + \frac{Z_x}{Z}\right] \\
&= -\varepsilon\frac{Z_x}{Z}.
\end{aligned} \tag{3.3.54}$$

Explicit solutions of the boundary value problem (3.3.49)–(3.3.51) for the heat equation is well known (see Carslaw and Jaeger 1959):

$$Z(x,t) = \frac{1}{(2\pi\varepsilon t)^{1/2}}\int_0^\infty \left[\exp\left(-\frac{(x-y)^2}{2t\varepsilon}\right) - \exp\left(-\frac{(x+y)^2}{2t\varepsilon}\right)\right] Z_0(y)dy$$

$$+ \frac{x}{(2\pi\varepsilon)^{1/2}}\int_0^t Z_b(s)\frac{\exp\left(-\dfrac{x^2}{2\varepsilon(t-s)}\right)}{(t-s)^{3/2}}ds. \tag{3.3.55}$$

When (3.3.55), with $Z_0(y)$ and $Z_b(y)$ given by (3.3.52) and (3.3.53), respectively, are substituted into (3.3.54), we get an explicit solution $u^\varepsilon(x,t)$ of (3.3.23)–(3.3.25).

To find the asymptotic form of $u^\varepsilon(x,t)$ as $t \to \infty$, we let $\varepsilon = 1$ for convenience. Thus, $u^1(x,t) \equiv u(x,t)$. We first write (3.3.55) in a more convenient form:

$$Z(x,t) = \frac{2}{\pi^{1/2}} \int_a^\infty Z_b\left(t - \frac{x^2}{2s^2}\right) e^{-s^2} ds + \frac{1}{\pi^{1/2}} \int_{-a}^\infty e^{-z^2} Z_0(x + (2t)^{1/2}z) dz$$

$$- \frac{1}{\pi^{1/2}} \int_a^\infty e^{-z^2} Z_0(-x + (2t)^{1/2}z) dz, \tag{3.3.56}$$

where $a = x/(2t)^{1/2}$.

Differentiating (3.3.56) with respect to x, we have

$$Z_x = -\frac{2}{\sqrt{\pi}} \int_a^\infty \exp(-s^2) Z_b'\left(t - \frac{x^2}{2s^2}\right) \frac{x}{s^2} ds$$

$$+ \frac{1}{\sqrt{\pi}} \int_a^\infty \exp(-z^2) Z_0'\left(z\sqrt{2t} - x\right) dz$$

$$+ \frac{1}{\sqrt{\pi}} \int_{-a}^\infty \exp(-z^2) Z_0'\left(z\sqrt{2t} + x\right) dz. \tag{3.3.57}$$

For $t \to \infty$, we may write

$$Z_b\left(t - \frac{x^2}{2s^2}\right) \approx \exp\left[\bar{\lambda} + \bar{u} + \lambda_\infty\left(t - \frac{x^2}{2s^2}\right)\right],$$

$$\frac{dZ_b}{dt}\left(t - \frac{x^2}{2s^2}\right) \approx \lambda_\infty \exp\left[\bar{\lambda} + \bar{u} + \lambda_\infty\left(t - \frac{x^2}{2s^2}\right)\right],$$

$$Z_0\left((2t)^{1/2}z \pm x\right) \approx \exp\left\{-u_\infty\left[(2t)^{1/2}z \pm x\right]\right\}; \tag{3.3.58}$$

see (3.3.52) and (3.3.53). Using (3.3.56)–(3.3.58) in (3.3.54), we have

$$u(x,t) \approx$$

$$2\lambda_\infty \int_a^\infty \frac{xe^{-s^2}}{s^2} \exp\left\{\bar{\lambda} + \bar{u} + \lambda_\infty\left(t - \frac{x^2}{2s^2}\right)\right\} ds$$

$$+ u_\infty \int_a^\infty \exp\left\{-[z^2 + u_\infty(\sqrt{2t}z - x)]\right\} dz + u_\infty \int_{-a}^\infty \exp\left\{-[z^2 + u_\infty(\sqrt{2t}z + x)]\right\} dz$$

$$\overline{2 \int_a^\infty e^{-s^2} \exp\left\{\bar{\lambda} + \bar{u} + \lambda_\infty\left(t - \frac{x^2}{2s^2}\right)\right\} ds - \int_a^\infty \exp\left\{-[z^2 + u_\infty(\sqrt{2t}z - x)]\right\} dz}$$

$$+ \int_{-a}^\infty \exp\left\{-[z^2 + u_\infty(\sqrt{2t}z + x)]\right\} dz.$$

$$\tag{3.3.59}$$

We restrict ourselves to the case $\lambda_\infty > 0$ (see (3.3.27)). To simplify (3.3.59), we observe the following identities,

$$e^{2(\lambda_\infty)^{1/2}x}\int_{a+b}^{\infty} e^{-y^2}\,dy = -\lambda_\infty^{1/2}\int_a^\infty \frac{x}{s^2}\exp\left[-\left(s^2+\frac{\lambda_\infty x^2}{s^2}\right)\right]ds + \int_a^\infty \exp\left[-\left(s^2+\frac{\lambda_\infty x^2}{s^2}\right)\right]ds, \tag{3.3.60}$$

$$e^{-2(\lambda_\infty)^{1/2}x}\int_{a-b}^{\infty} e^{-y^2}\,dy = \lambda_\infty^{1/2}\int_a^\infty \frac{x}{s^2}\exp\left[-\left(s^2+\frac{\lambda_\infty x^2}{s^2}\right)\right]ds + \int_a^\infty \exp\left[-\left(s^2+\frac{\lambda_\infty x^2}{s^2}\right)\right]ds, \tag{3.3.61}$$

where $b = (2\lambda_\infty t)^{1/2}$. From (3.3.60) and (3.3.61) we also find that

$$\int_a^\infty \exp\left[-\left(s^2+\lambda_\infty\frac{x^2}{s^2}\right)\right]ds = \frac{1}{2}\left[\exp\left[2(\lambda_\infty)^{1/2}x\right]\int_{a+b}^\infty e^{-y^2}\,dy + \exp\left[-2(\lambda_\infty)^{1/2}x\right]\int_{a-b}^\infty e^{-y^2}\,dy\right], \tag{3.3.62}$$

$$\int_a^\infty \frac{x}{s^2}\exp\left[-\left(s^2+\lambda_\infty\frac{x^2}{s^2}\right)\right]ds = \frac{1}{2(\lambda_\infty)^{1/2}}\left[\exp\left[-2(\lambda_\infty)^{1/2}x\right]\int_{a-b}^\infty e^{-y^2}\,dy - \exp\left[2(\lambda_\infty)^{1/2}x\right]\int_{a+b}^\infty e^{-y^2}\,dy\right]. \tag{3.3.63}$$

Moreover, by a simple change of variable, we have

$$\int_{\pm a}^\infty \exp\left\{-\left[z^2+u_\infty\left[(2t)^{1/2}z\mp x\right]\right]\right\}dz = \exp\left(\pm u_\infty x+\frac{u_\infty^2 t}{2}\right)\int_{\pm a+c}^\infty e^{-y^2}\,dy, \tag{3.3.64}$$

where $c = (t/2)^{1/2}u_\infty$.

Now, using (3.3.60)–(3.3.64) in (3.3.59), we get the asymptotic form

$$u(x,t) \approx \frac{\sqrt{2\lambda_\infty}\,e^{(\bar\lambda+\bar u+\lambda_\infty t)}\left\{\exp\left(-\sqrt{2\lambda_\infty}x\right)\int_{a-b_1}^\infty e^{-y^2}dy - \exp\left(\sqrt{2\lambda_\infty}x\right)\int_{a+b_1}^\infty e^{-y^2}dy\right\} + u_\infty\exp\left(\frac{tu_\infty^2}{2}+xu_\infty\right)\int_{a+c}^\infty e^{-y^2}dy + u_\infty\exp\left(\frac{tu_\infty^2}{2}-xu_\infty\right)\int_{-a+c}^\infty e^{-y^2}dy}{e^{(\bar\lambda+\bar u+\lambda_\infty t)}\left\{\exp\left(-\sqrt{2\lambda_\infty}x\right)\int_{a-b_1}^\infty e^{-y^2}dy + \exp\left(\sqrt{2\lambda_\infty}x\right)\int_{a+b_1}^\infty e^{-y^2}dy\right\} + \exp\left(\frac{tu_\infty^2}{2}-xu_\infty\right)\int_{-a+c}^\infty e^{-y^2}dy - \exp\left(\frac{tu_\infty^2}{2}+xu_\infty\right)\int_{a+c}^\infty e^{-y^2}dy}; \tag{3.3.65}$$

here $b_1 = (\lambda_\infty t)^{1/2}$. Three cases arise depending on whether $u_\infty = 0$, $u_\infty < 0$, or $u_\infty > 0$. Here we consider the case $u_\infty = 0$. In this case, (3.3.65) simplifies to

$$u(x,t) \approx$$

$$\frac{\sqrt{2\lambda_\infty}\, e^{(\bar{\lambda}+\bar{u}+\lambda_\infty t)} \left\{ e^{(-\sqrt{2\lambda_\infty}x)} \int_{a-b_1}^\infty e^{-y^2} dy - e^{(\sqrt{2\lambda_\infty}x)} \int_{a+b_1}^\infty e^{-y^2} dy \right\}}{e^{(\bar{\lambda}+\bar{u}+\lambda_\infty t)} \left\{ e^{(-\sqrt{2\lambda_\infty}x)} \int_{a-b_1}^\infty e^{-y^2} dy + e^{(\sqrt{2\lambda_\infty}x)} \int_{a+b_1}^\infty e^{-y^2} dy \right\} + 2 \int_0^a e^{-y^2} dy}.$$

$$(3.3.66)$$

Setting

$$\xi = \sqrt{2\lambda_\infty}x \quad \text{and} \quad \tau = \lambda_\infty t \qquad (3.3.67)$$

so that

$$\frac{x}{(2t)^{1/2}} \pm (\lambda_\infty t)^{1/2} = \frac{\xi}{2\tau^{1/2}} \pm \tau^{1/2} \qquad (3.3.68)$$

and observing the asymptotic properties of the complementary error function, namely,

$$\int_p^\infty e^{-y^2} dy \approx \frac{1}{2p} e^{-p^2} \quad \text{as } p \to \infty,$$

$$\int_p^\infty e^{-y^2} dy \approx \pi^{1/2} - \frac{1}{2p} e^{-p^2} \quad \text{as } p \to -\infty, \qquad (3.3.69)$$

(3.3.66) may be simplified to yield

$$u(x,t) \approx (2\lambda_\infty)^{1/2} \frac{e^{(\bar{\lambda}+\bar{u}+\tau)} \left\{ e^{-\xi} \pi^{1/2} + o(1) \right\}}{e^{(\bar{\lambda}+\bar{u}+\tau)} \left\{ e^{-\xi} \pi^{1/2} + o(1) \right\} + \pi^{1/2} + o(1)}; \qquad (3.3.70)$$

that is,

$$u(x,t) \approx \frac{(2\lambda_\infty)^{1/2}}{1 + e^{(\xi - \tau - \bar{\lambda} - \bar{u})}} \qquad (3.3.71)$$

as $\tau \to \infty$, where $\xi = (2\lambda_\infty)^{1/2}x$, $\tau = \lambda_\infty t$, and $\bar{\lambda}$ and $\bar{u}$ are given by (3.3.29). Similar expressions may be obtained for the cases $u_\infty < 0$ and $u_\infty > 0$ (see Joseph and Sachdev 1993). If we put $\lambda_\infty = 1/2$, $\bar{\lambda} = 0$, $\bar{u} = 0$ in (3.3.71), we recover the simple profile at infinity (3.3.18), obtained earlier by Clothier et al. (1981).

We may also refer to some related studies here. Weidman (1976) considered spin-up and spin-down of a rotating fluid. He studied (essentially) the Burgers equation in the form

$$v_\tau + (1 - v)v_x = E_\Omega^{1/2} v_{xx}, \qquad (3.3.72)$$

subject to the initial boundary conditions $v(x,0) = 0, v(0,\tau) = 1$. Here, E_Ω is a constant. The solution could be found explicitly in terms of complementary error functions.

Calogero and De Lillo (1991) attempted to solve the Burgers equation

$$u_t = u_{xx} + 2uu_x, \tag{3.3.73}$$

subject to the conditions

$$u(x,0) = u_0(x), \quad 0 \leq x < \infty, \tag{3.3.74}$$

$$H[u(0,t), u_x(0,t); t] = 0, \quad t \geq 0. \tag{3.3.75}$$

Using a generalised Cole–Hopf transformation, the problem (3.3.73)–(3.3.75) was first reduced to that for the heat equation, subject to two boundary conditions at $x = 0$. Further analysis led to a nonlinear integrodifferential equation which for the special case, $H[u(0,t), u_x(0,t); t] \equiv a(t)u(0,t) + b(t)(u_x(0,t) + u^2(0,t)) - F(t)$, reduces to a linear integral equation of Volterra type. This equation can be solved by quadratures if $a(t)/F(t) = c_1$ and $b(t)/F(t) = c_2$, where c_1 and c_2 are constants. Asymptotic behaviour of these solutions may be obtained in the manner of Joseph and Sachdev (1993).

3.4 Travelling waves describing flow in a porous medium

In this section we consider a more general model describing infiltration of water into a homogeneous soil, pose an initial boundary value problem for the same, and show rigorously how a travelling wave emerges as an intermediate asymptotic. In the present case, unlike for problems in Sections 3.2 and 3.3, it does not seem possible to construct an explicit solution of the initial boundary value problem; the asymptotic character of the travelling wave is demonstrated in a rigorous qualitative manner. Large time numerical solution of the problem is also obtained to confirm the analytic results. We follow here the work of Vanaja and Sachdev (1992). Let u denote the volume of water per unit volume of porous medium and q, the flux, that is, the volume of water flowing across the unit area per unit time. Assuming the density of water to be constant, the equation of continuity is

$$u_t - q_x = 0, \tag{3.4.1}$$

where the flux q is given by Darcy's law,

$$q = K(u)G_x. \tag{3.4.2}$$

x and t denote the space coordinate, measured positive downward from the surface, and time, respectively. $K(u)$ is the hydraulic conductivity. Under certain conditions applicable to unsaturated flows, the potential G is written out as the sum of a gravitational potential and a potential $H(u)$ due to capillary suction. Thus, we have

$$G = H(u) - x \tag{3.4.3}$$

so that (3.4.1) becomes

$$u_t = (K(u)H_x)_x - (K(u))_x. \tag{3.4.4}$$

With

$$D(u) = K(u)H_u,$$

equation (3.4.4) becomes

$$u_t = (D(u)u_x)_x - (K(u))_x \tag{3.4.5}$$

(see Section 3.8). We consider the special power law case (see Bear 1972, Philip 1970) for which

$$D(u) = D_0 u^{m-1}, \quad K(u) = K_0 u^n, \tag{3.4.6}$$

where D_0, K_0, m, and n are positive constants and $n \geq m > 1$. Scaling the variables suitably, we may now write (3.4.5) as

$$u_t = (u^m)_{xx} - (u^n)_x, \quad n \geq m > 1. \tag{3.4.7}$$

This is a standard porous medium equation. Taking into account the initial moisture distribution in the soil and infiltration on the surface of the ground, we may impose the following initial and boundary conditions for $u(x,t)$.

$$u(0,t) = u_1, \quad \text{a constant, for } 0 \leq t < \infty,$$
$$u(x,0) = u_0(x) \text{ for } 0 \leq x < \infty, \tag{3.4.8}$$
$$u_0 \leq u_0(x) \leq u_1, \quad \lim_{x \to \infty} u_0(x) = u_0.$$

It is also assumed that there is no water at a large depth beneath the ground. $u_1 = 1$ corresponds to full saturation of the soil on the surface of the ground.

We now show that the travelling wave solutions of (3.4.5) are intermediate asymptotics for the initial boundary value problem (3.4.5) and (3.4.8). We assume that the functions $D(u)$, $D'(u)$, $K(u)$, $K'(u)$, and $K''(u) > 0$ exist and are continuous and bounded for $u \geq u_0 > 0$; here u_0 is a constant. Let $E = (0, \infty) \times (0, \infty)$.

We may observe that, although equation (3.4.5) is more general than the Burgers equation, the boundary conditions considered by Joseph and Sachdev (1993) were more general than those in (3.4.8) (see Section 3.3). Besides, the intermediate asymptotic being studied here is a travelling wave, different from that in Section 3.3.

We first consider travelling wave solution $u = U(\eta)$ of (3.4.5), which depends only on $\eta = x - At + c$, where $A > 0$ and c are constants. Thus, (3.4.5) reduces to the ODE

$$(D(U)U_\eta)_\eta - (K(U))_\eta = -AU_\eta. \tag{3.4.9}$$

We impose the following end conditions on U at $\pm\infty$,

$$U(-\infty) = u_1, \quad U(+\infty) = u_0. \tag{3.4.10}$$

Integrating (3.4.9) and using the end conditions (3.4.10), we have

$$D(U)U_\eta - K(U) + K(u_0) = -A(U - u_0), \tag{3.4.11}$$

where

$$A = \frac{K(u_1) - K(u_0)}{(u_1 - u_0)}. \tag{3.4.12}$$

Integration of (3.4.11) yields

$$x - At + c = \int_{u_2}^{U} \frac{D(u)}{[K'(u_0 + \theta(u - u_0)) - A][u - u_0]} du, \tag{3.4.13}$$

where $u_0 < U(0) = u_2 < u_1$ and $0 < \theta(u) < 1$. Thus, (3.4.13) gives the travelling wave solution of (3.4.5) subject to end conditions (3.4.10). From (3.4.13), we observe that

$$\frac{dU}{d\eta} = \frac{[K'(u_0 + \theta(U - u_0)) - A](U - u_0)}{D(U)}. \tag{3.4.14}$$

If we expand $K(U)$ about u_0, we find that $K'(u_0 + \theta(U - u_0)) \leq A$ (see (3.4.12)). Because $D(U) \geq 0, U - u_0 \geq 0$, we conclude that $dU/d\eta \leq 0$ and the travelling wave solution $U(\eta)$ is a monotonically decreasing function of η. For the special choice $D(u) = mu^{m-1}$, $K(u) = u^n$, $u_0 < u_2 < u_1$, and $A > 0$, (3.4.13) becomes

$$x - At + c = \int_{u_2}^{U} \frac{mu^{m-1}[u_1 - u_0]}{[(u_1 - u_0)(u^n - u_0^n) - (u - u_0)(u_1^n - u_0^n)]} du. \tag{3.4.15}$$

Here, $U = u_2$ when $\eta = 0$. Denoting by E^+ the domain bounded by $x = 0$, $t = 0$, $t = T$, where T is a finite positive number, we have the following minimum principle due to Krzyżański (1959). Let $z(x,t)$ be a bounded solution of the differential inequality

$$a(x,t)z_{xx} + b(x,t)z_x + c(x,t)z - z_t \leq 0 \tag{3.4.16}$$

in E^+. Let $a(x,t), b(x,t)$, and $c(x,t)$ be bounded continuous functions of x and t and $a(x,t) > 0$. If $z \geq 0$ on $x = 0$ and $t = 0$, then $z \geq 0$ in E^+.

To prove the main theorem regarding the approach of the solution of (3.4.5) and (3.4.8) to the travelling wave, we need several intermediate steps.

First we prove that if u and v are two solutions of (3.4.5) and (3.4.8) in E and $u \leq v$ on $x = 0$ and $t = 0$, then $u \leq v$ in E. Let $\bar{u} = \int_0^u D(s)ds$ and $\bar{v} = \int_0^v D(s)ds$, then $\bar{u}$ and $\bar{v}$ satisfy

$$\bar{u}_t = D(u)\bar{u}_{xx} - K'(u)\bar{u}_x, \quad \bar{v}_t = D(v)\bar{v}_{xx} - K'(v)\bar{v}_x. \tag{3.4.17}$$

Setting $w = \bar{v} - \bar{u}$, we find that w satisfies

$$w_t = D(v)w_{xx} - K'(v)w_x + \bar{u}_{xx}[D(v) - D(u)] - \bar{u}_x[K'(v) - K'(u)]. \tag{3.4.18}$$

Because $D(v) - D(u) = (v-u)D'(\theta_1)$ and $K'(v) - K'(u) = (v-u)K''(\theta_2)$, where $\theta_1 = \theta_1(x,t)$ and $\theta_2 = \theta_2(x,t)$ lie between u and v, and because we may also write $\bar{v} - \bar{u} = (v-u)D(\theta_3)$, where again $u < \theta_3(x,t) < v$, we may put (3.4.18) as

$$w_t = D(v)w_{xx} - K'(v)w_x + w\left[\frac{\bar{u}_{xx}(v-u)D'(\theta_1)}{\bar{v}-\bar{u}} - \frac{\bar{u}_x(v-u)K''(\theta_2)}{\bar{v}-\bar{u}}\right]$$

or

$$D(v)w_{xx} - K'(v)w_x + \xi(x,t)w - w_t = 0, \qquad (3.4.19)$$

where

$$\xi(x,t) = \left[\frac{\bar{u}_{xx}D'(\theta_1)}{D(\theta_3)} - \frac{\bar{u}_x K''(\theta_2)}{D(\theta_3)}\right]. \qquad (3.4.20)$$

We observe that D, K', and $\xi(x,t)$ are bounded in E. Because $D > 0$ and $w = (v-u)$ $D(\theta_3) \geq 0$ on $x = 0$ and $t = 0$, w satisfies the conditions of the minimum principle (see (3.4.16) etc.). We infer that $w \geq 0$ in E^+ (see below (3.4.15)). The above argument is independent of the choice of T, therefore we infer that $u \leq v$ in E for all $t \geq 0$.

As a corollary, we easily check that if $u(x,t)$ is the solution of (3.4.5) and (3.4.8), then $u_0 \leq u(x,t) \leq u_1$ for each (x,t) in E. This is accomplished by applying the above result first to the functions $u(x,t)$ and u_1, leading to $u(x,t) \leq u_1$. The other inequality can be obtained in a similar fashion.

Now we show that $u(x,t)$ is bounded between two travelling wave solutions (see (3.4.22) below). Let $u(x,t)$ be the solution of (3.4.5) and (3.4.8) and let the initial condition $u_0(x)$ satisfy the inequality

$$u_0(x) - u_0 \leq M_1 e^{-\gamma_1 x}, \qquad (3.4.21)$$

where $\gamma_1 > |K'(u_0) - A|/D(u_0)$ is a positive constant; M_1 is another positive constant specified later. Then there exist travelling wave solutions $U_1(x - \lambda_1 t + c_1)$ and $U_2(x - \lambda_2 t - c_2)$ of (3.4.5) with the constants $\lambda_1, \lambda_2, c_1, c_2 > 0$, satisfying the conditions $U_1(-\infty) = u_1$, $U_1(+\infty) = u_0 - \varepsilon(\varepsilon > 0)$, $U_2(-\infty) = u_1 + \varepsilon$, and $U_2(+\infty) = u_0$ such that the following inequalities hold,

$$U_1(x - \lambda_1 t + c_1) \leq u(x,t) \leq U_2(x - \lambda_2 t - c_2) \qquad (3.4.22)$$

for every (x,t) in E.

First we observe that $\lambda_2 = [K(u_1 + \varepsilon) - K(u_0)]/(u_1 + \varepsilon - u_0) > A$ (see (3.4.12)). Let $m_2 = |K'(u_0) - \lambda_2|/D(u_0)$. We then choose $\varepsilon > 0$ sufficiently small such that $\gamma_1 \geq m_2 > |K'(u_0) - A|/D(u_0)$. This is possible in view of the inequality following (3.4.21) for γ_1.

The travelling wave solution U and the function $u_0(x)$ both decrease with x, thus we may find travelling wave solutions U_2 and U_1 satisfying the end conditions in the statement below (3.4.21) such that

$$u_0(x) \leq U_2(x - c_2) \quad \text{for each } x > 0, \qquad (3.4.23)$$

$$u_0(x) \geq U_1(x+c_1) \quad \text{for each } x > 0. \tag{3.4.24}$$

Also, we have $U_1 \leq u_1 \leq U_2$ at $x = 0$. The result (3.4.22) for the region E now follows from the statement made above (3.4.17).

Recalling the behaviour of U_1 and U_2 as $x \to \infty$, we have $\lim_{x \to \infty} u(x,t) = u_0$ for any finite t.

Now we proceed to prove the main result regarding the rate at which the solution of the IBVP (3.4.5) and (3.4.8) approaches the travelling wave solution (Vanaja and Sachdev 1992).

Theorem 3.4.1 *Let $u(x,t)$ be the solution of (3.4.5) and (3.4.8) and let $U(x-At+c)$ be a travelling wave solution of (3.4.5). Moreover, let the initial condition $u_0(x)$ satisfy the inequality (3.4.21). Then there exist constants $M, l > 0$, and c such that*

$$|u(x,t) - U(x-At+c)| \leq Me^{-lt} \tag{3.4.25}$$

for each (x,t) in E.

First we choose the constant c such that $K(u_0) - Au_0 + c = K(u_1) - Au_1 + c = 0$. We also write $\bar{u} = \int_0^u D(s)ds, \bar{U} = \int_0^U D(s)ds$ and let $y = \bar{u} - \bar{U}$. Then, y satisfies the equation

$$L(y) = D(u)y_{xx} - K'(u)y_x + \beta(x,t)y - y_t = 0 \tag{3.4.26}$$

(cf. (3.4.19)), where

$$\beta(x,t) = \left[\frac{\bar{U}_{xx}D'(\theta_4)}{D(\theta_6)} - \frac{\bar{U}_x K''(\theta_5)}{D(\theta_6)} \right] \tag{3.4.27}$$

and $\theta_4(x,t), \theta_5(x,t)$, and $\theta_6(x,t)$ all lie between u and U. We now show that $\beta(x,t)$ is bounded. We write $U_\eta = dU/d\eta$. Because $\eta = x - At + c$, and $\bar{U}_x = D(U)U_\eta$, we have

$$\bar{U}_{xx} = \frac{d}{d\eta}[D(U)U_\eta] = \frac{d}{d\eta}[K(U)] - AU_\eta = [K'(U) - A]U_\eta, \tag{3.4.28}$$

where we have used (3.4.9). Thus, (3.4.27) becomes

$$\beta(x,t) = (-U_\eta) \left[\frac{(A - K'(U))D'(\theta_4)}{D(\theta_6)} + \frac{D(U)K''(\theta_5)}{D(\theta_6)} \right]. \tag{3.4.29}$$

Because $U_\eta = [K'(u_0 + \theta(U - u_0)) - A](U - u_0)/D(U)$, where $K'(u_0 + \theta(U - u_0)) \leq A$, $U \leq u_1$, and $D(U) \geq D(u_0)$ for all U, we get $|U_\eta| \leq 2A(u_1 - u_0)/D(u_0)$. Thus, we get from (3.4.29) the inequality

$$|\beta| \leq \left[\frac{2A(u_1 - u_0)}{D(u_0)} \right] \frac{[2AD' + D(u_1)K'']}{D(u_0)}. \tag{3.4.30}$$

We choose the comparison function

$$z(x,t) = e^{-lt}\omega(\eta) \tag{3.4.31}$$

in the manner of Peletier (1970), where $\omega(\eta)$ is a positive, continuous, and piece-wise differentiable function to be chosen such that the inequality $L(z) \le 0$ holds for all η (see (3.4.26)). We first observe that

$$L(z) = z \left[\frac{D(u)\omega''}{\omega} + \frac{(A - K'(u))\omega'}{\omega} + \beta(x,t) + l \right]. \tag{3.4.32}$$

As in the original work of Khusnytdinova (1967), we let

$$\omega(\eta) = \begin{cases} e^{-\alpha \exp(\lambda \eta)}, & |\eta| < N \\ e^{-\alpha \eta}, & |\eta| \ge N, \end{cases} \tag{3.4.33}$$

where $\alpha = (1/k)\exp(-\lambda N), k \ge 1$; λ and N are specified presently. It is clear that

$$\frac{\omega'}{\omega} = \begin{cases} -\alpha \lambda e^{\lambda \eta} & \text{for } |\eta| < N, \\ -\alpha & \text{for } |\eta| > N \end{cases} \tag{3.4.34}$$

and

$$\frac{\omega''}{\omega} = \begin{cases} \alpha^2 \lambda^2 e^{2\lambda \eta} - \alpha \lambda^2 e^{\lambda \eta} & \text{for } |\eta| < N, \\ \alpha^2 & \text{for } |\eta| > N. \end{cases} \tag{3.4.35}$$

Now we show that with the above choice of $\omega(\eta), L(z) \le 0$ for all $|\eta| \ne N$. Referring to (3.4.14), we have

$$U_\eta = \frac{[K'(u_0 + \theta(U - u_0)) - A](U - u_0)}{D(U)}. \tag{3.4.36}$$

Also $U \to u_0$ as $\eta \to \infty$ and $[K'(u_0 + \theta(U - u_0)) - A] \to 0$ as $\eta \to -\infty$. We choose N so large that, for $|\eta| \ge N, \beta$ given by (3.4.29) is small.

We consider $L(z)$ for $|\eta| < N$ and $|\eta| > N$ separately. For the former we have

$$D(u)\frac{\omega''}{\omega} + (A - K'(u))\frac{\omega'}{\omega} + \beta + l = D(u)\lambda^2 \left[\frac{1}{k^2 e^{2\lambda(N-\eta)}} - \frac{1}{k e^{\lambda(N-\eta)}} \right]$$
$$- \frac{[A - K'(u)]\lambda}{k e^{\lambda(N-\eta)}} + \beta + l. \tag{3.4.37}$$

Because $D(u) \ge D(u_0)$ and $K'(u) \le K'(u_1)$, by assumption, and β is bounded, we may choose λ so large that

$$l_1 = D(u_0)\lambda^2 \left[\frac{1}{k e^{\lambda(N-\eta)}} - \frac{1}{k^2 e^{2\lambda(N-\eta)}} \right] + \frac{[A - K'(u_1)]\lambda}{k e^{\lambda(N-\eta)}} - \beta \ge 0. \tag{3.4.38}$$

Choosing $l \le l_1$, we have from (3.4.37) and (3.4.38) the inequality

$$L(z) \le 0 \quad \text{for } |\eta| < N. \tag{3.4.39}$$

Now we consider the interval $|\eta| > N$. Here,

$$D(u)\frac{\omega''}{\omega} + (A - K'(u))\frac{\omega'}{\omega} + \beta + l = \frac{D(u)}{k^2 e^{2\lambda N}} - \frac{(A - K'(u))}{K e^{\lambda N}} + \beta + l. \quad (3.4.40)$$

Because $K'(u) \leq A$, we let $\sup K'(u)$ be equal to s. We have already shown that β can be made arbitrarily small for $|\eta| \geq N$. Moreover, $D(u) \leq D(u_1)$. We may, therefore, choose k so large that

$$\frac{1}{k^2 e^{2\lambda N}} + \frac{\beta}{D(u_1)} \leq \frac{|A - s|}{D(u_1) k e^{\lambda N}}. \quad (3.4.41)$$

This would, in view of (3.4.40), ensure that

$$l_2 = \frac{(A - s)k e^{\lambda N} - D(u_1)}{k^2 e^{2\lambda N}} - \beta \geq 0. \quad (3.4.42)$$

The inequality (3.4.41), together with that following (3.4.21), imply that

$$\frac{1}{k e^{\lambda N}} < \frac{|A - s|}{D(u_1)} \leq \frac{|A - K'(u_0)|}{D(u_0)} < \gamma_1, \text{ say.} \quad (3.4.43)$$

Now we let $l \leq l_2$. For this choice of l, we have from (3.4.40)–(3.4.43) the inequality

$$L(z) \leq 0 \text{ for } |\eta| > N. \quad (3.4.44)$$

With $l = \min(l_1, l_2)$ in (3.4.31), we have shown that $L(z) \leq 0$ for all $\eta, |\eta| \neq N$.

For the final part of the proof, we let

$$\phi(x, t) = N_1 z(x, t) - y(x, t), \quad (3.4.45)$$

where the constant N_1 is chosen presently. We observe from (3.4.31) and (3.4.33) that $z(0, t) \geq 0$; moreover, $y(0, t) = \int_{U(0,t)}^{u_1} D(s)ds \geq 0$ because $U(0, t) \leq u_1$. We also have $\phi(x, 0) = N_1 z(x, 0) - y(x, 0)$, where $z(x, 0) > 0$. We may now choose N_1 sufficiently large that both $\phi(0, t)$ and $\phi(x, 0)$ are greater than or equal to zero. We observe that

$$L(\phi) = N_1 L(z) - L(y) = N_1 L(z) \leq 0 \text{ for all } \eta \quad (3.4.46)$$

in view of the fact that $L(y) = 0$ and $L(z) \leq 0$ for all η. Therefore the minimum principle (see (3.4.16) etc.) ensures that $\phi \geq 0$ everywhere in E^+, implying that

$$y(x, t) \leq N_1 z(x, t) \text{ in } E^+. \quad (3.4.47)$$

We may similarly show by using the function $\psi = N_2 z + y$, where N_2 is a constant, that

$$y(x, t) \geq -N_2 z(x, t) \text{ in } E^+. \quad (3.4.48)$$

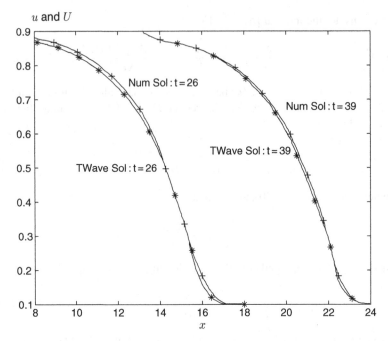

Fig. 3.3 Numerical solution of (3.4.5) subject to the initial condition (3.4.8) (+++) and travelling wave solution given by (3.4.9) and (3.4.10) (***) at $t = 26, 39$. Here, $D(u) = u$, $K(u) = u^2/2$, and $u_0(x) = 0.1 + (1.6/(1 + e^{10x}))$, $u_1 = 0.9$, $u_0 = 0.1$. (Vanaja and Sachdev 1992. Copyright © 1992 Brown University and American Mathematical Society. Reprinted with permission. All rights reserved.)

Combining (3.4.47) and (3.4.48) we have

$$-N_2 z(x,t) \leq y(x,t) \leq N_1 z(x,t) \text{ in } E^+. \tag{3.4.49}$$

We, therefore, infer that

$$|y| = |\bar{u} - \bar{U}| \leq |z| \max(N_1, N_2) = \left[\frac{|\omega(\eta)|}{e^{lt}}\right] \max(N_1, N_2). \tag{3.4.50}$$

We may check that

$$|u - U| = \frac{|\bar{u} - \bar{U}|}{|D(\theta)|}$$

$$\leq \left[\frac{|\omega(\eta)| \max(N_1, N_2)}{D(u_0)}\right] e^{-lt}$$

$$\leq \left[\frac{\max(N_1, N_2)}{D(u_0)} \right] e^{-lt}, \tag{3.4.51}$$

where we have used the results, $|\bar{u} - \bar{U}| = |(u - U)|D(\theta(x,t))$, θ lying between u and U, and $|\omega(\eta)| \leq 1$. Finally, if we write $M = \max(N_1, N_2)/D(u_0)$, we arrive at the estimate

$$|u - U| \leq M e^{-lt} \text{ in } E^+. \tag{3.4.52}$$

M does not depend on the choice of T, therefore (3.4.52) holds for all $t > 0$. We have thus vindicated the manner in which the solution u approaches the travelling wave U as $t \to \infty$. Vanaja and Sachdev (1992) carried out an extensive numerical study to confirm the above analytical results. The following cases were studied: (i) $D(u) = u$, $K(u) = u^2/2$; (ii) $D(u) = 2u$, $K(u) = u^3$; (iii) $D(u) = 2u$, $K(u) = u^4$. The far off conditions for all these cases were chosen to be $u_1 = 0.9$ and $u_0 = 0.1$. The initial condition for case (i), for example, was chosen to be $u_0(x) = 0.1 + (1.6/(1 + \exp(10x)))$. Figures 3.3–3.5 show numerical solutions for different times as well as the travelling wave solutions for different choices of $D(u)$ and $K(u)$. The numerical study of (3.4.5) and (3.4.8) clearly brings out the intermediate asymptotic character of the travelling wave solutions given by (3.4.9) and (3.4.10) (see Vanaja and Sachdev 1992 for details).

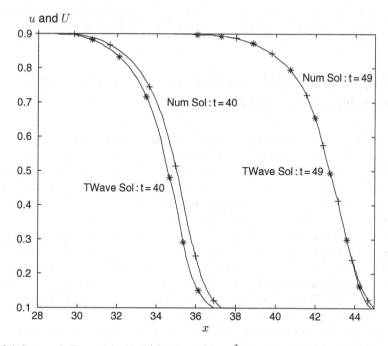

Fig. 3.4 Same as in Figure 3.3 with $D(u) = 2u$, $K(u) = u^3$. (Vanaja and Sachdev 1992. Copyright © 1992 Brown University and American Mathematical Society. Reprinted with permission. All rights reserved.)

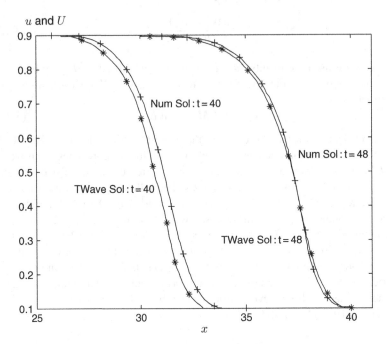

Fig. 3.5 Same as in Figure 3.3 with $D(u) = 2u$, $K(u) = u^4$. (Vanaja and Sachdev 1992. Copyright © 1992 Brown University and American Mathematical Society. Reprinted with permission. All rights reserved.)

3.5 Evolution of a stable profile describing cross-field diffusion in toroidal multiple plasma

In a series of papers, Berryman (1977) and Berryman and Holland (1978a, b, 1980, 1982) studied a class of nonlinear diffusion equations with fixed boundaries which describe particle diffusion across magnetic fields in the Wisconsin toroidal octupole plasma-containing device (see Drake 1973, Drake et al. 1977, Greenwood 1975). Their experiments involved a purely poloidal field and showed that after a few seconds the density profile evolves into a fixed shape, the so-called normal mode, and then simply decays with time. This suggested a separable form of the asymptotic solution which was later mathematically shown to be the case. It is the intermediate asymptotic character of this separable solution of an initial boundary value problem that we are concerned with here. First we briefly discuss the first paper in this series by Berryman (1977) and then detail a later study by Berryman and Holland (1982); the latter work has some interesting rigourous analysis.

In the standard form involving normalised quantities, we discuss the equation

$$F(x)n_t = (D(n)n_x)_x, \quad 0 \le x \le 1, \tag{3.5.1}$$

where n is the particle density, x is the spatial variable in one-dimension, and t is the time. The function $F(x)$ is positive and describes the geometry of the octupole. Assuming

$$D(n) = (1+\delta)n^\delta \text{ for } \delta > -1 \qquad (3.5.2)$$

and introducing the function

$$m(x,t) = n^{1+\delta} \text{ for } \delta > -1 \qquad (3.5.3)$$

in (3.5.1), we have

$$F(x)(m^{q-1})_t = m_{xx}, \qquad (3.5.4)$$

where $q = (2+\delta)/(1+\delta)$. It is clear that, for $\delta > -1$, $0 \le m(x,t) < \infty$ when $0 \le n(x,t) < \infty$. m may be referred to as the pseudo-density. The function $F(x)$, from physical considerations, is assumed to satisfy the conditions

$$F(x) > 0 \text{ for } 0 \le x \le 1 \qquad (3.5.5)$$

and

$$\int_0^1 F(x)dx < \infty. \qquad (3.5.6)$$

Thus, $F(x)$ may have a singularity at some $x = x_s$; it was found consistent and convenient to assume that $F'(x) \ge 0$ for $x < x_s$ and $F'(x) \le 0$ for $x > x_s$. The experimental results show that, for large times, the situation in the toroidal octupole is well represented by taking $n = 0$ at the boundaries so that

$$m(0,t) = m(1,t) = 0. \qquad (3.5.7)$$

For $\delta > -1$, (3.5.7) is consistent with the physical requirement of a finite flux.

Berryman (1977) assumed the solution of the BVP (3.5.4)–(3.5.7) in the form

$$m(x,t) = S(x)T(t) \qquad (3.5.8)$$

and showed that this separable solution evolves from an arbitrary initial distribution of particles. He then demonstrated, both analytically and numerically, the evolution and stability of this solution. Indeed, for the case for which initial particle distribution vanishes only at the boundaries, an approximate analysis showed that the perturbations decay exponentially causing a rapid evolution to the separable solution. We refer the reader to Berryman (1977) for details.

We discuss here in some detail another interesting paper by Berryman and Holland (1982) relating to the equation

$$n_t = D\left(\frac{1}{n}n_x\right)_x, \qquad (3.5.9)$$

where D is a constant. This equation also appears in other applications such as an expansion of a thermalised electron cloud. It is a special case of (3.5.1) with $F(x) =$

D^{-1}, a constant, and $D(n) = 1/n$. In the present case, the boundary conditions at the two ends of the interval, $0 \leq x \leq 1$, are assumed to be

$$n(0,t) = n(1,t) = n_0, \tag{3.5.10}$$

where $n_0 \neq 0$ is a small value of the density which may be thought of as the background value. Furthermore, the initial data $n(x,0) \geq n_0$. Introducing the variable

$$m(x,t) = \ln(n/n_0),$$

where m is nonnegative, one may write (3.5.9) and (3.5.10) as

$$(e^m)_t = m_{xx}, \quad 0 \leq x \leq 1, \tag{3.5.11}$$

and

$$m(0,t) = m(1,t) = 0. \tag{3.5.12}$$

The new time scale has a factor n_0/D. Berryman and Holland (1982), by using several inequalities, proved the result that

$$m(x,t) \to A\exp(-\pi^2 t)\phi_1(x) \tag{3.5.13}$$

where

$$\phi_k(x) = 2^{1/2}\sin(k\pi x), \quad k = 1,2,3,\ldots \tag{3.5.14}$$

are the normalised eigenfunctions satisfying the equation $\phi_{k,xx} + k^2\pi^2\phi_k = 0$ and the boundary conditions $\phi_k(0) = \phi_k(1) = 0$. The asymptotic amplitude A in (3.5.13) is a constant which depends on the initial data. Berryman and Holland (1982) derived other inequalities to determine the bounds for A.

First, a lower bound is found for the function

$$Q(t) = \frac{\pi}{2^{3/2}} \int_0^1 (\exp(m(x,t)) - 1)\phi_1(x)dx. \tag{3.5.15}$$

A simple calculation and use of (3.5.11) and (3.5.12) show that

$$-\frac{d}{dt}\int_0^1 (e^m - 1)\phi_1(x)dx = -\int_0^1 m_{xx}\phi_1 dx$$

$$= -\int_0^1 m\phi_{1xx}dx = \pi^2 \int_0^1 m\phi_1 dx$$

$$\leq \pi^2 \int_0^1 (e^m - 1)\phi_1 dx, \tag{3.5.16}$$

where we have also used the inequality $m \leq (e^m - 1)$. With the definition (3.5.15) of $Q(t)$, (3.5.16), on integration, shows that

$$Q(t) \geq Q(0)e^{-\pi^2 t} \equiv Q_0 e^{-\pi^2 t}, \tag{3.5.17}$$

giving a lower bound for $Q(t)$. To find an upper bound for m, we first observe that, for any differentiable function $f(x)$ such that $f(0) = f(1) = 0$,

$$f(x) = \int_0^x f_x(x)dx = -\int_x^1 f_x(x)dx. \qquad (3.5.18)$$

On using the Cauchy–Schwarz inequality, (3.5.18) yields

$$f^2(x) \le x \int_0^x f_x^2 dx \text{ and } f^2(x) \le (1-x) \int_x^1 f_x^2 dx. \qquad (3.5.19)$$

Adding the inequalities in (3.5.19), we have

$$\left(\frac{1}{x} + \frac{1}{1-x}\right) f^2 = \frac{f^2}{x(1-x)} \le \int_0^1 f_x^2 dx. \qquad (3.5.20)$$

With $f(x) = m(x,t)$, wherein t is treated as a parameter, we get

$$m^2(x,t) \le x(1-x) \int_0^1 m_x^2 dx \le \frac{1}{4} \int_0^1 m_x^2 dx \equiv z^2(t). \qquad (3.5.21)$$

We may now recall the standard inequality,

$$\pi^2 \le \frac{\int m_x^2 dx}{\int m^2 dx}. \qquad (3.5.22)$$

From (3.5.21), $\exp(m - z) \le 1$, therefore,

$$\int m^2 dx \ge e^{-z} \int m^2 e^m dx. \qquad (3.5.23)$$

Combining (3.5.22) and (3.5.23), we get

$$\pi^2 \le e^z \frac{\int m_x^2 dx}{\int m^2 e^m dx}. \qquad (3.5.24)$$

Furthermore, we have

$$\left(\int m_x^2 dx\right)^2 = \left(-\int m m_{xx} dx\right)^2$$
$$= \left[\int \left(m e^{m/2}\right) \left(m_{xx} e^{-m/2}\right) dx\right]^2$$
$$\le \int m^2 e^m dx \int m_{xx}^2 e^{-m} dx, \qquad (3.5.25)$$

where we have used integration by parts and the Cauchy–Schwartz inequality. Equations (3.5.24) and (3.5.25) yield

$$\pi^2 \le e^z \frac{\int m_{xx}^2 e^{-m} dx}{\int m_x^2 dx} = -\frac{e^{-z}}{z}\frac{dz}{dt}, \tag{3.5.26}$$

where we have used the result

$$\frac{d}{dt}z(t) = -\int m_{xx}^2 e^{-m} dx/4z \tag{3.5.27}$$

(see (3.5.21)). Integrating the inequality (3.5.26), we get

$$\mathrm{Ei}(z) \le \mathrm{Ei}(z_0) - \pi^2 t, \tag{3.5.28}$$

where $z_0 = z(0)$ and $\mathrm{Ei}(.)$ is the exponential integral (Abramowitz and Stegun (1972)). Using the expansion

$$\mathrm{Ei}(y) = \gamma + \ln|y| + \sum_{k=1}^{\infty} \frac{y^k}{k.k!}, \tag{3.5.29}$$

where γ is Euler's constant, and the fact that $z \ge 0$, it easily follows from (3.5.28) that

$$z \le \exp[\mathrm{Ei}(z) - \gamma] \le \exp[\mathrm{Ei}(z_0) - \gamma - \pi^2 t] \equiv z_B e^{-\pi^2 t}. \tag{3.5.30}$$

Thus, (3.5.30) gives an upper bound for z and, in view of (3.5.21), it bounds sup m. It may be observed that the bounds (3.5.17) and (3.5.30) both share exponential time dependence.

To find the actual asymptotic behaviour, we write

$$u(x,t) = e^{\pi^2 t} m(x,t) \tag{3.5.31}$$

so that (3.5.11) becomes

$$u_t = e^{-m} u_{xx} + \pi^2 u. \tag{3.5.32}$$

It is clear from (3.5.21) and (3.5.30) that $m \to 0$ as $t \to \infty$. We must now show that $|u_t| \to 0$ as $t \to \infty$. This will ensure that in some sense the solution approaches the solution of the steady equation $u_{xx} + \pi^2 u = 0$. To that end, we consider the functional

$$I(u) = \int_0^1 u_x^2 dx - \pi^2 \int_0^1 e^m u^2 dx. \tag{3.5.33}$$

Using (3.5.21), (3.5.30), and (3.5.31) we find that

$$0 \le \int u_x^2 dx = 4e^{2\pi^2 t} z^2 \le 4z_B^2 \tag{3.5.34}$$

and

$$0 \le \int e^m u^2 dx \le e^z \int u^2 dx \le \frac{e^z}{\pi^2} \int u_x^2 dx, \tag{3.5.35}$$

the last inequality following from (3.5.22). Therefore, both the integrals in (3.5.33) and hence $I(u)$ are bounded for all time. Differentiating (3.5.33) and using (3.5.31) and (3.5.32) we get

$$\frac{dI}{dt} = -2 \int e^m u_t^2 dx + 2\pi^2 e^{-\pi^2 t} \int u u_x^2 dx. \qquad (3.5.36)$$

To show that the second integral in (3.5.36) is bounded above by a positive constant, we employ the second inequality in (3.5.20):

$$0 \le \int u u_x^2 dx \le \frac{1}{2} \left(\int u_x^2 dx \right)^{3/2} \le 4z_B^3 \equiv C \qquad (3.5.37)$$

(see (3.5.34)). Now, for contradiction, suppose $|u_t|$ does not tend to zero as $t \to \infty$. Then there exists a constant ε such that

$$2 \int e^m u_t^2 dx \ge \varepsilon > 0. \qquad (3.5.38)$$

It, therefore, follows from (3.5.36), (3.5.37), and (3.5.38) that

$$\frac{dI}{dt} \le -\varepsilon + 2\pi^2 C e^{-\pi^2 t} \qquad (3.5.39)$$

which, on integration, yields

$$I \le 2C \left(1 - e^{-\pi^2 t} \right) - \varepsilon t + I(0) \le 2C - \varepsilon t + I(0). \qquad (3.5.40)$$

Equation (3.5.40) implies that $I(u)$ is not bounded from below as $t \to \infty$, contradicting the result proven earlier. Therefore, we must have a sequence t_i for which

$$\int e^m u_t^2 dx \to 0 \quad \text{as } t_i \to \infty. \qquad (3.5.41)$$

By an argument, detailed in Berryman and Holland (1980), there exists a function R such that $u(.,t_i) \to R(.)$ as $t_i \to \infty$. The final step is to show that $R(.)$ is, in fact, a solution of the linear equation satisfied by ϕ_1 (see below (3.5.14)).

Multiplying (3.5.32) by Pe^m, where P is a C^∞ function vanishing at the boundaries, and integrating we obtain

$$\int Pe^m u_t dx = - \int P_x u_x dx + \pi^2 \int Pe^m u dx. \qquad (3.5.42)$$

The Cauchy–Schwarz inequality gives

$$\left(\int Pe^m u_t dx \right)^2 \le \int e^m u_t^2 dx \int e^m P^2 dx. \qquad (3.5.43)$$

Inasmuch as P and m are bounded and the first integral on the RHS of (3.5.43) tends to zero as $t_i \to \infty$ (see (3.5.41)), the LHS of (3.5.43) tends to zero as the sequence $t_i \to \infty$. Furthermore, because $e^m \to 1$ as $t_i \to \infty$, it follows from (3.5.42) that $u \to R(x)$, where $R(x)$ satisfies the equation

$$-\int P_x R_x dx + \pi^2 \int PR dx = 0. \tag{3.5.44}$$

Equation (3.5.44) states that $R(x)$ is a weak solution of

$$u_{xx} + \pi^2 u = 0, \tag{3.5.45}$$

the steady form of (3.5.32) with $e^m = 1$. From the existence of the lower bound (3.5.17) for $Q(t)$ defined by (3.5.15), we infer that R is not identically zero. By an argument similar to that used by Berryman and Holland (1980), it may be shown that R is a classical solution of (3.5.45). To be able to show that $u \to R$ for all t, we consider

$$\frac{d}{dt}\int u^2 dx = 2\pi^2 \int u^2 dx + 2\int e^{-m} u u_{xx} dx = 2\pi^2 \int u^2 dx - 2\int (1-m)e^{-m} u_x^2 dx, \tag{3.5.46}$$

where we have used (3.5.31) and (3.5.32) as well as an integration by parts. Using the inequality (3.5.22) in (3.5.46), we have

$$\frac{d}{dt}\int u^2 dx \leq 2\int \left[1 - (1-m)e^{-m}\right] u_x^2 dx \leq 4\int m u_x^2 dx = 4e^{-\pi^2 t}\int u u_x^2 dx, \tag{3.5.47}$$

where we have employed the result that $[1 - (1-m)e^{-m}] \leq 2m$ for any m. Combining (3.5.37) and (3.5.47), we get

$$\frac{d}{dt}\int u^2 dx \leq 16 z_B^3 e^{-\pi^2 t}. \tag{3.5.48}$$

Again, for contradiction, assume that $\int u^2 dx$ does not converge to a constant as $t \to \infty$. Let s_i and t_i be two time sequences for which $\int u^2 dx$ converges to two distinct constants α and β, respectively, where $\alpha < \beta$. Then, using the inequality (3.5.48) and some standard arguments, one may show that both sequences must, in fact, converge to the lower constant α, contradicting the hypothesis. Therefore, the integral $\int u^2 dx$ must converge to a constant for all t. Again using the arguments of Berryman and Holland (1980), one may show that this limit is R for large t.

Thus, we have demonstrated that

$$u(x,t) \to R(x) = A\phi_1(x) \tag{3.5.49}$$

or

$$m(x,t) \to A e^{-\pi^2 t}\phi_1(x), \tag{3.5.50}$$

uniformly in x for large t. Bounds for the constant A may be obtained from (3.5.17) and (3.5.30):

$$\lim_{t \to \infty} e^{\pi^2 t}Q(t) = \frac{\pi}{2^{3/2}}\int_0^1 R(x)\phi_1(x)dx = \frac{\pi}{2^{3/2}}A \geq Q_0, \tag{3.5.51}$$

and

$$\lim_{t \to \infty} e^{\pi^2 t} z(t) = \frac{1}{2} \left[\int_0^1 R^2(x) dx \right]^{1/2}$$

$$= \frac{\pi}{2} A \le z_B. \tag{3.5.52}$$

Thus, for the amplitude A in (3.5.50), we have

$$\frac{2^{3/2}}{\pi} Q_0 \le A \le \left(\frac{2}{\pi} \right) z_B, \tag{3.5.53}$$

where Q_0 and z_B may be found from (3.5.17) and (3.5.30), respectively. To check the asymptotic nature of the solution (3.5.50), Berryman and Holland (1982) numerically integrated (3.5.11) and (3.5.12) subject to the initial conditions,

$$m(x,0) = \sum_{k=1}^{4} \alpha_k \sin(k\pi x), \tag{3.5.54}$$

for different sets of values for $\{\alpha_k\}$; they used a linear implicit three-level difference scheme for quasilinear parabolic equations (Lees 1966). The value of the parameter A was determined as the computation proceeded. The analytic bounds for A, however, were not found to be too close to those found numerically. Shenker and Roseman (1995) discussed a more general initial boundary value problem

$$\rho(x) \frac{\partial u}{\partial t} = (A(x,u,t)u_x)_x, \quad 0 < x < 1, \, 0 < t < \infty, \tag{3.5.55}$$

$$u(0,t) = 0 \quad \text{or} \quad \frac{\partial u}{\partial x}(0,t) = 0, \tag{3.5.56}$$

$$u(1,t) = 0 \quad \text{or} \quad \frac{\partial u}{\partial x}(1,t) = 0, \tag{3.5.57}$$

$$u(x,0) = f(x), \quad 0 < x < 1. \tag{3.5.58}$$

They showed that the solution of the IBVP (3.5.55)–(3.5.58) converges to a constant exponentially and the derivatives of the solution decay exponentially to zero as t becomes large.

3.6 Asymptotic solutions describing fast nonlinear diffusion

Friedman and Kamin (1980), Esteban et al. (1988), and King (1993) considered asymptotic behaviour of the nonlinear parabolic equation

$$u_t = \nabla \cdot \left(u^{-n} \nabla u \right), \tag{3.6.1}$$

where $n > 0$ and $N \geq 1$ is the dimension of space being considered. We have already considered (3.6.1) with $n = 1$ and $N = 1$ in Section 3.5. Equation (3.6.1) appears in many contexts including spreading of microscopic droplets and plasma physics. King (1993) considered (3.6.1) in great detail and analysed specific cases for which the solution with a finite mass either extinguishes in a finite time or decays over an infinite time; this depends strongly on the choice of the parameters n and N. Here, we are concerned with the radially symmetric form of (3.6.1) (see (3.6.5) below) and consider only those parameters for which the solution decays over an infinite time.

We may first observe that the solution of (3.6.1) subject to delta function initial conditions,

$$u(x,0) = M^* \delta(x) \text{ for } x \in \mathbb{R}^n, \tag{3.6.2}$$

has the similarity form

$$u(x,t) = t^{-N/(2-nN)} f\left(\frac{|x|}{t^{1/(2-nN)}}\right), \tag{3.6.3}$$

where, for $n > 0$,

$$f(\eta) = \left(\frac{n}{2(2-nN)}(a^2 + \eta^2)\right)^{-1/n}. \tag{3.6.4}$$

This solution holds for $n < \min(1, 2/N)$ and neatly illustrates the typical behaviour of similarity solutions as they arise from delta function initial conditions. The (unknown) constant a may be obtained from the conservation of mass condition

$$\int_{\mathbb{R}^N} u(x,t)dV = M^*,$$

which is assumed to hold for all time in the above parametric range (see Zel'dovich and Barenblatt 1958).

King (1993) considered the case $n \geq 2/N$ for which the solution (3.6.3) is not applicable and restricted himself to the radially symmetric form of (3.6.1), namely,

$$u_t = \frac{1}{r^{N-1}} \left(r^{N-1} u^{-n} u_r\right)_r. \tag{3.6.5}$$

Equation (3.6.5) was solved subject to the initial boundary conditions

$$u = I(r) \text{ at } t = 0, \tag{3.6.6}$$

$$r^{N-1} u^{-n} u_r = 0 \text{ at } r = 0, \tag{3.6.7}$$

and

$$u \to 0 \text{ as } r \to \infty. \tag{3.6.8}$$

In addition, it was assumed that the mass

$$M = \int_0^\infty r^{N-1} I(r) dr \tag{3.6.9}$$

under the profile is finite. As pointed out earlier, we are concerned here only with the case $N > 2$ and $n = 2/N$, which guarantees the finiteness of the mass M. It is shown that the large time solution of the initial boundary value problem in this case is given by an unusual form of a similarity solution. We also summarise the results for other cases for which either the solution vanishes after a finite time and/or the finite mass condition is not satisfied. This study demonstrates how the asymptotic form may depend crucially on the large distance $(r \to \infty)$ behaviour of the solution.

It is clear from (3.6.4) that this similarity solution breaks down when $n = 2/N$. It may also be observed that, for $n < 2/N$, it behaves as does the separable solution

$$u \sim \left(\frac{nr^2}{2(2-nN)t} \right)^{-1/n} \quad \text{as } r \to \infty \qquad (3.6.10)$$

when the initial profile decays faster than $r^{-2/N}$ as $r \to \infty$. Seeking a comparable product form of the solution of

$$u_t = \frac{1}{r^{N-1}} \left(r^{N-1} u^{-2/N} u_r \right)_r \qquad (3.6.11)$$

with $n = 2/N, N > 2$ in (3.6.5) we write

$$u \sim t^{N/2} F(r) \quad \text{as } r \to \infty, \qquad (3.6.12)$$

where $F(r)$ satisfies

$$\frac{N}{2} r^{N-1} F = \frac{d}{dr} \left(r^{N-1} F^{-2/N} \frac{dF}{dr} \right). \qquad (3.6.13)$$

In the manner of Grundy et al. (1994) (see Section 3.8), a balancing argument for (3.6.13) shows that, for $N > 2$ and $r \to \infty$,

$$F \sim \left[(N-2)^{-1} \left(r^2 \ln r \right) \right]^{-N/2}. \qquad (3.6.14)$$

The mass M defined by (3.6.9) with u given by (3.6.12) and (3.6.14) may be shown to be bounded; the corresponding flux

$$-r^{N-1} u^{-2/N} u_r \sim N \left(\frac{\ln r}{(N-2)t} \right)^{-(N-2)/2} \quad \text{as } r \to \infty \qquad (3.6.15)$$

decays logarithmically for large r.

King (1993) specifically considered the initial data with compact support:

$$I(r) = 0 \quad \text{for } r \geq r_0,$$
$$I(r) = A(r_0 - r)^b \quad \text{as } r \to r_0^-, \qquad (3.6.16)$$

where A and b are positive constants. The solution of (3.6.11) and (3.6.16) for $t \to 0^+$ may be found to be

$$u \sim I(r), \quad r < r_0, \tag{3.6.17}$$

$$u \sim t^{Nb/2(N+b)} \phi(w), \quad r = r_0 + O\left(t^{N/2(N+b)}\right), \tag{3.6.18}$$

where $w = (r - r_0)t^{-N/2(N+b)}$; $\phi(w)$ satisfies the equation

$$\frac{N}{2(N+b)} \left(b\phi - w\frac{d\phi}{dw} \right) = \frac{d}{dw} \left(\phi^{-2/N} \frac{d\phi}{dw} \right) \tag{3.6.19}$$

and the conditions $\phi \sim A(-w)^b$ as $w \to -\infty$ and $\phi \to 0$ as $w \to +\infty$. This initial behaviour may again be found by using the balancing argument. The dominant term in the solution of (3.6.19) for any value of b may be found to be

$$\phi \sim \left(\frac{w^2}{2(N-1)} \right)^{-N/2} \quad \text{as } w \to +\infty. \tag{3.6.20}$$

This yields the following useful result for $r > r_0, t \ll 1$,

$$u \sim \left(\frac{(r - r_0)^2}{2(N-1)t} \right)^{-N/2} \quad \text{for } t^{N/2(N+b)} \ll (r - r_0) \ll 1. \tag{3.6.21}$$

This approximation is independent of the behaviour of $I(r)$ in the neighbourhood of $r = r_0$. Now referring to (3.6.12) and (3.6.21), we find that F satisfies (3.6.13) with

$$F \sim \left(\frac{(r - r_0)^2}{2(N-1)} \right)^{-N/2} \quad \text{as } r \to r_0^+, \tag{3.6.22}$$

$$F \to 0 \quad \text{as } r \to +\infty.$$

The next order approximation to the solution of (3.6.13) (by perturbation or otherwise) may be found to be

$$F^{-2/N} = \frac{r^2}{N-2} \left(\ln r - \frac{N}{2(N-2)} \ln \ln r - x_0 + o(1) \right) \quad \text{as } r \to \infty, \tag{3.6.23}$$

where x_0 is a constant which may be found by solving (3.6.13) subject to (3.6.22). In fact, equation (3.6.13) is invariant under the transformation $r \to r_0 r$, $F \to r_0^{-N}F$, therefore, one may write $x_0 = \ln r_0 + \gamma_N$, where the constant γ_N is independent of r_0. The form (3.6.23) suggests the transformation

$$u = r^{-N}c, \quad x = \ln r \tag{3.6.24}$$

so that (3.6.11) becomes

$$c_t = \left(c^{-2/N}(c_x - Nc)\right)_x. \tag{3.6.25}$$

Now if we write the boundary condition at $x = 0$ as

$$u(0,t) = U(t), \tag{3.6.26}$$

we may solve (3.6.25) subject to the conditions

$$c \sim U(t)e^{Nx} \text{ as } x \to -\infty$$

$$c \sim \left(\frac{x}{(N-2)t}\right)^{-N/2} \text{ as } x \to +\infty, \tag{3.6.27}$$

where the function $U(t)$ must be found as part of the solution. The second condition in (3.6.27) conforms to the far field behaviour (3.6.14). The mass conservation condition (3.6.9) in terms of $c(x,t)$ may be written as

$$\int_{-\infty}^{\infty} c(x,t)dx = M. \tag{3.6.28}$$

In terms of new variables, (3.6.23) becomes

$$c^{-2/N} \sim \frac{\left(x - \dfrac{N}{2(N-2)} \ln x - x_0\right)}{(N-2)t} \quad \text{for } t \ll 1,\ x \to \infty. \tag{3.6.29}$$

This result actually holds for all t as $x \to \infty$ as we presently show. Writing

$$c - t^{N/2}G(x) \sim C(x,t) \text{ as } x \to \infty, \tag{3.6.30}$$

where $G(x) = e^{Nx}F(e^x)$ so that $G(x) \sim (x/(N-2))^{-N/2}$ as $x \to \infty$, and substituting in (3.6.25), we have, to the leading order,

$$C_t = -\left(\frac{xC}{t}\right)_x. \tag{3.6.31}$$

The solution to (3.6.31) may be found in the form

$$C = t^{-1}P(x/t), \tag{3.6.32}$$

where P is an arbitrary function. Because, for fixed $x > \ln r_0$, u goes to zero more rapidly than $t^{N/2}$ as $t \to 0$ (see (3.6.12)), $P(\sigma)$ with $\sigma = x/t$ must tend to zero faster than $\sigma^{-(N+2)/2}$ as $\sigma \to \infty$. We infer that

$$c^{-2/N} = \frac{\left(\left(x - \left(\dfrac{N}{2(N-2)}\right)\right) \ln x - x_0 + o(1)\right)}{(N-2)t} \quad \text{as } x \to \infty. \tag{3.6.33}$$

Now, the large time behaviour of the solution of (3.6.25) subject to (3.6.27), (3.6.28), and (3.6.33) is found in some detail by using the balancing argument (Grundy 1988; see Section 3.8):

$$c \sim t^{-N/(N-2)} g_0(\eta) + t^{-2N/(N-2)} \ln t \, g_1(\eta) + t^{-2N/(N-2)} g_2(\eta) \quad \text{as } t \to \infty, \quad (3.6.34)$$

where $\eta = x/t^{N/(N-2)}$. Substituting (3.6.34) into (3.6.25) and so on, we find that g_0 is governed by

$$-\frac{N}{N-2}\left(g_0 + \eta \frac{dg_0}{d\eta}\right) = -N\frac{d}{d\eta}\left(g_0^{1-2/N}\right), \quad (3.6.35)$$

showing that, to the leading order, the convective effect dominates (see Section 3.8). The relevant solution of (3.6.35) is

$$g_0(\eta) = \left(\frac{\eta}{(N-2)}\right)^{-N/2}. \quad (3.6.36)$$

As usual, (3.6.36) must be corrected by introducing an inner region. The asymptotic structure is schematically shown in Figure 3.6. The representation (3.6.34) therefore gives the outer expansion valid in the domain

$$x = O\left(t^{N/(N-2)}\right) \quad \text{as } t \to \infty \text{ with } \frac{x}{t^{N/(N-2)}} > \eta_0, \quad (3.6.37)$$

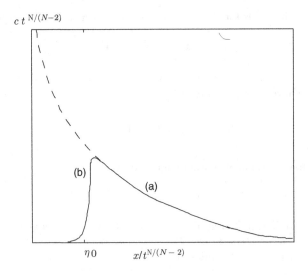

Fig. 3.6 Schematic diagram representing the asymptotic behaviour of the solution of (3.6.25) as $t \to \infty$. g_0 and h_0 are given by (a) and (b), respectively; see equations (3.6.36) and (3.6.52). (King 1993. Copyright © 1993 The Royal Society, Great Britain. Reprinted with permission. All rights reserved.)

where the constant η_0 may be found from the conservation equation (3.6.28):

$$\int_{\eta_0}^{\infty} g_0(\eta) = M$$

or

$$\eta_0 = (N-2)\left(\frac{M}{2}\right)^{-2/(N-2)}. \tag{3.6.38}$$

One may obtain the ODEs governing the correction terms g_1 and g_2 in (3.6.34) as follows.

$$-\frac{N}{(N-2)}\left(2g_1 + \eta\frac{dg_1}{d\eta}\right) = -(N-2)\frac{d}{d\eta}\left(g_0^{-2/N}g_1\right) \tag{3.6.39}$$

and

$$g_1 - \frac{N}{N-2}\left(2g_2 + \eta\frac{dg_2}{d\eta}\right) = \frac{d}{d\eta}\left(g_0^{-2/N}\frac{dg_0}{d\eta}\right) - (N-2)\frac{d}{d\eta}\left(g_0^{-2/N}g_2\right). \tag{3.6.40}$$

Equations (3.6.39) and (3.6.40) can be solved to yield

$$g_1 = \left(\frac{\eta}{N-2}\right)^{-(N+2)/2} A_1, \tag{3.6.41}$$

$$g_2 = \left(\frac{\eta}{N-2}\right)^{-(N+2)/2}\left(A_2 + \left(\frac{(N-2)A_1}{2} - \frac{N^2}{8(N-2)}\right)\ln\eta\right), \tag{3.6.42}$$

where A_1 and A_2 are constants of integration. For the solution (3.6.34) to match the distant behaviour (3.6.29) we must choose these constants to be

$$A_1 = \frac{N^3}{4(N-2)^3}, \quad A_2 = \frac{Nx_0}{2(N-2)}. \tag{3.6.43}$$

Thus the solution (3.6.34), with $g_i(i=0,1,2)$ found above reproduces the far field behaviour to this order.

As we remarked earlier, we must introduce an inner layer to complete the solution. To that end we introduce the variable

$$z = x - \eta_0 t^{N/(N-2)} - \frac{N^2}{2(N-2)^2}\ln t, \tag{3.6.44}$$

where $z = O(1)$ (see (3.6.37)). The inner expansion has the form

$$c \sim t^{-N/(N-2)}h_0(z) + t^{-2N/(N-2)}h_1(z) \quad \text{as } t \to \infty. \tag{3.6.45}$$

From the expressions (3.6.36) and (3.6.41) for $g_0(\eta)$ and $g_1(\eta)$, respectively, we may find the matching conditions to be

$$h_0 \sim \left(\frac{\eta_0}{N-2}\right)^{-N/2} \quad \text{as } z \to +\infty, \tag{3.6.46}$$

$$h_1 \sim -\frac{N\left(\dfrac{\eta_0}{N-2}\right)^{-N/2}\left(z - \dfrac{N}{2(N-2)}\ln \eta_0 - x_0\right)}{2\eta_0} \quad \text{as } z \to +\infty. \tag{3.6.47}$$

Moreover, we must have

$$h_0, \; h_1 \to 0 \text{ as } z \to -\infty. \tag{3.6.48}$$

Writing (3.6.25) in terms of z and t we obtain

$$c_t = \left(\frac{\eta_0 N}{N-2}t^{2/(N-2)} + \frac{N^2}{2(N-2)^2}\frac{1}{t}\right)c_z + \left(c^{-2/N}(c_z - Nc)\right)_z. \tag{3.6.49}$$

Substituting (3.6.45) into (3.6.49) and so on, we obtain the following ODEs for $h_0(z)$ and $h_1(z)$.

$$-\frac{N}{N-2}\eta_0 \frac{dh_0}{dz} = \frac{d}{dz}\left(h_0^{-2/N}\left(\frac{dh_0}{dz} - Nh_0\right)\right), \tag{3.6.50}$$

$$-\frac{Nh_0}{N-2} - \frac{N^2}{2(N-2)^2}\frac{dh_0}{dz} - \frac{N\eta_0}{N-2}\frac{dh_1}{dz} = \frac{d^2}{dz^2}\left(h_0^{-2/N}h_1\right) - (N-2)\frac{d}{dz}\left(h_0^{-2/N}h_1\right). \tag{3.6.51}$$

Equation (3.6.50) may be solved subject to the conditions (3.6.46) and (3.6.48):

$$h_0(z) = \left(\frac{\eta_0}{N-2}\left(1 + e^{-2(z-z_0)}\right)\right)^{-N/2}; \tag{3.6.52}$$

z_0 is the constant of integration. Integrating (3.6.51) with (3.6.48), we get

$$-\frac{N}{N-2}\int_{-\infty}^{z} h_0(z')dz' - \frac{N^2}{2(N-2)^2}h_0 - \frac{N}{N-2}\eta_0 h_1$$
$$= \frac{d}{dz}\left(h_0^{-2/N}h_1\right) - (N-2)h_0^{-2/N}h_1 \tag{3.6.53}$$

which, in the limit $z \to +\infty$, may be written as

$$h_1 \sim -N\left(\frac{\eta_0}{N-2}\right)^{-N/2}\frac{\left(z + \dfrac{1}{N-2} + \kappa_N - z_0\right)}{2\eta_0}, \tag{3.6.54}$$

where

$$\kappa_N = \int_{-\infty}^{\infty} \left(\left(1 + e^{-2z'}\right)^{-N/2} - H(z') \right) dz'. \tag{3.6.55}$$

Here, $H(z')$ denotes the Heaviside step function. Comparing (3.6.54) with (3.6.47) we find that

$$z_0 = \frac{N}{2(N-2)} \ln \eta_0 + \frac{1}{N-2} + \kappa_N + x_0. \tag{3.6.56}$$

Making use of (3.6.52), we arrive at the asymptotic form for the function $U(t)$ in the boundary condition (3.6.26):

$$U(t) \sim a^{-N} \left(\frac{M}{2}\right)^{N/(N-2)+N^2/(N-2)^2} t^{-N/(N-2)-N^3/2(N-2)^2}$$
$$\times \exp\left(-N\eta_0 t^{N/(N-2)}\right) \quad \text{as } t \to \infty, \tag{3.6.57}$$

where the constant a is given by

$$a = (N-2)^{N/2(N-2)} \exp\left(\frac{1}{(N-2)} + \kappa_N + x_0\right). \tag{3.6.58}$$

Thus, the function $U(t)$ decays exponentially for large time. In terms of the original variables we may write the leading order behaviour of the solution for large t:

$$u \sim \left(\frac{M}{2t}\right)^{N/(N-2)} \left(a^2 \left(\frac{M}{2}\right)^{-2N/(N-2)^2} t^{N^2/(N-2)^2} \exp\left(2\eta_0 t^{N/(N-2)} + r^2\right)\right)^{-N/2} \tag{3.6.59}$$

for

$$r = O\left(t^{N^2/2(N-2)^2} \exp\left(\eta_0 t^{N/(N-2)}\right)\right),$$

where η_0 is given by (3.6.38). The solution for $\ln r/(t^{N/(N-2)}) > \eta_0$ and $t \to \infty$ is found with the help of (3.6.36) as

$$u \sim \left(\frac{r^2 \ln r}{(N-2)t}\right)^{-N/2}. \tag{3.6.60}$$

The above large time analysis assumes that the initial conditions have a compact support. If one considers other types of initial conditions such as $I(r) \sim Ar^{-b}$, analysis similar to the above shows that the asymptotic solution also depends on the constant b. This is in contrast to the results for $n < 2/N$ where the large time behaviour (3.6.3) depends on the initial conditions only through M.

King (1993) showed that for $N > 2$ with $2/N < n < 1$ and for $N = 2$ with $n = 1$, the solution of IBVP (3.6.5)–(3.6.8) extinguishes in a finite time. For the former case it was shown that the behaviour close to the extinction time is governed by a self-similar solution of the second kind. The limit $N \to \infty$ is also considered to demonstrate how various types of asymptotic behaviour arise from the evolution over earlier times.

We refer to Galaktionov et al. (2000) for a rigourous analysis of solutions of (3.6.1) for $N \geq 3$, $n = 2/N$ subject to nonnegative initial data $u(x,0) \in L^1(\mathbb{R}^N)$ for large time.

Bowen and King (2001) considered the fourth-order 'thin film equation'

$$u_t = -(u^n u_{xxx})_x, \quad n < 2 \tag{3.6.61}$$

with the conditions

$$u = u_x = 0 \text{ at } x = \pm 1 \tag{3.6.62}$$

and

$$u = u_0(x) \text{ at } t = 0, \tag{3.6.63}$$

where $u_0(x) \geq 0$ is assumed to have a finite mass. Here, also, 'appropriate' similarity solutions are identified to describe asymptotics which extinguish over finite or infinite time.

In a related study, King and McCabe (2003) studied a generalised form of the Fisher–KPP equation with fast nonlinear diffusion:

$$u_t = \nabla \cdot \left(u^{-n}\nabla u\right) + u(1-u), \quad x \in \mathbb{R}^N, t > 0 \tag{3.6.64}$$

where $N \geq 1$ is the spatial dimension and $0 < n < \max(1, 2/N)$. This equation attempts to explain some observations concerning the dispersal of early Palaeoindian peoples in North America; in this context the diffusivity is believed to be a decreasing function of the population density so that low concentrations disperse very rapidly. King (1993) considered (3.6.64) in the absence of the reaction term.

King and McCabe (2003) studied (3.6.64) subject to the initial condition

$$u(x,0) = I(x) \text{ for } x \in \mathbb{R}^n \tag{3.6.65}$$

where the total mass $\int_{\mathbb{R}^N} I(x)dx$ is bounded.

In fact equations (3.6.64) and (3.6.65) do not possess a permanent travelling waveform as a large time asymptotic solution. This can be easily seen for $N = 1$ and $0 < n < 2$. In this case, if we put $z = x - ct$ in (3.6.64) we have

$$\frac{d}{dz}\left(u^{-n}\frac{du}{dz}\right) + c\frac{du}{dz} + u(1-u) = 0. \tag{3.6.66}$$

We look for solutions of (3.6.66) subject to the boundary conditions

$$u \to 0 \text{ as } z \to \infty, \quad u \to 1 \text{ as } z \to -\infty. \tag{3.6.67}$$

Writing $d\zeta/dz = u^n$ in (3.6.66) we obtain

$$\frac{d^2u}{d\zeta^2} + c\frac{du}{d\zeta} + u^{1-n}(1-u) = 0,\qquad(3.6.68)$$

where $-1 < 1-n < 1$. The nonlinear BVP (3.6.68) and (3.6.67) in terms of ζ do not possess a classical solution for $-1 < 1-n \leq 0$ because $u = 0$ is not a zero of the nonlinear term. It was also shown by McCabe et al. (2002) that this problem does not have any monotonically decreasing classical solutions for $0 < 1-n < 1$. Thus the Cauchy problem for (3.6.64) with $0 < n < 2$ and $N = 1$ does not have permanent travelling waveform solutions as its asymptotics.

With this background, King and McCabe (2003) investigated other large time behaviours of the above initial value problem. They considered specifically the radially symmetric form of (3.6.64), namely,

$$u_t = r^{1-N}\frac{\partial}{\partial r}\left(r^{N-1}u^{-n}\frac{\partial u}{\partial r}\right) + u(1-u),\quad r, t > 0,\qquad(3.6.69)$$

which is likely to furnish the relevant asymptotic behaviour of broader classes of solutions, subject to the initial and boundary conditions

$$u = I(r) \text{ at } t = 0,\qquad(3.6.70)$$
$$r^{N-1}u^{-n}u_r = 0 \text{ at } r = 0,\qquad(3.6.71)$$
$$u(r,t) \to 0 \text{ as } r \to \infty,\qquad(3.6.72)$$

where

$$M = \int_0^\infty r^{N-1}I(r)dr\qquad(3.6.73)$$

is finite and $0 \leq I(r) \leq 1$ is a smooth monotone decreasing function for all $r \geq 0$.

King and McCabe (2003) essentially used the balancing argument and distinguished a large number of asymptotic behaviours often the similarity solutions of simpler equation(s) containing the dominant terms would yield large time behaviour. These solutions, however, would in general not be analytic near $r = 0$. To complete the description of the large time behaviour of other solutions in the region $r = O(t^{1/2})$, $t \to \infty$ would have to be found and appropriately matched with the similarity solutions referred to above.

As pointed out earlier, similarity solutions of the truncated equations arising from the balancing argument play an important role in the asymptotic forms. It would be interesting to carry out the analysis of King and McCabe (2003) in a more formal manner, possibly combining distinct behaviours in a single form (see Section 3.8). These authors also considered the asymptotic behaviour for the more general quasi-linear reaction diffusion equation

$$u_t = (D(u)u_x)_x + f(u),\quad x \in \mathbb{R}, t > 0,$$

where

$$D(u) \sim u^m, \quad f(u) \sim u^p \text{ as } u \to 0^+,$$
$$f(u) > 0 \text{ for } 0 < u < 1, \ f(1) = 0,$$

for initial data $I(x)$ satisfying $0 < I < 1$ for all x. They assumed that $I = 0$ for $|x| > a$ where $a > 0$ is a constant. Several distinct cases were identified for different values of m and p and their asymptotic forms reviewed. It was observed that even semi-linear equations can exhibit accelerating wavefronts for compactly supported initial data provided these equations were 'nearly linear' for small u.

3.7 Large time asymptotic behaviour of periodic solutions of some generalised Burgers equations

The Burgers equation has been much studied with respect to periodic initial conditions; its analysis is easier because it is exactly linearisable to the heat equation via the Cole–Hopf transformation (see Section 3.2). The generalised Burgers equations (GBEs), which actually appear in applications, do not, in general, admit exact linearisation and must be treated directly. Sachdev et al. (2003) and Sachdev et al. (2005) have treated this class of equations with periodic initial conditions. They could directly find large time asymptotic behaviour of this class of equations. The GBEs studied by these authors include the nonplanar Burgers equation, the Burgers equation with linear damping, and the modified Burgers equation. We treat here the nonplanar Burgers equation in some detail and summarise the results for others; the analytic results thus obtained show excellent agreement with the numerical solution of the relevant initial/boundary value problems for large time. The analytic approach is similar to that of Bender and Orszag (1978) for nonlinear ordinary differential equations. The basic idea is to start with the solution of the linearised form of the equation and then include nonlinear effects. It sometimes becomes possible to write even a general term for the series form of the solution that is sought; however, the first few terms themselves give an excellent description of the asymptotic solution.

We seek the large time periodic solution of the nonplanar Burgers equation

$$u_t + uu_x + \frac{ju}{2t} = \frac{\delta}{2} u_{xx}, \tag{3.7.1}$$

where $\delta > 0$ is small. $j = 0, 1, 2$ for plane, cylindrical, and spherical geometry, respectively. The periodic solution of (3.7.1) should, for large time, tend to the periodic solution

$$u(x,t) = A_1 \exp(-kt) t^{(-j/2)} \sin(\pi x/l), \quad k = \frac{\delta \pi^2}{2l^2} \tag{3.7.2}$$

of the linearised form of (3.7.1), namely,

$$u_t + \frac{ju}{2t} = \frac{\delta}{2}u_{xx}.$$ (3.7.3)

This is true for the planar case with $j = 0$ (see Sachdev 1987). The linear solution (3.7.2) satisfies the initial condition

$$u(x,t_0) = A\sin\left(\frac{\pi x}{l}\right), \quad 0 \le x \le l,$$ (3.7.4)

and the boundary conditions

$$u(0,t) = u(l,t) = 0,$$ (3.7.5)

where l and A are positive constants. The idea here is to correct the solution (3.7.2) to take into account the effect of the nonlinear term in (3.7.1). This is done by first writing

$$u(x,t) = A_1 e^{-kt} t^{-j/2}\sin\left(\frac{\pi x}{l}\right) + \varepsilon(x,t),$$ (3.7.6)

where $\varepsilon(x,t)$ is small and $2l$ periodic in x; moreover, we must have $u(0,t) = 0 = u(l,t)$. Substituting (3.7.6) into (3.7.1) and retaining only the linear terms in ε and its derivatives, we have

$$\varepsilon_t + \frac{j}{2t}\varepsilon - \frac{\delta}{2}\varepsilon_{xx} \approx -\frac{\pi}{2l}A_1^2 e^{-2kt} t^{-j}\sin\left(\frac{2\pi x}{l}\right).$$ (3.7.7)

Writing the solution of (3.7.7) in the product form

$$\varepsilon(x,t) = T(t)X(x)$$ (3.7.8)

and solving the resulting equations for $T(t)$ and $X(x)$ and so on we obtain

$$\varepsilon(x,t) \sim -\frac{A_1^2 l}{2\delta\pi}e^{-2kt} t^{-j}\sin\left(\frac{2\pi x}{l}\right) + O\left(e^{-2kt} t^{-j-1}\right),$$ (3.7.9)

where we have imposed the periodicity condition in x on $\varepsilon(x,t)$. Thus, to this order, we have

$$u(x,t) = A_1 e^{-kt} t^{-j/2}\sin\left(\frac{\pi x}{l}\right) - \frac{A_1^2 l}{2\delta\pi}e^{-2kt} t^{-j}\sin\left(\frac{2\pi x}{l}\right) + O\left(e^{-2kt} t^{-j-1}\right)$$ (3.7.10)

as $t \to \infty$. The form (3.7.10) suggests that we seek solution of (3.7.1) in the form

$$u(x,t) = A_1 e^{-kt} f_1(x,t) + e^{-2kt} f_2(x,t) + e^{-3kt} f_3(x,t) + \ldots.$$ (3.7.11)

Substituting (3.7.11) into (3.7.1) and equating to zero the coefficients of e^{-nkt}, $n = 1,2,\ldots$, we obtain the following system of linear PDEs for $f_i(x,t)$, $i = 1,2,\ldots.$

$$f_{1,t} + \left(\frac{j}{2t} - k\right) f_1 - \frac{\delta}{2} f_{1,xx} = 0, \tag{3.7.12}$$

$$f_{2,t} + \left(\frac{j}{2t} - 2k\right) f_2 - \frac{\delta}{2} f_{2,xx} = -A_1^2 f_1 f_{1,x}, \tag{3.7.13}$$

$$f_{3,t} + \left(\frac{j}{2t} - 3k\right) f_3 - \frac{\delta}{2} f_{3,xx} = -A_1 \left(f_1 f_{2,x} + f_2 f_{1,x}\right), \tag{3.7.14}$$

$$\ldots$$

$$f_{n,t} + \left(\frac{j}{2t} - nk\right) f_n - \frac{\delta}{2} f_{n,xx} = -A_1 \left(f_1 f_{n-1,x} + \ldots + f_{n-1} f_{1,x}\right). \tag{3.7.15}$$

Using the initial and boundary conditions (3.7.4) and (3.7.5) and the periodicity condition with respect to x, the solution of (3.7.12) is found to be

$$f_1(x,t) = t^{-j/2} \sin\left(\frac{\pi x}{l}\right). \tag{3.7.16}$$

Using (3.7.16) in (3.7.13), we have

$$f_{2,t} + \left(\frac{j}{2t} - 2k\right) f_2 - \frac{\delta}{2} f_{2,xx} = -\frac{A_1^2 \pi}{2l} t^{-j} \sin\left(\frac{2\pi x}{l}\right). \tag{3.7.17}$$

Motivated by (3.7.10), we let

$$f_2(x,t) = t^{-j} \left(b_0 + \frac{b_1}{t} + \frac{b_2}{t^2} + \ldots\right) \sin\left(\frac{2\pi x}{l}\right). \tag{3.7.18}$$

Substituting (3.7.18) into (3.7.17) and equating coefficients of $t^{-j}, t^{-j-1}, \ldots$ on both sides we get, after some simplification, the coefficients b_i:

$$b_0 = -\frac{A_1^2 l}{2\pi\delta}, \tag{3.7.19}$$

$$4k b_{n+1} = (j+2n) b_n, \quad n \geq 0. \tag{3.7.20}$$

Using (3.7.16) and (3.7.18) for f_1 and f_2 in (3.7.14) and employing the perturbative approach as for $f_2(x,t)$, we find that $f_3(x,t)$ may be sought in the form

$$f_3(x,t) = t^{-3j/2} f_{31}(t) \sin\left(\frac{3\pi x}{l}\right) + t^{-3j/2} f_{32}(t) \sin\left(\frac{\pi x}{l}\right). \tag{3.7.21}$$

Now putting the functions f_1, f_2, and f_3 from (3.7.16), (3.7.18), and (3.7.21) into (3.7.14) and equating the coefficients of $\sin(\pi x/l)$ and $\sin(3\pi x/l)$ on both sides, we get the following ODEs for $f_{31}(t)$ and $f_{32}(t)$.

$$f'_{31}(t) + \left(-\frac{j}{t} + \frac{3\delta\pi^2}{l^2}\right) f_{31}(t) = -\frac{3\pi}{2l}A_1\left(b_0 + \frac{b_1}{t} + \frac{b_2}{t^2} + \ldots\right), \quad (3.7.22)$$

$$f'_{32}(t) + \left(-\frac{j}{t} - \frac{\delta\pi^2}{l^2}\right) f_{32}(t) = \frac{A_1\pi}{2l}\left(b_0 + \frac{b_1}{t} + \frac{b_2}{t^2} + \ldots\right). \quad (3.7.23)$$

We may now write

$$f_{31}(t) = c_0 + \frac{c_1}{t} + \frac{c_2}{t^2} + \ldots, \quad (3.7.24)$$

$$f_{32}(t) = d_0 + \frac{d_1}{t} + \frac{d_2}{t^2} + \ldots. \quad (3.7.25)$$

Putting (3.7.24) and (3.7.25) into (3.7.22) and (3.7.23), respectively, we get

$$c_0 = -\frac{A_1 b_0 l}{2\delta\pi},$$

$$6kc_{n+1} = (j+n)c_n - \frac{3\pi}{2l}A_1 b_{n+1}, \quad n \geq 0 \quad (3.7.26)$$

and

$$d_0 = -\frac{A_1 l}{2\delta\pi}b_0,$$

$$2kd_{n+1} = -(j+n)d_n - \frac{A_1\pi}{2l}b_{n+1}, \quad n \geq 0, \quad (3.7.27)$$

where b_i are given by (3.7.19) and (3.7.20). Thus, we obtain an elegant form of the periodic solution of (3.7.1):

$$u(x,t) = A_1 e^{-kt} t^{-j/2} \sin\left(\frac{\pi x}{l}\right)$$

$$+ e^{-2kt} t^{-j}\left(\sum_{n=0}^{\infty} b_n t^{-n}\right) \sin\left(\frac{2\pi x}{l}\right)$$

$$+ e^{-3kt} t^{-3j/2}\left(\sum_{n=0}^{\infty} c_n t^{-n} \sin\left(\frac{3\pi x}{l}\right) + \sum_{n=0}^{\infty} d_n t^{-n} \sin\left(\frac{\pi x}{l}\right)\right)$$

$$+ \ldots, \quad (3.7.28)$$

where $k = \delta\pi^2/(2l^2)$, and b_n, c_n, and d_n ($n = 1,2,3,\ldots$) are given by (3.7.19), (3.7.20), (3.7.26), and (3.7.27), respectively. If we put $j = 0$ in (3.7.28), we may verify, after considerable simplification, that it reduces to the exact periodic solution

$$u = \frac{2\pi v}{l} \frac{\left\{\sum_{n=1}^{\infty} e^{-vn^2\pi^2 t/l^2} n A_n \sin\left(\frac{n\pi x}{l}\right)\right\}}{A_0 + \sum_{n=1}^{\infty} e^{-vn^2\pi^2 t/l^2} A_n \cos\left(\frac{n\pi x}{l}\right)}, \quad v = \frac{\delta}{2} \quad (3.7.29)$$

of the Burgers equation. The coefficients A_n in (3.7.29) are defined in terms of Bessel functions.

We now summarise the results for some other GBEs. We consider the GBE with linear damping,

$$u_t + uu_x + \lambda u = \frac{\delta}{2}u_{xx},$$ (3.7.30)

where $\lambda > 0$, and $\delta > 0$ is small. This equation is again solved subject to (3.7.4)–(3.7.5). The balancing argument (see Section 3.8) leads to three distinct cases: (i) $\lambda < \delta\pi^2/l^2$, (ii) $\lambda = \delta\pi^2/l^2$, (iii) $\lambda > \delta\pi^2/l^2$. For case (i) and λ sufficiently small, the solution comes out to be

$$u(x,t) = e^{-kt}\left[B_1 \sin\left(\frac{\pi x}{l}\right)\right] + e^{-2kt}\left[B_2 \sin\left(\frac{2\pi x}{l}\right)\right]$$

$$+ e^{-3kt}\left[B_3 \sin\left(\frac{\pi x}{l}\right) + B_4 \sin\left(\frac{3\pi x}{l}\right)\right] + \dots \quad \text{as } t \to \infty, \ (3.7.31)$$

where

$$B_2 = \frac{B_1^2\pi}{2l(3\lambda - 2k)}, \quad B_3 = -\frac{B_1B_2\pi}{4kl}, \quad B_4 = \frac{3B_1B_2\pi}{4l(4\lambda - 3k)}, \quad \dots;$$ (3.7.32)

here, B_1 is an arbitrary constant. For case (ii), namely $\lambda = \delta\pi^2/l^2$, the asymptotic solution is found to be

$$u(x,t) \approx Ae^{-kt}\sin\left(\frac{\pi x}{l}\right) + \left(-\frac{A^2\pi}{2l}t\right)e^{-2kt}\sin\left(\frac{2\pi x}{l}\right)$$

$$+ e^{-3kt}\left(\frac{A^3t}{12\delta}\sin\left(\frac{\pi x}{l}\right) + \frac{3A^3t}{4\delta}\sin\left(\frac{3\pi x}{l}\right)\right) + \dots \quad (3.7.33)$$

as $t \to \infty$. The case $\lambda > \delta\pi^2/l^2$ leads to a more complicated form of the solution (see Srinivasa Rao and Satyanarayana (2008a) for details). In the appendix of the paper by Sachdev et al. (2003), the solution of (3.7.30) was found in a different manner: we used part of the Cole–Hopf transformation, namely, $u(x,t) = -2vG_x(x,t)$, $v = \delta/2$ in (3.7.30), and obtained, after an integration, the equation for $G(x,t)$:

$$G_t - vG_{xx} - vG_x^2 + \lambda G = 0.$$ (3.7.34)

The form

$$G(x,t) = \sum_{n=1}^{\infty} \Phi_n(x)e^{-n(\lambda+a)t}$$ (3.7.35)

of the solution with $u(x,0) = u_0 \sin(\pi x/l)$ was found such that it coincided with the (exact) Fay solution of the plane Burgers equation in the limit $\lambda \to 0$; here $a = v(\pi^2/l^2)$.

The numerical study of (3.7.1), subject to (3.7.4) and (3.7.5), confirmed the accuracy of the analytical solution as $t \to \infty$. We also compared our analytical solution with the (approximate) analytical results of Parker (1981) who considered the GBE

$$\frac{\partial V}{\partial t} + \beta(t)V - \gamma(t)V\frac{\partial V}{\partial x} = \nu(t)\frac{\partial^2 V}{\partial x^2}. \tag{3.7.36}$$

Parker (1981) used a generalised Cole–Hopf transformation and, under some further assumptions, could reduce (3.7.36) to the heat equation. The solution, thus found, agrees, under appropriate conditions, with the large time asymptotic solution that we have obtained.

Sachdev et al. (2005) also sought large time periodic solutions of the modified Burgers equation

$$u_t + u^n u_x = \frac{\delta}{2}u_{xx}, \tag{3.7.37}$$

where δ is greater than 0 and n is an integer greater than or equal to 2. The case $n = 1$ corresponds to the Burgers equation. Equation (3.7.37) displays analytically distinct solutions depending on whether n is odd or n even: for the former it enjoys the antisymmetry property $u(-x,t) = -u(x,t)$. N-wave solutions for the latter have a more complicated structure (see Sachdev and Srinivasa Rao 2000).

Now we summarise the results for (3.7.37) with $n = 2$ and $n = 3$. For the former we have

$$u_t + u^2 u_x = \frac{\delta}{2}u_{xx}, \tag{3.7.38}$$

where $\delta > 0$. Equation (3.7.38) has, for large time, the periodic solution

$$u(x,t) = A_1 e^{-(\delta/2)t} \sin(x - x_0); \tag{3.7.39}$$

it exactly satisfies the linearised form of (3.7.38), namely, the heat equation

$$u_t = \frac{\delta}{2}u_{xx}, \tag{3.7.40}$$

and the conditions

$$u(x,0) = A\sin(x - x_0), \quad -\infty < x < \infty; \quad u(x,t) = u(x + 2\pi, t), \quad t > 0. \tag{3.7.41}$$

Here, x_0 is an arbitrary constant. Our numerical study of (3.7.38), subject to (3.7.41) with $x_0 = 0$, showed that the zeros $x = 0, x = 2\pi$ of the initial profile $u(x,0) = A\sin x$ move as this profile evolves under (3.7.38). So the solution of (3.7.38) and (3.7.41) was sought in the form

$$u(x,t) = A_1 e^{-\varepsilon t} \sin(x - \tilde{x}_0(t)) + U_1(x,t); \quad \varepsilon = \frac{\delta}{2}, \tag{3.7.42}$$

where $\tilde{x}_0(t) \to x_0$ as $t \to \infty$ and $U_1(x,t) \ll A_1 e^{-\varepsilon t} \sin(x - \tilde{x}_0(t))$. The shift of the zero, $\tilde{x}_0(t)$, was assumed in the form

$$\tilde{x}_0(t) = x_0 + x_1 e^{-2\varepsilon t} + x_2 e^{-4\varepsilon t} + x_3 e^{-6\varepsilon t} + \dots. \tag{3.7.43}$$

Following the same procedure as for the nonplanar GBE, we found that

$$U_1(x,t) \sim e^{-3\varepsilon t} \left[c_1 \cos(x - \tilde{x}_0(t)) + c_2 \cos 3(x - \tilde{x}_0(t)) \right], \tag{3.7.44}$$

where

$$\tilde{x}_0(t) = x_0 - \frac{A_1^2}{6\varepsilon} e^{-2\varepsilon t} + O(e^{-4\varepsilon t}), \tag{3.7.45}$$

and

$$c_1 = -\frac{A_1^3}{24\varepsilon}, \quad c_2 = \frac{A_1^3}{24\varepsilon}, \quad x_1 = -\frac{A_1^2}{6\varepsilon}. \tag{3.7.46}$$

Continuing as for the nonplanar Burgers equation the solution with the first three terms was found to be

$$u(x,t) = A_1 e^{-\varepsilon t} \sin y + e^{-3\varepsilon t} \left[c_1 \cos y + c_2 \cos 3y \right]$$
$$+ e^{-5\varepsilon t} \left[c_4 \sin y + c_5 \sin 3y + c_6 \sin 5y \right] + \dots, \tag{3.7.47}$$

where $y = x - \tilde{x}_0(t)$, $\tilde{x}_0(t)$ and c_1, c_2 are given by (3.7.45) and (3.7.46). We observe that the solution (3.7.47) together with (3.7.45) and (3.7.46) involves two arbitrary constants, A_1 and x_0. This is in contrast to the nonplanar Burgers equation which involves only the initial amplitude A_0. This is possibly due to the lack of antisymmetry in the modified Burgers equation (3.7.37) with $n = 2$. For the case $n = 3$, we have the GBE

$$u_t + u^3 u_x = \frac{\delta}{2} u_{xx}. \tag{3.7.48}$$

The large time periodic solution of (3.7.48) subject to (3.7.4) and (3.7.5) was obtained in the same manner as for the nonplanar GBE (3.7.1); here, the transformation $x \to -x, u \to -u$ leaves (3.7.48) invariant. The periodic solution for (3.7.48) was found in the form

$$u(x,t) = e^{-kt} f_0(x,t) + e^{-4kt} f_1(x,t) + e^{-7kt} f_2(x,t) + \dots, \tag{3.7.49}$$

where $f_0, f_1,$ and f_2 are given by

$$f_0(x,t) \equiv f_0(x) = A_1 \sin\left(\frac{\pi x}{l}\right), \tag{3.7.50}$$

$$f_1(x,t) \approx -\frac{A_1^4 \pi}{4l} t \sin\left(\frac{2\pi x}{l}\right) + \frac{A_1^4 l}{48\delta\pi} \sin\left(\frac{4\pi x}{l}\right)$$
$$\equiv B_1 t \sin\left(\frac{2\pi x}{l}\right) + B_2 \sin\left(\frac{4\pi x}{l}\right), \tag{3.7.51}$$

$$f_2(x,t) = g_3(t) \sin\left(\frac{\pi x}{l}\right) + g_4(t) \sin\left(\frac{3\pi x}{l}\right) + g_5(t) \sin\left(\frac{5\pi x}{l}\right) + g_6(t) \sin\left(\frac{7\pi x}{l}\right), \tag{3.7.52}$$

where

$$g_3(t) = -\frac{l^2}{3\delta\pi^2} \left[D_1 t + E_1 + \frac{l^2 D_1}{3\delta\pi^2} \right], \tag{3.7.53}$$

$$g_4(t) = \frac{l^2}{\delta \pi^2} \left[D_2 t + E_2 - \frac{l^2 D_2}{\delta \pi^2} \right],$$

$$g_5(t) = \frac{l^2}{9\delta \pi^2} \left[D_3 t + E_3 - \frac{l^2 D_3}{9\delta \pi^2} \right],$$

$$g_6(t) = \frac{l^2 E_4}{21\delta \pi^2},$$

and

$$D_1 = \frac{A_1^3 B_1 \pi}{4l}, \quad D_2 = -\frac{9 A_1^3 B_1 \pi}{8l}, \quad D_3 = \frac{5 A_1^3 B_1 \pi}{8l},$$

$$E_1 = -\frac{A_1^3 B_2 \pi}{8l}, \quad E_2 = \frac{9 A_1^3 B_2 \pi}{8l},$$

$$E_3 = -\frac{15 A_1^3 B_2 \pi}{8l}, \quad E_4 = \frac{7 A_1^3 B_2 \pi}{8l}. \tag{3.7.54}$$

Table 3.1 Comparison of numerical and analytical solutions at $t = 250$. u_{anal} is calculated from (3.7.47) with three terms with $\delta = 0.01$; u_{num} is the numerical solution of (3.7.38) satisfying $u(x,0) = \sin x$; old age constant $A_1 = -0.3129$, $x_0 = 1.8069$. $\tilde{x}_0(t)$ is calculated from the expression (3.7.45). (Sachdev et al. 2005. Copyright © 2005 MIT and Blackwell Publishing, USA. Reprinted with permission. All rights reserved.)

x	u_{num}	u_{anal}
0.0000	0.0884	0.0869
0.0628	0.0895	0.0882
0.2670	0.0909	0.0906
0.4744	0.0882	0.0883
0.7320	0.0774	0.0770
1.1624	0.0408	0.0396
1.4074	0.0154	0.0143
1.6211	−0.0056	−0.0065
1.8441	−0.0250	−0.0258
2.0672	−0.0415	−0.0420
2.2682	−0.0541	−0.0541
2.4787	−0.0653	−0.0645
2.6892	−0.0747	−0.0730
3.0976	−0.0876	−0.0858
3.3301	−0.0907	−0.0901
3.5531	−0.0895	−0.0896
3.7762	−0.0826	−0.0824
4.0118	−0.0678	−0.0670
4.2066	−0.0505	−0.0494
4.4202	−0.0287	−0.0276
4.6496	−0.0053	−0.0043
4.8632	0.0147	0.0156
5.5323	0.0609	0.0604
5.7554	0.0716	0.0701
5.9502	0.0792	0.0772
6.2832	0.0884	0.0869

As we observed earlier, the modified Burgers equation (3.7.37) possesses distinct forms of the solution depending on whether n is odd or even. For $n = 2$, for example, the asymptotic solution involves two arbitrary constants whereas for $n = 3$ it involves only one arbitrary constant. These constants represent memory of the initial conditions and must be found by matching the asymptotic solution with the numerical solution. The constants A_1 and x_0 in the solution (3.7.47) for $n = 2$ were obtained by matching the numerical solution with the linear form of the solution for large t when the nonlinear terms become negligible. With these values of A_1 and x_0, our analytic solution gives a good description of the asymptotic behaviour; this is clearly brought out by comparison of this solution with the numerical solution in Table 3.1 For $n = 3$, there is only one unknown constant, A_1, in the asymptotic form of the solution. The asymptotic solution again agreed very well with the numerical results.

3.8 Asymptotic behaviour of some generalised Burgers equations via balancing argument

As we have observed earlier, the self-similar solutions (when they exist) describe behaviour of the original nonlinear PDEs subject to some singular initial conditions. These special solutions arise only if the basic system of nonlinear PDEs enjoys certain symmetries (Sachdev 2000, Mayil Vaganan 1994). That is generally not the case. Other avenues such as exact linearisation or transformation to simpler PDEs better amenable to analysis are rather limited. One must therefore look for other means to find asymptotic solutions which may subsequently be improvised or may even be rendered 'exact' in some series form. This is what was accomplished in the context of generalised Burgers equations by Grundy and his collaborators. Here we describe in detail the work of Grundy et al. (1994) for the nonplanar Burgers equation with a more general convective term:

$$u_t + u^\alpha u_x = \frac{\delta}{2} u_{xx} - \frac{ju}{2t}, \quad \alpha > 0, \; j > 0. \tag{3.8.1}$$

Here, $j = 0, 1, 2$ refer to plane, cylindrical, and spherical symmetry, respectively. The case $j = 0, \alpha = 1$ corresponds to the plane Burgers equation which can be exactly linearised to the heat equation via Cole–Hopf transformation and hence fully analysed (see Sachdev 1987). We later summarise the results (via this method) for other GBEs such as one with nonlinear damping. Sachdev and Nair (1987) first sought similarity solutions of (3.8.1) in the form

$$u = t^{-1/2\alpha} f(\eta), \quad \eta = \frac{x}{(2\delta t)^{1/2}}. \tag{3.8.2}$$

Thus, (3.8.1) reduces to the ODE

$$f'' + 2\eta f' + \frac{2(1-\alpha j)}{\alpha} f - 2^{3/2}\delta^{-1/2}f^\alpha f' = 0. \qquad (3.8.3)$$

Sachdev and Nair (1987) solved the connection problem for (3.8.3) subject to the conditions

$$f \sim A\exp\left(-\eta^2\right)H_\gamma(\eta) \text{ as } \eta \to \infty, \qquad (3.8.4)$$
$$f \to 0 \text{ as } \eta \to -\infty, \qquad (3.8.5)$$
$$|f| < \infty, \qquad (3.8.6)$$

where $\gamma = (1/\alpha) - (j+1)$ and A is a positive constant; $H_\gamma(\eta)$ is the Hermite function. The condition (3.8.4) states that the solution of (3.8.3) tends to the solution of its linear form as $\eta \to +\infty$. The numerical solution of (3.8.3) subject to (3.8.4) revealed the following features of the solution for special values of $j = 0, 1, 2$. (i) If $1/(j+2) < \alpha < 1/(j+1)$, the solution of (3.8.3) and (3.8.4) vanishes at a finite point. (ii) For $1/(j+1) \le \alpha < 1/j$, the solution of (3.8.3) and (3.8.4) is positive over $(-\infty, \infty)$ and tends to zero as $\eta \to -\infty$. (iii) For $\alpha j = 1$, the solution of (3.8.3) and (3.8.4) is positive on $(-\infty, \infty)$ and tends monotonically to a nonzero constant value as $\eta \to -\infty$. (iv) For $\alpha j > 1$, the solution of (3.8.3) and (3.8.4) is positive and diverges to $+\infty$ as $\eta \to -\infty$.

Sachdev and Nair (1987) also numerically demonstrated the intermediate asymptotic character of the self-similar solution for $\alpha = 1/(j+1)$.

In a more recent study, Srinivasa Rao et al. (2002) studied both the connection problem (3.8.3) and (3.8.4) and an initial value problem for (3.8.3). For the former they showed that (i) f has a finite zero when $0 < \alpha < 1/(j+1)$, (ii) f is positive on $(-\infty, \infty)$ and decays algebraically to zero as $\eta \to -\infty$ provided that $1/(j+1) < \alpha < 1/j$, (iii) f is monotonic and tends to a nonzero positive constant as $\eta \to -\infty$ if $\alpha j = 1$, and (iv) f is monotonic and becomes unbounded as $\eta \to -\infty$ if $\alpha j > 1$. It was further shown that, for $1/(j+1) < \alpha < 1/j$, there exists a positive solution of (3.8.3) which satisfies (3.8.4)–(3.8.6) and decays algebraically to zero as $\eta \to -\infty$.

The IVP for (3.8.3) was solved subject to the conditions

$$f(0) = v, \quad f'(0) = 0. \qquad (3.8.7)$$

A variety of solutions of the IVP (3.8.3) and (3.8.7) exists depending on the parameters α and j; there is also a strong dependence on the amplitude parameter v. These solutions include all the behaviours at $\eta = \pm\infty$, summarised above in the context of the connection problem. Reference may be made to Srinivasa Rao et al. (2002) for further details. The (possible) intermediate asymptotic character of these solutions has not yet been explored analytically or numerically.

Numerical study of Sachdev and Nair (1987) shows that the self-similar solution of (3.8.1) for $\alpha > 1/(j+1)$ does not constitute intermediate asymptotics for solutions of (3.8.1) subject to initial data with compact support. Therefore, we seek more general bounded solutions of (3.8.1) in the entire (α, j)-plane by relaxing the requirement of self-similarity. Thus, we may seek different dominant balances of the terms in (3.8.1) in the lowest order approximation (of the solution) and improve upon them by including the effect of the terms which were ignored earlier. The problem, following this procedure, does not reduce to solving ODEs but to simpler PDEs accruing from different (possible) dominant balances. This is indeed the approach adopted by Grundy et al. (1994). They sought large time asymptotic behaviour of solutions of a slightly different form of (3.8.1) which have nonnegative initial data and decay sufficiently rapidly as $x \to \pm\infty$ to ensure that the area under the profile,

$$M(t) = \int_{-\infty}^{\infty} u(x,t)dx, \qquad (3.8.8)$$

remains finite for all time. It readily follows from the new version of (3.8.1), namely,

$$u_t = u_{xx} - (u^{\alpha+1})_x - \frac{ju}{2t}, \quad j > 0, \ \alpha > 0, \qquad (3.8.9)$$

that

$$\frac{dM}{dt} = \int_{-\infty}^{\infty} \left[u_{xx} - (u^{\alpha+1})_x - \frac{ju}{2t} \right] dx \qquad (3.8.10)$$

$$= [u_x]_{-\infty}^{\infty} - \left[u^{\alpha+1} \right]_{-\infty}^{\infty} - \frac{jM}{2t}. \qquad (3.8.11)$$

Therefore, provided u and u_x tend to zero as $|x| \to \infty$, we have

$$Mt^{j/2} = M(1) = M_1, \ \text{say},$$

or

$$M(t) = M_1 t^{-j/2}. \qquad (3.8.12)$$

Thus, we seek solution of (3.8.9) with the bounded initial data

$$u(x,1) = u_1(x) \qquad (3.8.13)$$

which has either finite support or vanishes sufficiently rapidly as $|x| \to \infty$. It seems natural, and numerical results seem to suggest, that we may generalise the self-similar concept and introduce the form

$$u(x,t) = t^{-a}v(\eta,t), \quad \eta = xt^{-\delta}, \qquad (3.8.14)$$

where the constants $a > 0$ and $\delta > 0$ may be chosen to meet other requirements regarding the solutions. Substituting (3.8.14) into (3.8.9) we obtain the following PDE for $v = v(\eta,t)$,

$$tv_t + \left(\frac{j}{2} - a \right)v - \delta\eta v_\eta = t^{1-2\delta}v_{\eta\eta} - t^{1-\delta-a\alpha}\left(v^{\alpha+1} \right)_\eta. \qquad (3.8.15)$$

The so-called balancing argument compares the relative importance, as $t \to \infty$, of the terms on the RHS of (3.8.15) involving t explicitly. The simplest situation relates to the case when diffusion dominates nonlinear convection. Thus, assuming that $tv_t = o(1), \eta = O(1)$, and all the η derivatives are bounded as $t \to \infty$, the case for which diffusion is dominant, requires that the first term on the RHS must (asymptotically) balance the terms on the left, the second term being less important. Thus we find that

$$1 - 2\delta = 0 \quad \text{and} \quad 1 - \delta - a\alpha < 0, \tag{3.8.16}$$

implying that

$$\delta = \frac{1}{2}, \quad a > \frac{1}{2\alpha}. \tag{3.8.17}$$

Expressing the mass M in (3.8.8) in terms of the variables v and η, we have

$$M = M_1 t^{-j/2} = \int_{-\infty}^{\infty} u(x,t)dx = t^{\delta - a} \int_{-\infty}^{\infty} v(\eta,t)d\eta. \tag{3.8.18}$$

Because, to the lowest order, v is assumed to be a function of η alone, one may write the large time expansion for $v(\eta,t)$ as

$$v(\eta,t) = v_0(\eta) + o(1), \quad \eta = O(1). \tag{3.8.19}$$

Therefore, (3.8.18) becomes

$$M_1 = t^{\delta - a + j/2} \left[\int_{-\infty}^{\infty} v_0(\eta)d\eta + o(1) \right]. \tag{3.8.20}$$

Inasmuch as M_1 is a constant and $\delta = 1/2$, we must have

$$a = \frac{j+1}{2} \tag{3.8.21}$$

and hence the inequality in (3.8.17) becomes

$$\alpha > \frac{1}{j+1}. \tag{3.8.22}$$

This is the region of the (α, j)-plane where diffusion dominates nonlinear convection. Putting (3.8.19) into (3.8.15) we find the equation satisfied by $v_0(\eta)$:

$$v_0'' + \frac{\eta}{2}v_0' + \frac{v_0}{2} = 0. \tag{3.8.23}$$

The general solution of (3.8.23) is

$$v_0 = A\exp(-\eta^2/4)\int_0^{\eta} \exp(r^2/4)dr + B\exp(-\eta^2/4), \tag{3.8.24}$$

where A and B are constants. For the solution v_0, given by (3.8.24), to converge as $\eta \to \infty$, we must require that $A = 0$. Substituting (3.8.24) with $A = 0$ into (3.8.20), we evaluate the constant B and hence obtain the asymptotic behaviour

$$v_0 = \frac{M_1}{2\sqrt{\pi}} \exp\left(-\frac{\eta^2}{4}\right) \tag{3.8.25}$$

to the lowest order. Thus, the asymptotic solution u (as $t \to \infty$) of (3.8.9) for $\alpha > 1/(j+1)$ may be written as

$$u(x,t) = \frac{M_1}{2\sqrt{\pi}} t^{-(j+1)/2} \exp\left(-\frac{x^2}{4t}\right) \{1 + o(1)\}. \tag{3.8.26}$$

The result (3.8.26) holds uniformly in $-\infty < x < \infty$. Substituting

$$v(\eta,t) = v_0(\eta) + \varepsilon(\eta,t), \tag{3.8.27}$$

into (3.8.15) and solving the equation for the perturbation term ε and so on, one may check for the special case $\alpha = 2/(j+1)$ that

$$v(\eta,t) = v_0(\eta) - 2\left(\frac{M_1}{2\sqrt{\pi}}\right)^{2/(j+1)} \sqrt{\frac{j+1}{j+3}} v_0' t^{-1/2} \log t + O\left(t^{-1/2}\right) \tag{3.8.28}$$

which may conveniently be written as

$$v(\eta,t) = v_0(\eta_1) + O\left(t^{-1/2}\right), \tag{3.8.29}$$

where

$$\eta_1 = \eta - 2\left(\frac{M_1}{2\sqrt{\pi}}\right)^{2/(j+1)} \sqrt{\frac{j+1}{j+3}} t^{-1/2} \log t. \tag{3.8.30}$$

Figure 3.7 shows the solution (3.8.14), (3.8.28)–(3.8.30) for $\alpha = 2/3$, $j = 2$ for different values of $t = 500, 1500$. The initial condition for the numerical solution for this set of parameters was chosen to be

$$u_0(x) = \frac{\{H(x+1) - H(x-1)\}}{2}, \tag{3.8.31}$$

where H denotes the Heaviside function. This gives $M_1 = 1$ (see (3.8.18)). Figure 3.8 shows $v = ut^{3/2}$ versus η for different times. The agreement of the numerical solution with the analytic asymptotic results is remarkable.

Next we consider the case for which both nonlinear convection and diffusion terms in (3.8.15) balance those on the left-hand side as $t \to \infty$. This requires that

$$1 - 2\delta = 1 - \delta - a\alpha = 0; \tag{3.8.32}$$

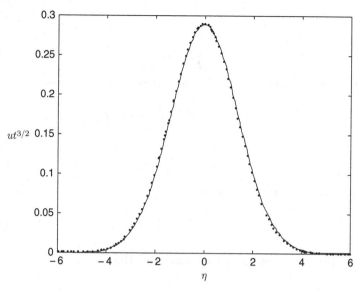

Fig. 3.7 Convergence of the numerical solution of (3.8.9) and (3.8.31) to the large time solution (3.8.14) and (3.8.28) for $\alpha = 2/3, j = 2$ at $t = 500$ (—), 1500 (...). (Grundy et al. 1994. Copyright © 1994 Narosa Publishing House, New Delhi. Reprinted with permission. All rights reserved.)

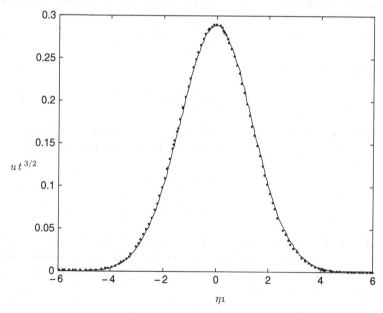

Fig. 3.8 Same as in Figure 3.7 with η replaced by the scaled coordinate η_1 (see (3.8.30)). (Grundy et al. 1994. Copyright © 1994 Narosa Publishing House, New Delhi. Reprinted with permission. All rights reserved.)

that is,

$$\delta = \frac{1}{2}, \quad a = \frac{1}{2\alpha}. \tag{3.8.33}$$

The mass invariance condition (3.8.18) in the present case becomes

$$M_1 = t^{[\alpha(j+1)-1]/2\alpha} \int_{-\infty}^{\infty} v(\eta,t)d\eta. \tag{3.8.34}$$

If we write

$$v(\eta,t) = v_0(\eta) + o(1) \tag{3.8.35}$$

in (3.8.34), as $t \to \infty$, $\eta = O(1)$, we obtain

$$M_1 = t^{[\alpha(j+1)-1]/2\alpha} \int_{-\infty}^{\infty} v_0(\eta)d\eta. \tag{3.8.36}$$

We assume that the integral in (3.8.36) converges. Then,

$$\alpha = \frac{1}{j+1} \tag{3.8.37}$$

and hence the mass

$$M_1 = \int_{-\infty}^{\infty} v_0(\eta)d\eta. \tag{3.8.38}$$

is constant. Continuing as before, we find that, to leading order terms, v_0 is governed by

$$v_0'' - (v_0^{\alpha+1})' + \frac{\eta}{2}v_0' + \frac{v_0}{2} = 0 \tag{3.8.39}$$

which, on integration, yields

$$v_0' - v_0^{\alpha+1} - \frac{\eta}{2}v_0 = A, \tag{3.8.40}$$

where A is a constant. For the integral in (3.8.38) to converge we require that $v_0 = o(\eta^{-1})$ as $|\eta| \to \infty$; therefore, $A = 0$ in (3.8.40). An integration of (3.8.40) with $A = 0$ yields

$$v_0 = \frac{\exp(-\eta^2/4)}{\left\{B + \sqrt{\alpha\pi} \operatorname{erfc}\left(\frac{\eta\sqrt{\alpha}}{2}\right)\right\}^{1/\alpha}}, \tag{3.8.41}$$

where B is an arbitrary constant which may be found from (3.8.38) in terms of M_1. In fact (3.8.41) is the exact similarity solution of (3.8.9) which was first found by Sachdev and Nair (1987) for $\alpha = 1/(j+1)$. In the present context we write the solution for $\alpha = 1/(j+1)$ in the form

$$u = t^{-1/2\alpha}v_0(\eta)\{1 + o(1)\} \tag{3.8.42}$$

as $t \to \infty$, where $v_0(\eta)$ is given by (3.8.41) in terms of $\eta = x/t^{1/2}$. This result is also uniform in $-\infty < x < \infty$. The numerical solution of (3.8.9) for $\alpha = 1/3, j = 2$ (satisfying $\alpha = 1/(j+1)$) with (3.8.31) as the initial condition as well as the asymptotic solution (3.8.41) is depicted in Figure 3.9. This plot shows $ut^{3/2}$ versus η. The excellent agreement of the large time asymptotic solution with the numerical solution is apparent. We show later how to write an 'exact solution' of (3.8.9) for $\alpha > 1/(j+1)$ as an infinite series which, in the limit $\alpha \to 1/(j+1)$, tends to the closed form solution (3.8.41).

The case $\alpha < 1/(j+1)$ presents more difficulties. To understand it better, we briefly review the exact single hump solution of the Burgers equation, the case $j = 0, \alpha = 1$ in (3.8.1). By using the Cole–Hopf transformation, the solution of the Burgers equation subject to the δ function initial condition

$$u(x,0) = A\delta(x)$$

may be explicitly written as

$$u(x,t) = \sqrt{\frac{\delta}{2t}} \frac{(e^R - 1)e^{-x^2/2\delta t}}{\sqrt{\pi} + (e^R - 1)\int_{x/\sqrt{2\delta t}}^{\infty} e^{-r^2} dr} \tag{3.8.43}$$

(see Whitham 1974, Sachdev 1987); here $R = A/\delta$ is a constant in the present case and referred to as the Reynolds number. The Reynolds number R (a dimensionless

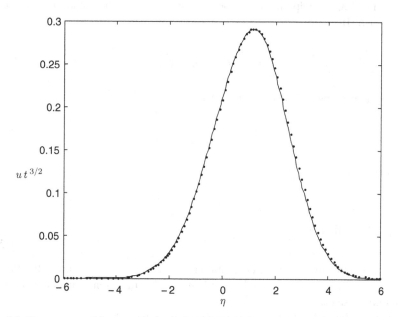

Fig. 3.9 Convergence of the numerical solution of (3.8.9) and (3.8.31) to the large time solution (3.8.14) and (3.8.41) for $\alpha = 1/3, j = 2$ at $t = 500$ (—), 1500 (...). (Grundy et al. 1994. Copyright © 1994 Narosa Publishing House, New Delhi. Reprinted with permission. All rights reserved.)

number) represents the ratio of nonlinear effects to diffusive effects and facilitates the discussion of the solution for large or small viscosity.

For $R \ll 1$, diffusion dominates convection and (3.8.43) may be approximated by

$$u(x,t) \approx \sqrt{\frac{\delta}{2\pi t}} R \exp\left(-\frac{x^2}{2\delta t}\right)$$

$$= \frac{A}{\sqrt{2\pi \delta t}} \exp\left(-\frac{x^2}{2\delta t}\right). \tag{3.8.44}$$

This, in fact, is the exact solution of the heat equation, $u_t = (\delta/2)u_{xx}$. On the other hand, if convection dominates diffusion, we may write (3.8.43) as

$$u = \sqrt{\frac{2A}{t}} v(\eta, R), \quad \eta = \frac{x}{(2At)^{1/2}}, \tag{3.8.45}$$

where

$$v(\eta, R) = \frac{e^R - 1}{2\sqrt{R}} \frac{e^{-\eta^2 R}}{\sqrt{\pi} + (e^R - 1)\int_{\eta\sqrt{R}}^{\infty} \exp(-r^2)dr}$$

$$\sim \frac{1}{2\sqrt{R}} \frac{e^{R(1-\eta^2)}}{\sqrt{\pi} + e^R \int_{\eta\sqrt{R}}^{\infty} \exp(-r^2)dr} \quad \text{as } R \to \infty. \tag{3.8.46}$$

The form (3.8.46) helps to write approximate forms of the solution in different ranges of η. Because

$$\int_{\zeta}^{\infty} e^{-r^2} dr \sim \frac{e^{-\zeta^2}}{2\zeta} \quad \text{as } \zeta \to \infty, \tag{3.8.47}$$

we may approximate (3.8.46) for $\eta > 0$ as

$$v \sim \frac{\eta}{1 + 2\eta\sqrt{\pi R} e^{R(\eta^2 - 1)}} \tag{3.8.48}$$

and, therefore, for $0 < \eta < 1$, we have

$$v \sim \eta \quad \text{as } R \to \infty; \tag{3.8.49}$$

for $\eta > 1$, we find that

$$v \to 0 \quad \text{as } R \to \infty. \tag{3.8.50}$$

Thus, we get the inviscid or outer solution $v \sim \eta$ in $0 < \eta < 1$ and zero solution outside. This inviscid solution must be supplemented by transition layers in the leading and trailing edges.

In the leading edge $\eta \gtrsim 1$, we may approximate (3.8.48) by

$$v \approx \frac{1}{1 + 2\sqrt{\pi R} e^{R(\eta^2 - 1)}} \tag{3.8.51}$$

which shows that the transition layer near $\eta \approx 1$ is $O(R^{-1})$. There is another (weaker) transition layer about $\eta = 0$ which smooths out the discontinuity in the derivative at $\eta = 0$: $v = 0$ in $\eta < 0$ and $v = \eta$ in $0 < \eta < 1$. It may be observed from the denominator of (3.8.46) that this transition layer occurs when $\eta = O(R^{-1/2})$. We may approximate (3.8.46) in this layer as

$$v \approx \frac{e^{-R\eta^2}}{2\sqrt{R}\int_{\eta\sqrt{R}}^{\infty} e^{-r^2}dr}. \tag{3.8.52}$$

We infer from (3.8.46), (3.8.51), and (3.8.52) that for some fixed time and $R \to \infty$, the leading edge becomes a discontinuity in u and the trailing edge displays a discontinuity in u_x.

We observe the same behaviour for the nonplanar Burgers equation for which we have no Cole–Hopf transformation; we must deal with this equation directly using the balancing argument or otherwise.

We return now to (3.8.15) and consider the third possibility that convection dominates diffusion for large time when $\eta = O(1)$. Then (3.8.15) shows that, for this to happen, we must have

$$1 - \delta - a\alpha = 0 \quad \text{and} \quad \delta > \frac{1}{2} \tag{3.8.53}$$

so that convection balances the terms on the left. Here, also, we assume that $v_t = o(1)$. The relations (3.8.53), together with the mass conservation condition (3.8.20), give

$$a = \frac{(2+j)}{2(1+\alpha)}, \tag{3.8.54}$$

$$\delta = \frac{(2-\alpha j)}{2(1+\alpha)} \tag{3.8.55}$$

and

$$\alpha < \frac{1}{j+1}. \tag{3.8.56}$$

The condition (3.8.56), in conjunction with (3.8.22) and (3.8.37), completes (α, j) parameter space. Using the constraints (3.8.54) and (3.8.55) in (3.8.15) we get

$$tv_t - \frac{(2+j)}{2(\alpha+1)}v - \frac{(2-\alpha j)}{2(1+\alpha)}\eta v_\eta + \frac{jv}{2} + (v^{\alpha+1})_\eta = t^{[\alpha(j+1)-1]/(\alpha+1)}v_{\eta\eta}. \tag{3.8.57}$$

Now, for $t \to \infty, \eta = O(1)$, we substitute

$$v(\eta,t) = v_0(\eta) + o(1)$$

into (3.8.57) and obtain, to leading order,

$$(v_0^{\alpha+1})' - \frac{(2-\alpha j)}{2(1+\alpha)}\eta v_0' + \frac{(j\alpha-2)}{2(\alpha+1)}v_0 = 0, \tag{3.8.58}$$

which may be integrated to yield

$$v_0\left\{(2-\alpha j)\eta - 2(\alpha+1)v_0^\alpha)\right\} = C, \tag{3.8.59}$$

where C is the constant of integration. We must choose C negative to avoid multi-valuedness of $v_0(\eta)$. Therefore, as $\eta \to -\infty$, (3.8.59) gives

$$v_0 \sim \frac{C}{(2-\alpha j)\eta} \tag{3.8.60}$$

which, however, would make the integral $\int_{-\infty}^\infty v_0(\eta)d\eta$ diverge. We must therefore choose $C = 0$. Let the solution be zero for $\eta < 0$ and for $\eta > \eta_1$, a constant, so that the above integral converges. Thus, we have

$$v_0 = \begin{cases} 0, & \eta < 0, \\ \left\{\dfrac{(2-\alpha j)\eta}{2(\alpha+1)}\right\}^{1/\alpha}, & 0 \le \eta \le \eta_1, \\ 0, & \eta > \eta_1. \end{cases} \tag{3.8.61}$$

The conservation of mass (3.8.20) gives

$$M_1 = \left\{\frac{(2-\alpha j)}{2(\alpha+1)}\right\}^{1/\alpha} \int_0^{\eta_1} \eta^{1/\alpha}d\eta$$

or

$$\eta_1 = \left\{\frac{M_1(1+\alpha)}{\alpha}\right\}^{\alpha/(\alpha+1)} \left\{\frac{2(\alpha+1)}{(2-\alpha j)}\right\}^{1/(\alpha+1)}, \tag{3.8.62}$$

to leading order. If M_1 is known, (3.8.62) gives the value of η_1.

The numerical solution for $ut^{(2+j)/2(\alpha+1)}$ as a function of η for the parametric values $j = 1$ and $\alpha = 1/4$ is shown in Figure 3.10. Here, the value of the unscaled diffusion coefficient δ was chosen to be 10^{-2}; this, however, does not affect the outer solution because the particular combination of variables chosen in the analysis eliminates δ so that η_1 in (3.8.62) is independent of δ. The numerical results confirm the veracity of the outer solution. It is clear from (3.8.61) that the solution to this order is not uniformly valid for all η. As we showed earlier for the Burgers equation (see (3.8.46) and (3.8.52)), it must be supplemented by leading and trailing edges to make it continuous for all x; it would thus have an infinite support. The discontinuities at $\eta = 0$ and $\eta = \eta_1$ must be smoothed out by taking into account the effect of the diffusion term. To that end we first introduce the scaled variable

$$y = (\eta - \eta_1)t^\beta, \quad \beta > 0 \tag{3.8.63}$$

near $\eta = \eta_1$ and transform (3.8.57) in terms of $v = v(y,t)$. We get

$$\left[tv_t + \left\{\beta - \frac{(2-\alpha j)}{2(1+\alpha)}\right\}yv_y + \frac{(j\alpha-2)}{2(\alpha+1)}v\right]t^{-\beta} - \frac{(2-\alpha j)}{2(\alpha+1)}\eta_1 v_y + \left(v^{\alpha+1}\right)_y$$
$$= t^{\beta-[1-\alpha(j+1)]/(\alpha+1)}v_{yy}. \tag{3.8.64}$$

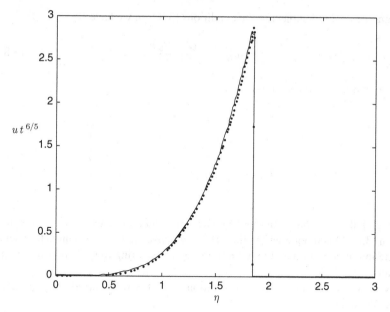

Fig. 3.10 Convergence of the numerical solution of (3.8.9) and (3.8.31) to the large time solution (3.8.81) for $\alpha = 1/4, j = 1$ at $t = 500$ (—), 1500 (...). (Grundy et al. 1994. Copyright © 1994 Narosa Publishing House, New Delhi. Reprinted with permission. All rights reserved.)

To include the effect of the diffusion term on the RHS of (3.8.64) as $t \to \infty$, we must choose

$$\beta = \frac{\{1 - \alpha(j+1)\}}{(\alpha + 1)}. \tag{3.8.65}$$

Now letting

$$v(y,t) = w_0(y) + o(1) \tag{3.8.66}$$

as $t \to \infty$, $y = O(1)$, and substituting it into (3.8.64), we get, to the leading order, the ODE

$$w_0'' - \left(w_0^{\alpha+1}\right)' + \frac{(2 - \alpha j)}{2(\alpha + 1)} \eta_1 w_0' = 0, \tag{3.8.67}$$

where prime now denotes the derivative with respect to y. Equation (3.8.67) must be solved subject to the matching conditions arising from the outer solution (3.8.61) for $\eta < \eta_1, t \to \infty$. Thus, we require that

$$w_0 \to \left\{ \frac{(2 - \alpha j)\eta_1}{2(\alpha + 1)} \right\}^{1/\alpha} \quad \text{as } y \to -\infty. \tag{3.8.68}$$

In addition, we must satisfy the vanishing conditions as $y \to \infty$:

$$w_0, w_0' \to 0 \quad \text{as } y \to +\infty. \tag{3.8.69}$$

Integrating (3.8.67) subject to the conditions (3.8.69), we have

$$w_0' = w_0^{\alpha+1} - \frac{(2-\alpha j)}{2\alpha+1} \eta_1 w_0 \tag{3.8.70}$$

which, in turn, has the solution

$$w_0 = \left\{ \frac{\eta_1(2-\alpha j)}{2(\alpha+1)\left[1+\exp\left(\alpha\eta_1 \dfrac{(2-\alpha j)}{2(\alpha+1)}(y-y_0)\right)\right]} \right\}^{1/\alpha}, \tag{3.8.71}$$

satisfying the end condition (3.8.68). Here, y_0 is an arbitrary constant. Thus the solution (3.8.71) is unique within this arbitrary constant. This constant is the memory of the initial condition and may be found by considering higher-order terms in the expansion (3.8.66).

Now we correct the large time solution near $\eta = 0$ by introducing the trailing edge. We write

$$\zeta = \eta t^b, \quad b > 0. \tag{3.8.72}$$

The (nonzero) outer solution (3.8.61) in terms of ζ becomes

$$v_0 = \left\{ \frac{(2-\alpha j)}{2(\alpha+1)} \right\}^{1/\alpha} \zeta^{1/\alpha} t^{-b/\alpha}. \tag{3.8.73}$$

Motivated by the form (3.8.73) we write

$$v = t^{-b/\alpha} \tau(\zeta, t), \tag{3.8.74}$$

where $\zeta = O(1)$, corresponding to $t \to \infty$, $\eta \to 0$. Equation (3.8.57) now becomes

$$t\tau_t + \left\{ b - \frac{(2-\alpha j)}{2(\alpha+1)} \right\} \zeta \tau_\zeta + \left\{ \frac{\alpha j-2}{2(\alpha+1)} - \frac{b}{\alpha} \right\} \tau + \left(\tau^{\alpha+1}\right)_\zeta$$
$$= t^{2b-[1-\alpha(j+1)]/(\alpha+1)} \tau_{\zeta\zeta}. \tag{3.8.75}$$

To make the diffusion term on the RHS of (3.8.75) comparable to those on the left as $t \to \infty$, we must choose

$$b = \frac{[1-\alpha(j+1)]}{2(\alpha+1)}. \tag{3.8.76}$$

Now seeking the solution of (3.8.75) in the form

$$\tau(\zeta, t) = \tau_0(\zeta) + o(1) \tag{3.8.77}$$

as $t \to \infty$, $\zeta = O(1)$, we get, to leading order,

$$\tau_0'' - (\tau_0^{\alpha+1})' + \frac{\zeta \tau_0'}{2} + \frac{(1-j\alpha)}{2\alpha}\tau_0 = 0. \tag{3.8.78}$$

The boundary condition as $x \to -\infty$ now becomes

$$\tau_0 \to 0, \quad \zeta \to -\infty. \tag{3.8.79}$$

The matching condition (3.8.73), relating (3.8.78) to the outer solution, becomes

$$\tau_0 \sim \left\{ \frac{2-\alpha j}{2(\alpha+1)} \right\}^{1/\alpha} \zeta^{1/\alpha}, \quad \zeta \to +\infty. \tag{3.8.80}$$

Thus, one must solve the nonlinear ODE (3.8.78) for τ_0 subject to the end conditions (3.8.79) and (3.8.80). Cazenave and Escobedo (1994) showed that a 'similar' boundary value problem for $j = 0$ posed by Grundy et al. (1994) has a unique solution with exponential decay as $\zeta \to \infty$.

We may now summarise the asymptotic results in terms of the original variables for the case $\alpha < 1/(j+1)$ in the language of singular perturbation theory in the three layers we have discussed above.

In the outer region, we have $\eta = x/t^\delta = O(1)$, $0 < \eta < \eta_1$,

$$u(x,t) = t^{-a} \left\{ \frac{(2-\alpha j)x}{2(\alpha+1)t^\delta} \right\}^{1/\alpha} \{1+o(1)\} \tag{3.8.81}$$

$$= \left\{ \frac{(2-\alpha j)x}{2(\alpha+1)t} \right\}^{1/\alpha} \{1+o(1)\} \text{ as } t \to \infty.$$

In the leading edge described by

$$y = \left\{ \frac{x-\eta_1 t^\delta}{t^\delta} \right\} t^\beta = \left\{ x - \eta_1 t^\delta \right\} t^{\alpha(2+\alpha j)/2(\alpha+1)} = O(1), \tag{3.8.82}$$

the solution is given by

$$u(x,t) = t^{-a} w_0(y) \{1+o(1)\}$$

$$= t^{-a} \left\{ \frac{\eta_1(2-\alpha j)}{2(\alpha+1)\left[1+e^{\alpha\eta_1(2-\alpha j)(x-\eta_1 t^\delta - y_0 t^{\delta_1})/2(\alpha+1)}t^{\delta_1}\right]} \right\}^{1/\alpha} \{1+o(1)\} \tag{3.8.83}$$

as $t \to \infty$, where $\delta_1 = \alpha(2+\alpha j)/2(\alpha+1)$ and $a = (2+j)/2(\alpha+1)$. Finally, in the trailing edge,

$$\zeta = \eta t^{[1-\alpha(j+1)]/2(\alpha+1)} = xt^{-1/2}, \tag{3.8.84}$$

we have

$$u = t^{-1/2\alpha} \tau_0 \left(\frac{x}{t^{1/2}} \right) \{1+o(1)\} \tag{3.8.85}$$

as $t \to \infty$ where $\tau_0(\zeta)$ is given by the solution of the boundary value problem (3.8.78)–(3.8.80).

We now summarise the asymptotic results obtained by the balancing argument for a related GBE,

$$u_t = au_{xx} - b\left(u^n\right)_x - cu^p, \quad p \geq 1, \, n \geq 1, \tag{3.8.86}$$

where a, b, and c are positive constants (Grundy 1988). This equation has several interesting special cases of physical interest. For $b = c = 0$, we have the heat equation. The special case $b = 0$ represents the reaction–diffusion equation which has been extensively studied. The case $c = 0$ is a direct generalisation of the Burgers equation with higher-order nonlinearity; it is often referred to as the modified Burgers equation.

Equation (3.8.86) with nonnegative initial data on a finite support was first considered by Sachdev et al. (1986). Supported by extensive computational work, these authors proposed that, for $p = 2n - 1$, $p > 3$, the solution of (3.8.86) for large time tends to the linear solution,

$$u(x,t) \to At^{-1/2} \exp\left(-x^2/4t\right), \tag{3.8.87}$$

which vanishes as $x \to \pm\infty$; the constant A depends only on the initial data. The result (3.8.87) is obtained by retaining only the linear terms in (3.8.86). For $p < 3$, Sachdev et al. (1986) conjectured that

$$u(x,t) \to t^{-1/(p-1)} f(\eta), \quad \eta = xt^{-1/2} \tag{3.8.88}$$

as $t \to \infty$, where $f(\eta)$ is the solution of the boundary value problem

$$af'' + \frac{\eta f'}{2} + \frac{f}{p-1} - b\left(f^{(p+1)/2}\right) - cf^p = 0, \tag{3.8.89}$$

$$f(\pm\infty) = 0. \tag{3.8.90}$$

A detailed numerical solution of (3.8.89) and (3.8.90) showed that it behaves like

$$f \sim A\eta^{(3-p)/(p-1)} \exp\left(-\eta^2/4\right) \quad \text{as } \eta \to +\infty; \tag{3.8.91}$$

the vanishing condition at $\eta = -\infty$ could only be satisfied for a certain range of values of A.

Grundy (1988) studied this problem in great detail in the manner of the nonplanar GBE (3.8.9) above. The parameters a, b, and c in (3.8.86) may be scaled out except when $p = 2n - 1$; in the latter case, the parameter c still persists. Thus, without loss of generality, one may write (3.8.86) as

$$u_t = u_{xx} - \left(u^n\right)_x - cu^p, \quad p \geq 1, \, n \geq 1, \tag{3.8.92}$$

where $c = 1$ when $p \neq 2n - 1$. The initial and boundary conditions for (3.8.92) may be chosen to be

$$u(x,0) = u_0(x) \tag{3.8.93}$$

and

$$u(x,t) \to 0, \quad x \to \pm\infty. \tag{3.8.94}$$

Here, $u_0(x) \geq 0$ has either a compact support or vanishes sufficiently rapidly as $x \to \pm\infty$. Grundy (1988) again introduced the 'mass'

$$M(t) = \int_{-\infty}^{\infty} u(x,t)dx \tag{3.8.95}$$

and, by using (3.8.92)–(3.8.94), concluded that

$$\frac{dM}{dt} = -c \int_{-\infty}^{\infty} u^p dx. \tag{3.8.96}$$

Clearly, if the absorption term is absent, M is invariant in time; if the absorption term is small for large time, M may approach a constant as $t \to \infty$. Here again, Grundy (1988) introduced the independent variables $\eta = xt^{-\delta}$, $\delta > 0$, and t and transformed (3.8.92) to

$$\underbrace{tv_t + \alpha v - \delta\eta v_\eta}_{A} = \underbrace{t^{1-2\delta}v_{\eta\eta}}_{B} - \underbrace{t^{\alpha(n-1)+1-\delta}(v^n)_\eta}_{C} - \underbrace{t^{\alpha(p-1)+1}v^p}_{D} \tag{3.8.97}$$

via the transformation

$$u = t^\alpha v(\eta,t), \tag{3.8.98}$$

where $\alpha < 0$. The respective terms in (3.8.97) were denominated as A, B, C, and D as indicated above. Putting (3.8.98) into (3.8.95) and (3.8.96), we obtain

$$M(t) = t^{\alpha+\delta} \int_{-\infty}^{\infty} v(\eta,t)d\eta \tag{3.8.99}$$

and

$$\frac{dM}{dt} = -ct^{\alpha p+\delta} \int_{-\infty}^{\infty} v^p(\eta,t)d\eta. \tag{3.8.100}$$

Depending on which set of terms in (3.8.97) balances, the following asymptotic behaviours of solutions of (3.8.92) were discovered.

(i)   The linear balance with $n > 2$ and $p > 3$. In this case

$$u(x,t) \to ct^{-1/2}\exp\left(-x^2/4t\right) \tag{3.8.101}$$

as $t \to \infty$ uniformly in x. Here, $M(t) \sim 2c\sqrt{\pi}$.

(ii)   The linear absorption balance with $p < 2n - 1$, $p < 3$. Here the solution was found in the form

$$u(x,t) \sim t^{-1/(p-1)}v_0\left(xt^{-1/2}\right)$$

as $t \to \infty$ uniformly in x, where $v_0(\eta)$ is the solution of the BVP

$$v_0'' + \frac{\eta v_0'}{2} + \frac{v_0}{p-1} - v_0^p = 0, \quad v_0(\pm\infty) = 0. \tag{3.8.102}$$

The asymptotic behaviour of the solution of the BVP (3.8.102) was found to be

$$v_0(\eta) \sim A|\eta|^{(3-p)/(p-1)} \exp\left(-\eta^2/4\right) \left\{1 + O(\eta^{-2})\right\}, \quad \eta \to \pm\infty$$

by Brezis et al. (1986). Here,

$$M(t) \sim t^{(p-3)/2(p-1)} \int_{-\infty}^{\infty} v_0(\eta) d\eta. \tag{3.8.103}$$

(iii) $p = 2n - 1$, $p < 3$. In this case a similarity solution exists (Sachdev et al. (1986)) and is governed by (3.8.89)–(3.8.91) with $a = 1$, $b = 1$. Here,

$$M(t) \sim O(t^{(p-3)/2(p-1)}), \quad t \to \infty. \tag{3.8.104}$$

(iv) $p > n+1$, $n < 2$. In this case, the convection term dominates. The dominant behaviour is given by

$$v_0 = \left(\frac{\eta}{n}\right)^{1/(n-1)} \tag{3.8.105}$$

and must be supplemented by leading and trailing edge solutions. Specifically, for $1 < n < 2$ and $p > n+1$, we have

$$u(x,t) \sim t^{-1/n} \left(\frac{\eta}{n}\right)^{1/(n-1)} \tag{3.8.106}$$

as $t \to \infty$, $\eta = xt^{-1/n} = O(1)$. For $t \to \infty$, $y = (\eta - \eta_1)^{(2-n)/n} = O(1)$, we have

$$u(x,t) \sim t^{-1/n} \left\{\frac{\eta_1}{n\left[1 + \mu^2 e^{\eta_1(n-1)y/n}\right]}\right\}^{1/(n-1)}. \tag{3.8.107}$$

For $t \to \infty$, $\zeta = xt^{-1/2} = O(1)$,

$$u(x,t) \to t^{-1/2(n-1)} W_0(\zeta), \tag{3.8.108}$$

where $W_0(\zeta)$ is governed by the boundary value problem

$$W_0'' + \frac{\zeta}{2} W_0' + \frac{W_0}{2(n-1)} - (W_0^n)' = 0, \tag{3.8.109}$$

$$W_0 \to 0 \text{ as } \zeta \to -\infty \text{ and } W_0 \sim \left(\frac{\zeta}{n}\right)^{1/(n-1)} \text{ as } \zeta \to +\infty. \tag{3.8.110}$$

Here,

$$M(t) \sim \int_0^{\eta_1} v_0(\eta) d\eta = \frac{(n-1)}{n} \left(\frac{1}{n}\right)^{1/(n-1)} \eta_1^{n/(n-1)}. \tag{3.8.111}$$

(v) $2n - 1 < p < n + 1$, $1 < n < 2$. This case refers to the situation when the terms A, C, and D in (3.8.97) balance and B is small, implying that diffusion is subdominant. Here,

$$u(x,t) \to t^{-1/(p-1)} v_0(\eta) \qquad (3.8.112)$$

as $t \to \infty$, $\eta = xt^{-(p-n)/(p-1)} = O(1)$. $v_0(\eta)$ is governed by

$$\frac{v_0}{(p-1)} + \frac{(p-n)}{(p-1)} \eta v_0' = (v_0^n)' + v_0^p$$

with the implicit solution

$$\eta = \frac{n v_0^{n-p}}{(p-n)} \left\{ 1 - \left[1 - (p-1) v_0^{p-1} \right]^{(p-n)/(p-1)} \right\}.$$

Leading and trailing edge solutions must be found as for case (iv) satisfying $v_0(0) = 0$.

(vi) $n = 2$, $p > 3$. In this case the terms A, B, and C balance and $D = O\left(t^{-(p-3)/2}\right)$ is small. The solution here is found to be

$$u(x,t) \sim t^{-1/2} v_0(\eta) \qquad (3.8.113)$$

as $t \to \infty$, $\eta = xt^{-1/2} = O(1)$, where

$$v_0(\eta) = \frac{e^{-\eta^2/4}}{\left[C - \int_0^\eta \exp\left(-s^2/4\right) ds \right]} \qquad (3.8.114)$$

and $C > \sqrt{\pi}$ is an arbitrary constant. Here, $v_0 \to 0$ as $\eta \to \infty$. The asymptotic form (3.8.113) holds uniformly in x. Here,

$$M(t) \to \int_{-\infty}^{\infty} v_0(\eta) d\eta \qquad (3.8.115)$$

as $t \to \infty$.

Grundy (1988) considered two other interesting 'singular' cases which need introduction of complicated similarity variables different from those used for the cases (i)–(vi).

(vii) $p = 3$, $n \geq 2$. In this case the terms D, A, and B balance and C is small, requiring that $\alpha = -\delta = -1/2$; with this choice, $M(t)$ is asymptotically invariant even when the absorption term is present. This is clearly untenable. So Grundy (1988) introduced more general variables involving $\log t$ terms:

$$u = t^{-1/2} (\log t)^\gamma v(\eta, t), \eta = xt^{-1/2} (\log t)^\beta.$$

Proceeding with the balancing argument as in the previous cases, he arrived at the lowest-order solution

$$u(x,t) = \sqrt{\frac{3^{1/2}}{2c}} t^{-1/2} (\log t)^{-1/2} e^{-x^2/4t} \qquad (3.8.116)$$

as $t \to \infty$, uniformly in x. Here,

$$M(t) \sim 2\sqrt{\frac{\pi 3^{1/2}}{2c}} (\log t)^{-1/2}. \qquad (3.8.117)$$

(viii)   For the case $p = n+1$, $1 < n < 2$, Grundy (1988) proceeded in the manner of case (vii) and concluded that, for $t \to \infty$, $y = (\eta - \eta_1)^{(2-n)/n} (\log t)^{-2(n-1)/n} = O(1)$,

$$u(x,t) \to t^{-1/n} (\log t)^{-1/n} \left\{ \frac{\eta_1}{n[1 + \mu^2 e^{\eta_1 (n-1)y/n}]} \right\}^{1/(n-1)}. \qquad (3.8.118)$$

For $t \to \infty$, $\zeta = x t^{-1/2} = O(1)$,

$$u(x,t) \to t^{-1/2(n-1)} W_0(\zeta), \qquad (3.8.119)$$

where the function $W_0(\zeta)$ is the same as for case (iv) (see (3.8.109)–(3.8.110)).

(ix)   For the case $n = 1$, $p \geq 1$, Grundy (1988) introduced the variable $X = x - t$ into (3.8.92) to obatain

$$u_t = u_{XX} - u^p, \qquad (3.8.120)$$

with $u(X,0) = u_0(X)$ and $u(X,t) \to 0$ as $X \to \pm\infty$.

He then carried out the balancing argument for $(3.8.120)_1$ by introducing the usual transformation $u = t^\alpha v(\eta,t)$, $\eta = X t^{-\delta}$, and so on. He finally arrived at the following result for the special case $p = 3$, $n = 1$;

$$u(x,t) \to \left(\frac{\sqrt{3}}{2}\right)^{1/2} t^{-1/2} (\log t)^{-1/2} e^{-x^2/4t} \qquad (3.8.121)$$

as $t \to \infty$, uniformly in x, and

$$M(t) \sim 2\sqrt{\pi} \left(\frac{\sqrt{3}}{2}\right)^{1/2} (\log t)^{-1/2}. \qquad (3.8.122)$$

Other cases with $p < 3$ and $p > 3$ were also treated.

(x)   The final case $p = 1$, $n > 1$ turns out to be singular (see (3.8.112)). One must introduce the transformation

$$u(x,t) = e^{-t} v(\eta,t), \quad \eta = x t^{-1/2} \qquad (3.8.123)$$

in (3.8.86). A balancing argument in the transformed equation leads to the asymptotic solution

$$u(x,t) \sim Ae^{-t-x^2/4t} \qquad (3.8.124)$$

as $t \to \infty$, uniformly in x, and

$$M(t) \sim 2A\sqrt{\pi}t^{1/2}e^{-t}. \qquad (3.8.125)$$

Observing the large diversity of asymptotic solutions of (3.8.86) summarised above, we may marvel at its rich structure. Clearly, further work is needed to fully understand the evolution of solutions of generalised Burgers equations such as (3.8.9) and (3.8.86).

We refer the reader to Sections 4.6–4.8 for a rigourous analysis of solutions for special cases of (3.8.92) and the references therein.

Grundy and his collaborators (Grundy et al. 1994, Dawson et al. 1996, Van Duijn et al. 1997) extended the balancing argument to the study of a related class of non-linear parabolic equations. We quote the main results of Dawson et al. (1996) for the (scaled) two-dimensional equation

$$\frac{\partial}{\partial t}(u+u^p) + \frac{\partial u}{\partial x} = \frac{\partial^2 u}{\partial x^2} + \frac{\partial^2 u}{\partial y^2} \text{ for } (x,y,t) \in Q, \qquad (3.8.126)$$

subject to the initial condition

$$u(x,y,0) = u_0(x,y), \quad (x,y) \in \mathbb{R}^2. \qquad (3.8.127)$$

Here $Q = \{(x,y,t) : -\infty < x,y < \infty, \ t > 0\}$ and $p > 0$. They investigated the large time behaviour of the nonnegative solution of the problem (3.8.126) and (3.8.127), where $u \geq 0$ is the (redefined) concentration which satisfies the mass conservation law

$$\int\int_{\mathbb{R}^2}(u+u^p)(x,y,t)dxdy = \int\int_{\mathbb{R}^2}(u_0+u_0^p)(x,y)dxdy := M \qquad (3.8.128)$$

for all $t \geq 0$. This requires that for all $t \geq 0$, $u(x,y,t) \to 0$ sufficiently fast as $|x|, |y| \to \infty$ so that the term $u+u^p$ is integrable (see (3.8.128)). Using the method of dominant balance discussed in detail earlier they obtained several reduced equations which depend crucially on the values of the parameters that appear in the problem. They first summarised the corresponding one-dimensional results for (3.8.126) and (3.8.127) due to Grundy et al. (1994a). In some cases the reduced equations via the balancing arguments could be solved explicitly whereas others posed intractable analytical or numerical difficulties. However, Dawson et al. (1996) could extract a number of global and local properties of the solution which helped them to form a reasonably complete picture of the kind of asymptotic profiles that may emerge. This work, just as the one-dimensional one, was much supported by the numerical solution of the original initial value problem to confirm the asymptotic nature of the approximate analytical solutions in different parametric domains.

We summarise here considerable work by Sachdev and his collaborators (1994, 1999, 1996) on the 'exact' asymptotic solutions of generalised Burgers equations with N-wave initial conditions. This work was initiated by Sachdev, Joseph and Nair (1994) who constructed an exact representation of the N-wave solution of the nonplanar Burgers equation (3.8.1) with $\alpha = 1$ for $0 < j < 2, j = p/q$, where p and q are positive integers with no common factor. This equation may be written explicitly as

$$u_t + u u_x + \frac{ju}{2t} = \frac{\delta}{2} u_{xx}. \tag{3.8.129}$$

The N-wave solution of (3.8.129) was sought in the form

$$u(x,t) = \frac{(2\delta)^{1/2}\xi}{V(\eta,\tau)}, \quad \xi = \frac{x}{(2\delta t)^{1/2}}, \quad \eta = \xi^2, \quad \tau = t^{1/2q}, \tag{3.8.130}$$

where

$$V(\eta,\tau) = \sum_{i=0}^{\infty} f_i(\tau) \frac{\eta^i}{i!}. \tag{3.8.131}$$

Substitution of (3.8.130) and (3.8.131) into (3.8.129) leads to an infinite system of coupled nonlinear ODEs for $f_i(\tau), i \geq 0$. To embed the inviscid ($\delta = 0$) and linear behaviours of solutions of (3.8.129) into the exact N-wave solution under appropriate limits, the functions $f_i(\tau), i = 0,1,2$ were sought in the polynomial form

$$f_i(\tau) = \tau^q \sum_{k=0}^{p+q} a_k^{(i)} \tau^k. \tag{3.8.132}$$

This approach was inspired by the exact N-wave solutions of the plane Burgers equation ($j = 0$ in (3.8.129)), namely,

$$u(x,t) = \frac{x/t^{1/2}}{t^{1/2} \left[1 + \left(t^{1/2}/c_0\right) \exp\left(x^2/2\delta t\right)\right]}, \quad c_0 \text{ a constant}, \tag{3.8.133}$$

which may be written in the form (3.8.130)–(3.8.132) with $q = 1$, $p = 0$ so that

$$f_0(\tau) = \tau + a\tau^2, \quad f_i(\tau) = a\tau^2, \quad i \geq 1 \tag{3.8.134}$$

(see Whitham 1974; Sachdev 1987). Here $a = 1/c_0$. The N-wave solution (3.8.133) of the plane Burgers equation tends to its inviscid solution as the lobe Reynolds number (area under one lobe of the N-wave divided by δ) tends to infinity. It also asymptotes to the linear solution of the Burgers equation as $t \to \infty$. Srinivasa Rao and Satyanarayana (2008b) constructed large-time asymptotic N-wave solutions for (3.8.129) for $j > 0$ following a perturbative approach similar to that in Section 3.7.

Sachdev et al. (1996) considered the more general equation

$$u_t + u^n u_x + \left(\frac{j}{2t} + \alpha\right)u + \left(\beta + \frac{\gamma}{x}\right)u^{n+1} = \frac{\delta}{2} u_{xx}, \tag{3.8.135}$$

where j, α, β, and γ are nonnegative constants and n is a positive integer. This equation contains a large number of PDEs of physical interest as special cases. For $j = \gamma = 0$, for example, it is the well known Fisher–Burgers equation; for $j = \gamma = \beta = 0$ it is GBE with linear damping (see also Sachdev and Joseph 1994). It may be observed that the N-wave solutions of (3.8.135) are not antisymmetric about $x = 0$ when n is an even integer or when $\beta \neq 0$ (we discuss this case later). These solutions are antisymmetric about $x = 0$ when n is an odd integer and $\beta = 0$. The work of Sachdev et al. (1996) writes out the solution of (3.8.135) such that it tends to the solution of its linearised form as $t \to \infty$. The linearised form of (3.8.135),

$$u_t + \left(\alpha + \frac{j}{2t} \right) u = \frac{\delta}{2} u_{xx}, \tag{3.8.136}$$

has the N-wave solution

$$u(x,t) = C \frac{x/t^{1/2}}{e^{\alpha t} t^{1+j/2}} \exp \left(\frac{-x^2}{2\delta t} \right), \tag{3.8.137}$$

where C is a constant. When n is an odd integer, two kinds of solutions were found for different sets of parameters. The solution is now written in the form

$$u(x,t) = \frac{(2\delta)^{1/2} \xi}{[V(\xi, \tau)]^{1/n}}, \quad \xi = \frac{x}{(2\delta t)^{1/2}}, \quad \tau = t^{1/2}, \tag{3.8.138}$$

where

$$V(\xi, \tau) = \sum_{i=0}^{\infty} f_i(\tau) \frac{\xi^i}{i!}. \tag{3.8.139}$$

(i) For $\beta = 0$ or $\beta \neq 0, \alpha \neq 0$, N-wave solutions of (3.8.135) were found in the form (3.8.138) and (3.8.139) such that

$$f_{2i}(\tau) \approx \frac{a n^i (2i)!}{i!} \tau^{nj+2n} e^{n\alpha \tau^2} \text{ as } \tau \to \infty, \ i \geq 0,$$

$$\frac{f_{2i+1}(\tau)}{\tau^{nj+2n} e^{n\alpha \tau^2}} \to 0 \text{ as } \tau \to \infty, \ i \geq 0. \tag{3.8.140}$$

a in (3.8.140) is related to the old age constant C by $a = 1/C^n$ (see (3.8.137)).

(ii) For $\alpha = \beta = \gamma = 0$, (3.8.135) reduces to

$$u_t + u^n u_x + \frac{ju}{2t} = \frac{\delta}{2} u_{xx}; \tag{3.8.141}$$

the solution is now sought in the form (3.8.138) and (3.8.139), where

$$f_i(\tau) = \tau^{nj+2n} \sum_{k=0}^{r} \frac{a_k^{(i)}}{\tau^k} \tag{3.8.142}$$

and r is a nonnegative integer. As $r \to \infty$, the solution of the system of ODEs governing $f_i(\tau)$ would, in general, be an infinite series involving negative powers of τ. The series (3.8.139) together with (3.8.142) embed the old age solution and extend its validity far back in time.

For the case for which n is even, we refer to the work of Sachdev and Srinivasa Rao (2000) who treated the so-called modified Burgers equation

$$u_t + u^n u_x = \frac{\delta}{2} u_{xx}, \qquad (3.8.143)$$

with N-wave initial conditions

$$u(x,0) = \begin{cases} -x, & |x| \leq 1, \\ 0, & \text{otherwise.} \end{cases} \qquad (3.8.144)$$

This problem was considered earlier by Lee-Bapty and Crighton (1987) and Harris (1996). When (3.8.143) and (3.8.144) were solved numerically with n even, it was observed that the node of the N-wave moves from its initial position as it evolves. The shift, however, becomes negligible after some initial time. The large time analytic solution of (3.8.143) and (3.8.144) was sought with the assumption that the node has come to a halt. The solution was written out in the form (3.8.138)–(3.8.140) where $\xi = (x - x_0)/(2\delta t)^{1/2}$; $x = x_0$ is the point where the node of the evolving N-wave finally comes to rest. The approach here is similar to that in Sachdev et al. (1996); the linear solution is embedded in the exact large time solution and is approached as $t \to \infty$. Here, the inviscid solution, unlike for the nonplanar case, cannot be recovered in the limit of the large Reynolds number. The numerical solution of (3.8.143) and (3.8.144) was carried out for $n = 2$ and $n = 4$. It was found that this solution begins to agree with the analytic one at $t \approx 300$ and then it merges smoothly with the old age solution as time increases. It may also be remarked that the series solution of (3.8.143) with n even involves two constants instead of one, manifesting the fact that this equation is not antisymmetric for even n. This is in contrast to the case when n is odd for which there is only one arbitrary constant (see Sachdev and Srinivasa Rao (2000) for further details). These constants represent the memory of initial conditions in some integral sense.

We must emphasise that, in the analysis summarised here, no initial value problem for (3.8.143) is solved to find its limiting asymptotic behaviour. Instead, the N-wave solution of the plane Burgers equation is mimicked and simulated such that the corresponding solution of the GBE embeds its old age solution and tends to it as t becomes large.

It is of some interest to summarise an approach complementary to the balancing argument of Grundy and his collaborators which gives the asymptotic behaviour of a certain class of parabolic equations. Here we discuss the simple model

$$u_t + (u^q)_x - (u^m)_{xx} = 0 \quad \text{in } \mathbb{R} \times (0, \infty), \qquad (3.8.145)$$

$$u(x,0) = u_0 \quad \text{in } \mathbb{R}, \qquad (3.8.146)$$

where

$$m > 1, \quad \text{and} \quad q > 1 \tag{3.8.147}$$

and

$$u_0 \in L^1(\mathbb{R}), \quad u_0 \geq 0 \text{ a.e.} \tag{3.8.148}$$

This model describes the transport of a solute through a porous medium under the assumption that it undergoes equilibrium adsorption with the porous matrix (Grundy et al. 1994). The relevant physical cases for m and q are $m = q > 1$ and $m = 1$, $q \geq 1$. We summarise here the long time behaviour of (3.8.145)–(3.8.148) due to Laurencot and Simondon (1998). The same problem was discussed earlier for $m = 1$ and $q > 1$ by Escobedo et al. (1993) and Escobedo and Zuazua (1991). Their analysis showed that there exists a critical value $q_1 = 2$ such that: if $q > q_1$, the solution to (3.8.145) behaves as $t \to \infty$ as the fundamental solution to the linear heat equation with the same mass as $u(0)$; that is, the large time behaviour is given by the diffusive part of (3.8.145) with $m = 1$; if $q = q_1$, the solution u to (3.8.145) tends, as $t \to \infty$, to its source type solution with initial data $M\delta$, where δ denotes the Dirac delta mass centered at zero and $M = \int u(x,0)dx$; if $q \in (1, q_1)$, the long time behaviour of (3.8.145)–(3.8.148) is dominated by the convective part of (3.8.145); that is, the solution of (3.8.145) behaves, as $t \to \infty$, as the unique nonnegative entropy solution of $A_t + (A^q)_x = 0$ with the initial data $M\delta$.

Now we summarise the result of Laurencot and Simondon (1998) for $m > 1$ and $q > 1$. To that end we introduce the following notations. For $m > 1$ and $M > 0$, let

$$e_M(x) = ((\beta_M^2 - c_m^2 x^2)_+)^{1/(m-1)}, \quad x \in \mathbb{R},$$

where β_M is a positive real number such that $|e_M|_{L^1} = M$ and

$$c_m^2 = (m-1)/(2m(m+1)).$$

Define

$$E_M(x,t) = t^{-1/(m+1)} e_M\left(xt^{-1/(m+1)}\right), \quad (x,t) \in \mathbb{R} \times (0,\infty);$$

E_M satisfies

$$E_{Mt} - (E_M^m)_{xx} = 0, \quad \mathbb{R} \times (0,\infty),$$

$$\lim_{t \to 0} \int E_M(x,t)\zeta(x)dx = M\zeta(0), \quad \forall \, \zeta \in \mathscr{C}_b(\mathbb{R});$$

here $\mathscr{C}_b(\mathbb{R})$ is the space of continuous bounded functions on $\mathbb{R}$. Laurencot and Simondon (1998) proved the following result.

Theorem 3.8.1 *Suppose that $m > 1$ and $q > m + 1$. Let $u_0 \in L^1(\mathbb{R})$ be a nonnegative function with $M = |u_0|_{L^1} > 0$. Then, for $p \geq 1$,*

$$\lim_{t \to +\infty} t^{(1-1/p)/(m+1)}|u(t) - E_M(t)|_{L^p} = 0;$$

here u is the mild solution of (3.8.145)–(3.8.146).

Laurencot and Simondon (1998) discussed the cases $q = m + 1$ and $1 < q < m + 1$. We refer the reader to the work of Laurencot and Simondon (1998) for more details.

3.9 Evolution of travelling waves in generalised Fisher's equations via matched asymptotic expansions

We have observed in Section 3.8 that exact similarity solutions may not always constitute large time asymptotics. Grundy and his collaborators (Dawson et al. 1996, Grundy et al. 1994, Van Duijn et al. 1997) attempted to go beyond similarity solutions by the so-called balancing argument. They introduced the similarity variable and time as new independent variables and then, by balancing different sets of terms corresponding to different dominant physical effects, they could go beyond the limited class of similarity solutions. Grundy et al. (1994) and Grundy (1988) imposed the requirement that the area under the evolving profile remained bounded in time. Here, explicit initial conditions were not imposed. So the asymptotic solutions corresponded to a class of initial conditions vanishing appropriately at $x = \pm\infty$ and having a given 'mass' of the initial profile (see Section 3.8).

In a series of papers, Merkin and Needham (1989), Needham (1992), and Leach and Needham (2001) addressed initial/initial boundary value problems for a class of nonlinear PDEs which may be termed generalised Fisher's equations. They used matched asymptotic expansions to study the evolution of travelling waves from some initial conditions in the large time limit. Thus, by using an (approximate) analytic approach, they studied different behaviours in different space and time regimes, starting from $t = 0$ till the wave assumed its travelling waveform.

In the present section we follow the work of Leach and Needham (2001) and summarise main results of other related studies.

The relevant reaction–diffusion equation and the corresponding initial boundary conditions in a nondimensional form may be stated as follows.

$$u_t = u_{xx} + F(u), \quad x, \, t > 0, \tag{3.9.1}$$

$$u(x,0) = \begin{cases} u_0 g(x), & 0 \le x \le \sigma \\ 0, & x > \sigma, \end{cases} \tag{3.9.2}$$

$$u_x(0,t) = 0, \quad t > 0, \tag{3.9.3}$$

$$u(x,t) \to 0 \quad \text{as } x \to \infty, \, t \ge 0, \tag{3.9.4}$$

where $F : (-\infty, \infty) \to \mathbb{R}$ satisfies the following (normalised) conditions: (i) $F(u)$ is continuous and differentiable for $u \in (-\infty, \infty)$; (ii) $F'(0) = 1$, $F'(1) < 0$; (iii) $F(u) > 0$ for $u \in (0,1)$; (iv) $F(u) < 0$ for $u \in (1,\infty)$; and (v) $F(0) = F(1) = 0$. It is known that the above IBVP has a unique global solution with $0 < u(x,t) < \max[1, u_0]$ for all $x, t > 0$ (Smoller (1989)). Before we take up the case of more general $F(u)$ satisfying the conditions (i)–(v) above, we summarise the results for the well known Fisher–Kolmogorov equation with $F(u) = u(1 - u)$ (see Larson 1978). If we write $z = x - vt$

in (3.9.1) with $v > 0$ and look for nonnegative travelling wave solutions $u = u(z)$ which join 'unreacted' state $u = 0$ ahead of the wavefront with the fully reacted state $u = 1$ at the rear of the wavefront, we get the following nonlinear eigenvalue problem to discover permanent travelling waves (PTW).

$$u_{zz} + vu_z + u(1-u) = 0, \quad -\infty < z < \infty,$$

$$u(z) \rightarrow \begin{cases} 1 & \text{as } z \rightarrow -\infty, \\ 0 & \text{as } z \rightarrow +\infty, \end{cases} \tag{3.9.5}$$

$$u(z) \geq 0, \quad -\infty < z < \infty.$$

We refer to (3.9.5) as BVP I. It is known that this problem has a unique (to a translation) solution if and only if the eigenvalue $v \geq 2$ (Kolmogorov et al. 1937). With reference to initial and boundary conditions (3.9.2)–(3.9.4), it was shown by Kolmogorov et al. (1937) and McKean (1975) that, with $g(x) \equiv 1$ and $u_0 = 1$, a PTW does evolve from the solution of the IBVP as $t \rightarrow \infty$ and propagates with the minimum possible speed $v = 2$. For this specific case, Bramson (1978) proved that the asymptotic estimate for the propagation speed is given by

$$\dot{s} = 2 - \frac{3}{2}t^{-1} + o(t^{-1}) \quad \text{as } t \rightarrow \infty. \tag{3.9.6}$$

Needham (1992) recovered the formula (3.9.6) by requiring $g(x)$ to have a finite support. His approach is the same as discussed below but he primarily used a linearised form of the IBVP. It is instructive to discuss this linear form of the IBVP (3.9.1)–(3.9.4) with $F(u) = u(1-u)$ and $u_0 \ll 1$. For the linearised form of (3.9.1),

$$u_t = u_{xx} + u, \quad x, t > 0 \tag{3.9.7}$$

with initial and boundary conditions (3.9.2)–(3.9.4), we have the solution

$$u(x,t) = e^t D(x,t), \quad x, t \geq 0, \tag{3.9.8}$$

where $D(x,t)$ is the relevant solution of the heat equation. It may be checked from (3.9.8) that, under the assumptions that $t \gg 1$, $x \gg O(t)$, and $u \ll 1$, we have

$$u(x,t) \sim u_0 t^{-1/2} \exp\left[-t\left(\frac{y^2}{4} - 1\right)\right], \tag{3.9.9}$$

where $y = x/t \gg O(1)$ and $t \gg 1$. The linear approximation (3.9.8) and (3.9.9) is valid only when u remains small, that is, when $t \gg 1$, $y \gg 2$; it fails when $y \leq 2$. Thus, for small initial data with finite support, the solution (3.9.9) with exponential decay undergoes a transition at $y \sim 2$; that is, $x \sim 2t$, $t \gg 1$. This transition from $u = O(1)$ to $u \ll 1$ when $x \sim 2t$ and $t \gg 1$ is indicative of the large t development of the PTW with minimum speed $v = 2$, starting from the given initial/boundary conditions. This simple argument is in agreement with the more rigourous results (see Needham 1992) that the development of a PTW from the initial conditions of

IBVP for $t \gg 1$ takes place by selecting the propagation speed from those available ($v \geq 2$) via the evolution in the far field, namely, $x \gg 1, t \geq 0$, where a linear solution holds. This may not be always true for the generalised Fisher's equation as we show later in this section.

Now we consider the more general problem (3.9.1)–(3.9.4) where $F(u)$ satisfies the conditions (i)–(v) listed therein. First we quote the results concerning the PTW for $F(u)$ satisfying (i)–(v) and then discuss in detail the evolution of given initial/boundary conditions (3.9.1)–(3.9.4) to the travelling waveform. Thus, if we substitute $z = x - vt$ in (3.9.1) and pose the boundary conditions as in (3.9.5), we have

$$u''(z) + vu'(z) + F(u) = 0, \quad -\infty < z < \infty, \tag{3.9.10}$$

$$u(z) \to 0 \text{ as } z \to \infty, \tag{3.9.11}$$

$$u(z) \to 1 \text{ as } z \to -\infty, \tag{3.9.12}$$

$$u(z) \geq 0, \quad -\infty < z < \infty. \tag{3.9.13}$$

This eigenvalue problem for the wavespeed $v > 0$ may be referred to as BVP2. Any solution to this problem may provide a permanent waveform to which the solution of initial value problem (3.9.1)–(3.9.4) may evolve as t tends to infinity. The problem (3.9.10)–(3.9.13) has been well studied (Fife 1979) and the following theorem summarises the results. BVP2 has a unique solution $u_T(z, v)$ for each $v \in [v^*, \infty)$ with $v^* \geq 2$. Furthermore, the following asymptotic forms result depending on the nature of the function F and the value of $v = v^*$.

(a) If $F(u) \leq u \ \forall u \in [0, 1]$, then $v^* = 2$ and

$$u_T(z, v) \sim \begin{cases} (A^* z + B^*) e^{-z} & \text{as } z \to \infty, \ v = v^*, \\ A e^{\lambda_+(v) z} & \text{as } z \to \infty, \ v > v^*. \end{cases} \tag{3.9.14}$$

(b) If $F(u) \nleq u \ \forall u \in [0, 1]$ and $v^* = 2$, then

$$u_T(z, v) \sim \begin{cases} (A^* z + B^*) e^{-z} & \text{as } z \to \infty, \ v = v^*, \\ A e^{\lambda_+(v) z} & \text{as } z \to \infty, \ v > v^*. \end{cases} \tag{3.9.15}$$

(c) If $F(u) \nleq u \ \forall u \in [0, 1]$ and $v^* > 2$, then

$$u_T(z, v) \sim \begin{cases} A^* e^{\lambda_-(v^*) z} & \text{as } z \to \infty, \ v = v^*, \\ A e^{\lambda_+(v) z} & \text{as } z \to \infty, \ v > v^*, \end{cases} \tag{3.9.16}$$

where

$$\lambda_\pm = -\frac{v}{2} \pm \frac{1}{2}(v^2 - 4)^{1/2}. \tag{3.9.17}$$

For each of the cases above we also have the following behaviour as $z \to -\infty$,

$$u_T(z, v) \sim 1 - c^* e^{\lambda_m(v) z} \text{ as } z \to -\infty, \tag{3.9.18}$$

where

$$\lambda_m(v) = -\frac{v}{2} + \frac{1}{2}(v^2 - 4F'(1))^{1/2} \quad (> 0). \tag{3.9.19}$$

Here, the constants A^*, B^*, and A are such that $A^* \geq 0$, with $B^* > 0$ when $A^* = 0$ and A is positive. These constants can be determined analytically or otherwise. Thus, for any $F(u)$ satisfying the conditions (i)–(v), there exists a travelling wave with minimum wavespeed $v = v^*$; faster travelling waves for each $v > v^*$ also exist. $v^* = 2$ or $v^* > 2$ depending on the curvature of the function $F(u)$ in $[0, 1]$.

Now consider specifically the more general Fisher's equation

$$u_t = u_{xx} + F(u, \sigma), \tag{3.9.20}$$

where

$$F(u, \sigma) = u(1 - u)(1 + \sigma u), \quad -\infty < u < \infty. \tag{3.9.21}$$

Here, $\sigma \in [0, \infty)$ is a dimensionless parameter. Equation (3.9.20) with $F(u, \sigma)$ defined by (3.9.21) describes migration of advantageous genes; it may also model mixed quadratic and cubic autocatalysis where σ measures the ratio of the quadratic to cubic reaction rates. Equation (3.9.20) reduces to the standard Fisher's equation when $\sigma = 0$; it also satisfies conditions (i)–(v). It is fortunate that (3.9.20) and (3.9.21) admit an exact travelling wave solution

$$u_e(z) = \left[1 + \exp\left(\frac{\sigma^{1/2}z}{\sqrt{2}}\right)\right]^{-1}, \quad -\infty < z < \infty, \tag{3.9.22}$$

for any $\sigma > 0$, with the propagation speed

$$v = v_e = \frac{\sqrt{2}}{\sigma^{1/2}} + \frac{\sigma^{1/2}}{\sqrt{2}}, \quad \sigma > 0. \tag{3.9.23}$$

It is clear from (3.9.23) that

$$v_e \begin{cases} > 2, & \sigma \in (0, 2) \cup (2, \infty), \\ = 2, & \sigma = 2. \end{cases} \tag{3.9.24}$$

Moreover,

$$u_e(z) \sim \exp\left(-\frac{\sigma^{1/2}z}{\sqrt{2}}\right) \quad \text{as } z \to \infty. \tag{3.9.25}$$

Therefore, we have

$$u_e(z) \sim \begin{cases} e^{\lambda_+(v_e)z}, & 0 < \sigma < 2, \\ e^{-z}, & \sigma = 2, \\ e^{\lambda_-(v_e)z}, & \sigma > 2 \end{cases} \tag{3.9.26}$$

as $z \to \infty$ (see (3.9.17)). In view of (3.9.14)–(3.9.16), we find that, for the present model with $\sigma > 0$, we have

$$2 \le v^* \le v_e, \tag{3.9.27}$$

which, together with (3.9.24), gives

$$v^* = 2 \quad \text{when } \sigma = 2. \tag{3.9.28}$$

Because, here, $F(u) \le u$ if and only if $\sigma \in [0,1]$, we infer from (3.9.14) that

$$v^* = 2 \quad \text{when } \sigma \in [0,1]. \tag{3.9.29}$$

Now, combining (3.9.26) with (3.9.14)–(3.9.16), we conclude that

$$v^* = v_e = \frac{\sqrt{2}}{\sigma^{1/2}} + \frac{\sigma^{1/2}}{\sqrt{2}} \; (> 2) \quad \text{when } \sigma \in (2,\infty). \tag{3.9.30}$$

It was independently proved by Hadeler and Rothe (1975) that

$$v^* = 2 \quad \text{when } \sigma \in (1,2). \tag{3.9.31}$$

Combining (3.9.29)–(3.9.31) we have propagation speed v^* of the permanent wave for all $\sigma \in (0,\infty)$. Referring to the cases (a)–(c) of the general theorem (see (3.9.14)–(3.9.16)) and (3.9.28)–(3.9.31), we have case (a) when $\sigma \in [0,1]$, case (b) when $\sigma \in (1,2]$, and case (c) when $\sigma \in (2,\infty)$.

Now, following Leach and Needham (2001), we discover how the (initial) waves starting from the initial and boundary conditions (3.9.2)–(3.9.4), actually evolve to assume their asymptotic form. We revert here to the general form of F satisfying conditions (i)–(v) and consider the case $v^* > 2$ ($F(u) \not\le u \; \forall \, u \in [0,1]$).

Because the initial function $u(x,0) > 0$ and is analytic in the region $0 \le x \le \sigma - O(1)$, to be called region I, and because $u = O(1)$ as $t \to 0$, we may easily write a power series solution in the neighbourhood of $t = 0$ with $0 \le x \le \sigma - O(1)$:

$$u(x,t) = u_0 g(x) + t \left[u_0 g''(x) + F(u_0 g(x)) \right] + O(t^2). \tag{3.9.32}$$

If we assume that $0 < (\sigma - x) \ll 1$, we may write (3.9.32) as

$$u(x,t) \sim u_0 g_\sigma \left[(\sigma - x)^r + \ldots \right] + t \left[u_0 (r-1) r g_\sigma (\sigma - x)^{r-2} + \ldots \right.$$
$$\left. + u_0 g_\sigma (\sigma - x)^r + \ldots \right] + \ldots \tag{3.9.33}$$

as $t \to 0$. This expansion obviously fails when $\sigma - x = O\left(t^{1/2}\right)$; here, $u = O\left(t^{r/2}\right)$. This requires introduction of a second region, called region II, where $x = \sigma \pm O\left(t^{1/2}\right)$ as $t \to 0$.

To study this region we introduce the scaled coordinate $\eta = (x - \sigma)t^{-1/2}$ and seek the asymptotic expansion (as $t \to 0$)

$$u(\eta,t) = t^{r/2}\breve{u}(\eta) + o\left(t^{r/2}\right) \tag{3.9.34}$$

with $\eta = O(1)$. Substituting (3.9.34) into (3.9.1), written in terms of η and t, we obtain

$$\breve{u}_{\eta\eta} + \frac{\eta}{2}\breve{u}_\eta - \frac{r}{2}\breve{u} = 0, \quad -\infty < \eta < \infty, \tag{3.9.35}$$

to leading order. The solution of (3.9.35) must satisfy the initial condition (3.9.2) as $t \to 0$ and match the solution (3.9.33) in region I as $\eta \to -\infty$ (because $x < \sigma$ and $t \to 0$). Thus we have the conditions

$$\breve{u}(\eta) \sim u_0 g_\sigma(-\eta)^r \text{ as } \eta \to -\infty, \tag{3.9.36}$$

$$\breve{u}(\eta) \sim o(\eta^r) \text{ as } \eta \to +\infty. \tag{3.9.37}$$

The solution of the boundary value problem (3.9.35)–(3.9.37) is unique and may explicitly be written out as

$$\breve{u}(\eta) = \begin{cases} \dfrac{u_0 g_\sigma r!}{\left(\dfrac{r}{2}\right)! k_1} A(\eta) \displaystyle\int_\eta^\infty \dfrac{e^{-s^2/4}}{A^2(s)}\,ds, & r \text{ even} \\[4ex] \dfrac{u_0 g_\sigma r!}{\left(\dfrac{1}{2}(r-1)\right)! k_2} \left[\dfrac{A(\eta)}{\eta} - A(\eta)\displaystyle\int_\eta^\infty \left\{\dfrac{1}{s^2} - \dfrac{e^{-s^2/4}}{A^2(s)}\right\}ds\right], & r \text{ odd,} \end{cases} \tag{3.9.38}$$

where

$$A(\eta) = \begin{cases} \displaystyle\sum_{p=0}^{r/2} \dfrac{\left(\dfrac{r}{2}\right)! \eta^{2p}}{(2p)!\left(\dfrac{r}{2}-p\right)!}, & r \text{ even,} \\[4ex] \displaystyle\sum_{p=0}^{(r-1)/2} \dfrac{\left(\dfrac{r-1}{2}\right)! \eta^{2p+1}}{(2p+1)!\left(\dfrac{r-1}{2}-p\right)!}, & r \text{ odd} \end{cases} \tag{3.9.39}$$

and

$$k_1 = \int_{-\infty}^\infty \frac{e^{-s^2/4}}{A^2(s)}\,ds, \quad k_2 = \int_{-\infty}^\infty \left(\frac{1}{s^2} - \frac{e^{s^2/4}}{A^2(s)}\right)ds. \tag{3.9.40}$$

It is clear from (3.9.38) that $\breve{u}(\eta)$ is positive and decreases monotonically for all $-\infty < \eta < \infty$. We find from (3.9.38) that

$$\breve{u}(\eta) \sim c_\infty \frac{1}{\eta^{r+1}} e^{-\eta^2/4} \text{ as } \eta \to +\infty \tag{3.9.41}$$

with

$$c_\infty = \begin{cases} \dfrac{2u_0 g_\sigma (r!)^2}{k_1 \left[(r/2)!\right]^2}, & r \text{ even}, \\[4mm] \dfrac{2u_0 g_\sigma (r!)^2}{k_2 \left[\left(\dfrac{r-1}{2}\right)!\right]^2}, & r \text{ odd}. \end{cases} \qquad (3.9.42)$$

In the above we have assumed that $\eta = O(1)$. To treat the region where $\eta \gg 1$, we introduce region III for which $x = \sigma + O(1)$.

In region III, $\check{u}$ is exponentially small (see (3.9.41)). The form (3.9.34) and (3.9.41) of the solution as $\eta \to \infty$ in region II suggests that we may write the solution in region III as

$$u(x,t) = e^{-\check{F}(x,t)/t} \text{ as } t \to 0, \qquad (3.9.43)$$

where

$$\check{F}(x,t) = F_0(x) + F_1(x)t \ln t + F_2(x)t + O(t^2) \qquad (3.9.44)$$

and $x = \sigma + O(1)$. $\check{F}(x,t) > 0$ for all $x > \sigma$. Substituting (3.9.43)–(3.9.44) into (3.9.1) and solving the resulting ODEs for F_0, F_1, and so on, we find that

$$u(x,t) = \exp\left\{ -\frac{(x+C)^2}{4t} - A \ln t - \left(\frac{1}{2} - A\right) \ln(x+C) - B + O(t) \right\} \qquad (3.9.45)$$

as $t \to 0$ with $x = \sigma + O(1)$. Here, A, B, and C are arbitrary constants. The solution (3.9.45) holding in region III satisfies the initial condition $u(x,t) \to 0$ as $t \to 0$. It also satisfies the boundary condition (3.9.4) as $x \to \infty$. Matching the solution (3.9.45) as $x \to \sigma^+$ with the solution (3.9.34) in region II as $\eta \to \infty$ leads to the determination of the constants:

$$A = -\left(r + \frac{1}{2}\right), \quad B = -\ln c_\infty, \quad C = -\sigma. \qquad (3.9.46)$$

Writing the exponential more explicitly one may check that the expansion of (3.9.45) remains uniform for $x \gg 1$ as $t \to 0$.

Now the solution in region III is generalised and extended to apply to the region IV where $x \to \infty$ and $t = O(1)$. The expansion (3.9.45) which holds for $x \gg 1$ as $t \to 0$ suggests that we write the solution in region IV in the form

$$u(x,t) = e^{-\check{H}(x,t)} \text{ as } x \to \infty, \qquad (3.9.47)$$

where $\check{H}$ is a series in decreasing power of x and $\ln x$ with coefficients functions of time:

$$\check{H}(x,t) = H_0(t)x^2 + H_1(t)x + H_2(t)\ln x + H_3(t) + H_4(t)x^{-1} + O(x^{-2}), \qquad (3.9.48)$$

where now $t = O(1)$ and $x \to \infty$ (the form (3.9.48) may be discovered by a balancing argument). Substituting (3.9.47) and (3.9.48) into (3.9.1) and solving the resulting ODEs for $H_i(t)$ and so on we get

$$u(x,t) = \exp\left\{-\frac{x^2}{4t} + \frac{\sigma x}{2t} - (r+1)\ln x\right.$$
$$\left. + \left[\left(r+\frac{1}{2}\right)\ln t + t + \ln c_\infty - \frac{\sigma^2}{4t}\right] + O(x^{-1})\right\}. \qquad (3.9.49)$$

Comparing the term $x^2/4t$ with t in the exponential, and so on, we find that (3.9.49) remains valid for $t \gg 1$ provided that $x \gg t$. It clearly fails when $x = O(t), t \to \infty$.

To extend the solution (3.9.49) such that it remains valid when $x = O(t)$ and $t \to \infty$, called region V, we introduce the variable $y = x/t$ and rewrite the form (3.9.49) as

$$u(y,t) = e^{-t\check{F}(y,t)} \quad \text{as } t \to \infty, \qquad (3.9.50)$$

where

$$\check{F}(y,t) = f_0(y) + f_1(y)\frac{\ln t}{t} + f_2(y)\frac{1}{t} + O(t^{-2}). \qquad (3.9.51)$$

Substituting (3.9.50)–(3.9.51) into (3.9.1) we have, to leading order, the problem

$$f_{0y}^2 - yf_{0y} + f_0 + 1 = 0, \quad y > 0, \qquad (3.9.52)$$
$$f_0(y) > 0, \quad y > 0, \qquad (3.9.53)$$
$$f_0(y) \sim \frac{1}{4}y^2 - 1 \quad \text{as } y \to \infty. \qquad (3.9.54)$$

The condition (3.9.54) arises from matching (3.9.50) with $y \gg 1$ and (3.9.49) where $x = O(t)$. Equation (3.9.52) has two solutions

$$f_0(y) = c_0(y - c_0) - 1, \quad y > 0, \qquad (3.9.55)$$

and

$$f_0(y) = \frac{1}{4}y^2 - 1, \quad y > 0, \qquad (3.9.56)$$

the latter being the envelope solution of the former linear solutions. However, the matching condition (3.9.54) selects the envelope solution

$$f_0(y) = \frac{1}{4}y^2 - 1, \quad y > 0$$

or

$$f_0(y) = \begin{cases} \frac{1}{4}y^2 - 1, & y > y_0, \\ \frac{1}{2}y_0\left(y - \frac{1}{2}y_0\right) - 1, & 0 < y \le y_0 \end{cases} \qquad (3.9.57)$$

for any $y_0 > 2$. However, neither of these solutions satisfies the condition $f_0(y) > 0, y > 0$ (see (3.9.53)); (3.9.56) vanishes as $y \to 2^+$ and (3.9.57) goes to zero as $y \to ((2/y_0) + (1/2)y_0)^+$ which is greater than 2 when $y_0 > 2$. We conclude that a nonuniformity arises in the expansion (3.9.51) when $y \to y_c(\ge 2)$, where

$$y_c = \frac{2}{y_0} + \frac{y_0}{2} \begin{cases} = 2, & y_0 = 2, \\ > 2, & y_0 > 2, \end{cases} \tag{3.9.58}$$

for some $y_0 \geq 2$. We observe that, for $y_0 = 2$, $f_0(y)$ is given by (3.9.56), whereas for $y_0 > 2$, $f_0(y)$ is defined by (3.9.57). If one computes more terms in the expansion (3.9.51), one may discover that a nonuniformity appears when

$$y = y_c + O(t^{-1}) \quad \text{with } u = O(1) \text{ as } t \to \infty. \tag{3.9.59}$$

It is therefore natural to introduce the independent variable

$$y = y_c + \frac{z}{t} \tag{3.9.60}$$

with $z = O(1)$ as $t \to \infty$ and write

$$u(z,t) = u_c(z) + o(1), \quad z = O(1). \tag{3.9.61}$$

We call this region TW. Substitution of (3.9.61) into (3.9.1) yields, to leading order, the problem

$$u_c'' + y_c u_c' + F(u_c) = 0, \quad -\infty < z < \infty, \tag{3.9.62}$$

$$u_c(z) > 0, \quad -\infty < z < \infty, \tag{3.9.63}$$

$$u_c(z) \to 0 \quad \text{as } z \to +\infty, \tag{3.9.64}$$

$$u_c(z) \text{ is bounded as } z \to -\infty. \tag{3.9.65}$$

The vanishing condition (3.9.64) arises from matching this solution with (3.9.50) in region V as $y \to y_c^+$. A phase plane analysis of (3.9.62) and use of the Poincaré–Bendixson theorem with the conditions (3.9.63) and (3.9.64) allows the boundedness condition (3.9.65) to be replaced by

$$u_c(z) \to 1 \quad \text{as } z \to -\infty. \tag{3.9.66}$$

All other possibilities in the phase plane lead to an unbounded solution $u_c(z)$ as $z \to -\infty$. Thus we arrive exactly at the BVP2 – (3.9.10)–(3.9.13) – which has permanent travelling wave solutions. Using (3.9.14)–(3.9.16) we infer that this problem has a unique solution $u_c(z) = u_T(z, y_c)$ for each $y_c \in [v^*, \infty)$. In the present case, $v^* > 2$, therefore, we have $y_c > 2$. It follows from (3.9.58) that $y_0 > 2$ in region V; $f_0(y)$ itself is given by (3.9.57).

We now match expansion (3.9.50) and (3.9.51) in region V to the expansion (3.9.61) in terms of $U = \log u$. We follow the matching principle of Van Dyke (1975). Writing (3.9.50) and (3.9.51) in terms of U and employing it in the TW region up to $O(1)$, we have

$$U_{t0} = -\frac{1}{2} y_0 z = \lambda_-(y_c) z, \tag{3.9.67}$$

where we have used (3.9.57) and (3.9.58). Conversely, expanding (3.9.61) for U in region V up to O(t) gives the expression

$$U_{0t} = \begin{cases} \lambda_-(y_c)z, & y_c = v^*, \\ \lambda_+(y_c)z, & y_c > v^*. \end{cases} \tag{3.9.68}$$

The matching principle requires $U_{0t} \equiv U_{t0}$. Thus matching (3.9.67) and (3.9.68) we obtain

$$y_c = v^*. \tag{3.9.69}$$

Thus the travelling wave solution of minimum speed $v^*(>2)$ is selected in the region TW. Now, combining (3.9.58) and (3.9.69), we find that

$$y_0 = v^* + ((v^*)^2 - 4)^{1/2} = -2\lambda_-(v^*)(>2). \tag{3.9.70}$$

This completes the expansions in both the regions V and TW to leading order.

The main objective of the present section is to show by using matched asymptotic expansions how the travelling waveform emerges for large times, starting the solution at the initial time. Leach and Needham (2001) analysed region V to make the solution smooth. It is clear from (3.9.56) and (3.9.57) that $f_0''(y)$ is discontinuous at the point $y = y_0(> v^*)$. The approach here is quite intuitive and requires considerable experience in matched asymptotic expansions.

3.10 Periodic travelling wave solutions in reaction–diffusion systems

In the present section, we consider systems of reaction–diffusion equations. Here we show, following the work of Sherratt (2003), how periodic travelling waves (PTWs) are generated by such a system with Dirichlet conditions at one edge of a semi-infinite domain: both the dependent variables are assumed to be zero at the edge $x = 0$. Sherratt (2003) was much motivated in his analysis by the results of his numerical study which we presently summarise. The 'oscillatory' reaction–diffusion system that we study is

$$u_t = \nabla^2 u + \left(1 - r^2\right) u - (w_0 - w_1 r^2)v, \tag{3.10.1}$$

$$v_t = \nabla^2 v + (w_0 - w_1 r^2)u + \left(1 - r^2\right) v, \tag{3.10.2}$$

where $r = \sqrt{u^2 + v^2}$. The system (3.10.1) and (3.10.2) is often said to belong to the '$\lambda - w$' class, first introduced by Kopell and Howard (1973), and is considered natural for studying generic behaviour in systems in which each variable has the same diffusion coefficient. Here, we restrict ourselves to the one-dimensional form of (3.10.1) and (3.10.2).

It is known (see Kopell and Howard 1973) that the above system has an unstable equilibrium at $u = v = 0$ and a stable circular limit cycle of radius 1, centred at this equilibrium. Kopell and Howard (1973) found a travelling wave periodic solution of (3.10.1)–(3.10.2) in the form

$$u = r^* \cos \left[\theta_0 \pm \sqrt{1 - r^{*2}} x + \left(w_0 - w_1 r^{*2} \right) t \right],$$ (3.10.3)

$$v = r^* \sin \left[\theta_0 \pm \sqrt{1 - r^{*2}} x + \left(w_0 - w_1 r^{*2} \right) t \right],$$ (3.10.4)

where r^* is a parameter and θ_0 is an arbitrary constant. They showed that the above wave solution is stable provided that

$$r^* > r_{stab} \equiv \left(\frac{2 + 2w_1^2}{3 + 2w_1^2} \right)^{1/2}.$$ (3.10.5)

The solution (3.10.3) and (3.10.4) suggests that we introduce the polar coordinates $r = \sqrt{u^2 + v^2}$ and $\theta = \tan^{-1}(v/u)$ in the (u, v) plane. We thus write (3.10.1) and (3.10.2) as

$$r_t = r_{xx} - r\theta_x^2 + r(1 - r^2),$$ (3.10.6)

$$\theta_t = \theta_{xx} + \frac{2r_x\theta_x}{r} + w_0 - w_1 r^2.$$ (3.10.7)

The PTW (3.10.3) and (3.10.4) now becomes

$$r = r^*, \quad \theta = \theta_0 \pm \sqrt{1 - r^{*2}} x + \left(w_0 - w_1 r^{*2} \right) t.$$ (3.10.8)

Indeed, it may be shown that any solution of (3.10.6) and (3.10.7) with $r = $ constant < 1 is a PTW.

Sherratt (2003) carried out a very interesting numerical study of (3.10.1) and (3.10.2) in the one-dimensional semi-infinite domain $x > 0$ with the boundary condition $u = v = 0$ at $x = 0$. He chose the numerical domain to be $0 < x < X_\infty$ where X_∞ is large. He assumed zero flux conditions $u_x = v_x = 0$ at $x = X_\infty$. The initial conditions in $0 < x < X_\infty$ were chosen to be random, obtained by a random number generator to calculate u and v values between $+1$ and -1 at equidistant points with $\Delta x = 5$ throughout the domain. These random values were joined by straight lines to give the initial conditions. This is a very useful numerical experiment which shows the 'independence' of the asymptotic solution of the details of the initial conditions.

It was found that, for a wide range of values of the parameters w_0 and w_1, the numerical solutions of this problem showed the same behaviour. Generally, the solution changes rapidly from the random initial conditions to spatially uniform oscillations everywhere away from the boundary $x = 0$. A transition wave then develops so that there is a homogeneous oscillation ahead of it and a PTW behind. For large time it is the PTW that prevails away from the boundary $x = 0$. This behaviour is strongly dependent on the specific boundary conditions $u = v = 0$ at $x = 0$. If this condition, for example, is changed to zero flux, namely, $u_x = v_x = 0$ at $x = 0$, the

PTW disappears and is replaced by spatially uniform oscillations. The intermediate asymptotic character of the PTW is confirmed by the fact that the speed/amplitude of this wave is independent of the 'seed' in the random number generator used for the initial conditions and, thus, the Dirichlet boundary conditions very carefully select a particular member of the PTW family.

Numerical results of Sherratt (2003) show distinct behaviours when $|w_1|$ is small and when it is large. For the former, as we remarked earlier, a transition front moves across the domain behind which PTWs develop and move in the positive x- or negative x-direction. For large $|w_1|$, the long-term behaviour does not exhibit PTW but irregular spatiotemporal oscillation. This behaviour arises when the PTW that is selected by the boundary conditions has an amplitude less than r_{stab} (see (3.10.5)). It is an unstable solution of the PDEs and hence the spatiotemporal oscillations. If the numerical results referred to above are shown in terms of r and θ_x rather than u and v (see Figure 3.11) then it is observed that the solution changes rapidly from its random initial behaviour until $r \approx 1$ and $\theta_x \approx 0$ everywhere away from the boundary $x = 0$; this defines the spatially homogeneous oscillations in u and v. This behaviour is followed by a transition front in r and θ_x which moves in the positive x-direction. The conditions ahead of this front are such that $r \to 1$ and $\theta_x \to 0$ whereas behind it r and θ_x have constant values, r_{PTW} and ψ_{PTW}, say, corresponding

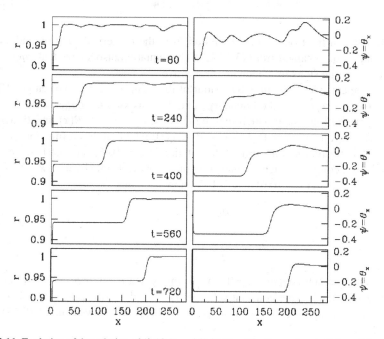

Fig. 3.11 Evolution of the solution of (3.10.1) and (3.10.2) subject to random initial conditions and the boundary conditions $u = v = 0$ at $x = 0$ and $u_x = v_x = 0$ at $x = 400$. (Sherratt 2003. Copyright © 2003 Society for Industrial and Applied Mathematics. Reprinted with permission. All rights reserved.)

to the PTW. The numerical solution suggests that this transition front moves with constant shape and speed. Therefore, we look for the travelling waveform of the solution of (3.10.1) and (3.10.2) or more specifically of (3.10.6) and (3.10.7) in the form

$$r(x,t) = \hat{r}(x-st) \text{ and } \theta_x(x,t) = \hat{\psi}(x-st) \text{ or } \theta(x,t) = \int^{z=(x-st)} \hat{\psi}(z)dz + f(t),$$

(3.10.9)

where $s > 0$ is the speed of the wavefront and $f(t)$ is a function of integration.

Substitution of (3.10.9) into (3.10.6) and (3.10.7) leads to the ODEs

$$\hat{r}'' + s\hat{r}' + \hat{r}\left(1 - \hat{r}^2 - \hat{\psi}^2\right) = 0,$$

(3.10.10)

$$\hat{\psi}' + s\hat{\psi} + w_0 - w_1\hat{r}^2 + 2\hat{\psi}\frac{\hat{r}'}{\hat{r}} = f'(t).$$

(3.10.11)

It is clear from (3.10.11) that $f'(t)$ must be a constant; moreover, because $\hat{r} \to 1$ and $\hat{\psi} \to 0$ as $x - st \to \infty$, $f'(t) = w_0 - w_1$. Putting the asymptotic values $\hat{r} = r_{PTW}$, $\hat{\psi} = \psi_{PTW}$ in (3.10.10) and (3.10.11) we check that

$$r_{PTW} = \sqrt{1 - \frac{s^2}{w_1^2}}, \quad \psi_{PTW} = -\frac{s}{w_1}.$$

(3.10.12)

The velocity of the front, s, may be found from the numerical solution and hence r_{PTW} and ψ_{PTW} obtained from (3.10.12). This equation also shows that ψ_{PTW} and w_1 have opposite signs inasmuch as s is positive.

Now we consider large time behaviour of the solutions of (3.10.6) and (3.10.7). The numerical results of the boundary value for this system suggest that this is a (steady) equilibrium state which may be denoted by $r(x,t) = R(x)$ and $\theta_x(x,t) = \psi(x)$. Thus, we may write $\theta = \int^x \psi(\bar{x})d\bar{x} + g(t)$, where $g(t)$ is a function of integration. By substituting this form of the solution into (3.10.6) and (3.10.7), we infer that $g'(t)$ is a constant which we may, for convenience, denote by $w_0 - k$, where k is arbitrary. The system (3.10.6) and (3.10.7) now becomes

$$R_{xx} + R(1 - R^2 - \Psi^2) = 0,$$

(3.10.13)

$$\Psi_x + \frac{2\Psi R_x}{R} + k - w_1R^2 = 0.$$

(3.10.14)

The boundary conditions $u = v = 0$ at $x = 0$ imply that $R = 0$ there. We also require that R and Ψ tend to the constant solution, r_{PTW} and ψ_{PTW}, as $x \to \infty$, where the sign of ψ_{PTW} is opposite to that of w_1 (see (3.10.12)). It follows from (3.10.13) and (3.10.14) that

$$r_{PTW} = \sqrt{\frac{k}{w_1}}, \quad \psi_{PTW} = -\text{sgn}(w_1)\sqrt{1 - \frac{k}{w_1}},$$

(3.10.15)

where we have used the fact that ψ_{PTW} and w_1 have opposite signs. k is related to the speed s by $k = w_1 - (s^2/w_1)$ (see (3.10.12)). Clearly, k and w_1 must have the same sign. Rescaling (3.10.13) and (3.10.14) according to

$$\phi = R\left(\frac{w_1}{k}\right)^{1/2}, \quad w = R_x\left(\frac{w_1 - k}{k}\right)^{1/2}\frac{\mathrm{sgn}(w_1)}{k},$$

$$\Gamma = -\frac{\Psi}{R}\left(\frac{k}{w_1 - k}\right)^{1/2}\mathrm{sgn}(w_1), \quad z = x\left(\frac{w_1}{w_1 - k}\right)^{1/2}k\,\mathrm{sgn}(w_1), \quad (3.10.16)$$

we obtain

$$\phi_z = w, \tag{3.10.17}$$

$$w_z = -\frac{\alpha}{k^2}\phi\left[1 - \phi^2 - \alpha\phi^2(\Gamma^2 - 1)\right], \tag{3.10.18}$$

$$\Gamma_z = \frac{1 - 3w\Gamma - \phi^2}{\phi}, \tag{3.10.19}$$

where $\alpha = 1 - k/w_1$ so that $0 \le \alpha \le 1$. The boundary conditions in terms of the new variables become

$$\phi = 0 \ \text{ at } z = 0 \text{ and } \phi = 1, \quad w = 0, \quad \Gamma = 1 \ \text{ at } z = \infty. \tag{3.10.20}$$

The constant k in (3.10.18) is arbitrary. Thus, one must enquire for what values of k the boundary value problem (3.10.17)–(3.10.20) has a solution. To solve this eigenvalue problem one may integrate (3.10.17)–(3.10.19) backward in z from $(1,0,1)$. Numerical solution of this problem shows that there is a unique stable eigenvector for given values of k and w_1 if ϕ becomes zero along one of these trajectories. There is a large discrete set of values of k for which this happens. These values of k are widely separated when $|k|$ is just below $|w_1|$ and become closer as $|k| \to 0$.

Sherratt (2003) used perturbation analysis to find the analytic character of the solution for small $|w_1|$. Here, we content ourselves with some other simple aspects of the solution. Numerical results for the one-dimensional form of (3.10.1) and (3.10.2) with $u = v = 0$ at $x = 0$ suggest that Ψ/R is constant in the observed solution. So, one may seek a solution of (3.10.17)–(3.10.19) under the assumption that $\Gamma \equiv 1$ (see (3.10.16)). This system now assumes the form

$$\phi_z = w, \tag{3.10.21}$$

$$w_z = -\frac{\alpha}{k^2}\phi\left(1 - \phi^2\right), \tag{3.10.22}$$

$$3w + \phi^2 = 1. \tag{3.10.23}$$

Combining (3.10.21) and (3.10.22) suitably and hence integrating with the condition $\phi = 1$ when $w = 0$, we obtain

$$w^2 = \frac{\alpha}{2k^2}\left(1 - \phi^2\right)^2. \tag{3.10.24}$$

Equations (3.10.23) and (3.10.24) are compatible only when $k^2 = 9\alpha/2$; here we have used the fact that k and w_1 have the same sign. Because $\alpha = 1 - (k/w_1)$, we get

$$k = k^* \equiv \frac{-9 + \left(81 + 72w_1^2\right)^{1/2}}{4w_1}. \tag{3.10.25}$$

The solution corresponding to this value of k is monotonic in w as well as ϕ. We may now explicitly solve (3.10.17)–(3.10.19) using (3.10.23). This solution, in terms of the original variables, is found to be

$$R(x) = r_{PTW} \tanh\left(\frac{x}{\sqrt{2}}\right), \quad \Psi(x) = \psi_{PTW} \tanh\left(\frac{x}{\sqrt{2}}\right), \tag{3.10.26}$$

where

$$r_{PTW} = \left\{\frac{1}{2}\left[1 + \left(1 + \frac{8}{9}w_1^2\right)^{1/2}\right]\right\}^{-1/2},$$

$$\psi_{PTW} = -\text{sgn}(w_1)\left\{\frac{\sqrt{1 + \frac{8}{9}w_1^2} - 1}{\sqrt{1 + \frac{8}{9}w_1^2} + 1}\right\}^{1/2}. \tag{3.10.27}$$

The solution (3.10.26) and (3.10.27) was shown by Sherratt (2003) to be in excellent agreement with the large time behaviour predicted by the numerical solution of (3.10.1) and (3.10.2) subject to the conditions $u = v = 0$ at $x = 0$. Substituting (3.10.27) into the stability condition (3.10.5) for the PTW, we find that

$$8w_1^6 + 16w_1^4 - 10w_1^2 - 27 < 0, \tag{3.10.28}$$

implying that $|w_1| < 1.110468$. Details of the numerical solution of the initial/boundary value problem indicate that irregular oscillations develop exactly when the values of w_1 are chosen to be above this critical value. One may also determine the direction of PTWs with the help of (3.10.27). The travelling wave (3.10.3) and (3.10.4) for u and v, in conjunction with (3.10.8) and (3.10.27), shows that it propagates in the positive x-direction if and only if

$$\psi_{PTW}\left(w_0 - w_1 r_{PTW}^2\right) < 0, \tag{3.10.29}$$

implying that w_0 and w_1 have the same sign and

$$|w_0| > \frac{2|w_1|}{1 + \sqrt{1 + \frac{8}{9}w_1^2}}. \tag{3.10.30}$$

The conditions (3.10.28)–(3.10.30) are shown graphically in Figure 3.12.

The main result of the present study is that the boundary conditions at $x = 0$ for the system (3.10.1) and (3.10.2) select a unique PTW amplitude which is

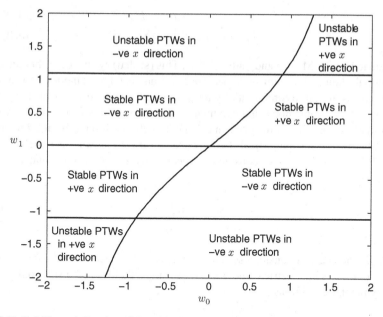

Fig. 3.12 Stability and direction of the periodic travelling wave solution when $u = v = 0$ at $x = 0$ (see (3.10.28)–(3.10.30)). (Sherratt 2003. Copyright © 2003 Society for Industrial and Applied Mathematics. Reprinted with permission. All rights reserved.)

independent of initial conditions. This amplitude is given by (3.10.27). If the conditions $u = 0, v = 0$ at $x = 0$, for example, are changed to a zero flux condition, namely, $u_x = v_x = 0$, PTWs disappear and are replaced by spatially uniform oscillations.

We may summarise some other work on travelling waves and their asymptotic nature. Sneyd et al. (1998) considered travelling waves in buffered systems with application to calcium waves which themselves result from the reaction and diffusion of calcium. Specifically, they studied buffered reaction–diffusion system

$$c_t = D_c c_{xx} + f(c) + k^- b - k^+ c(b_t - b), \qquad (3.10.31)$$

$$b_t = D_b b_{xx} - k^- b + k^+ c(b_t - b), \qquad (3.10.32)$$

where b is $[CaB]$ and $b_t = [B] + [CaB]$ and c is the concentration of free cytosolic Ca^{2+}. $f(c)$ in (3.10.31) is taken to be $c(1 - c)(c - a)$. D_c, D_b, k^+, and k^- are constants. Sneyd et al. (1998) derived an approximate expression for the wavespeed of the travelling wave which is numerically confirmed for both high and low affinity buffers. It was also shown that similar behaviour holds for the related FitzHugh–Nagumo equations.

A very detailed study of travelling wave solutions of the coupled system of non-linear diffusion equations

$$a_t = a_{xx} + 1 - a - \mu ab^2, \qquad\qquad (3.10.33)$$
$$b_t = b_{xx} + \mu ab^2 - \phi b \qquad\qquad (3.10.34)$$

was carried out by Merkin and Sadiq (1996). This (scaled) system describes an open isothermal chemical system governed by cubic autocatalytic kinetics. It was first shown to sustain up to three spatially uniform steady states: the (trivial) unreacted state, which is a stable node, and two nontrivial states one of which is always unstable, a saddle point. The third state can change its stability through Hopf bifurcation, both subcritical and supercritical. Different sets of travelling waves connecting different behaviour at $\pm\infty$ were considered. Specifically, it was assumed that at $t = 0$,

$$
\begin{aligned}
a &= 1, \quad -\infty < x < \infty \\
b &= \beta_0 g(x) \quad \text{if } |x| < \sigma \qquad\qquad (3.10.35)\\
&= 0 \quad \text{if } |x| > \sigma,
\end{aligned}
$$

where β_0 and σ are dimensionless constants. It was also assumed that the behaviour of the solution is symmetrical about $x = 0$; one may, therefore, consider only $x > 0$ with symmetry conditions

$$a_x = b_x = 0 \ \text{ at } x = 0, \ t > 0. \qquad\qquad (3.10.36)$$

Different kinds of exact solutions of (3.10.33) and (3.10.34) were considered: spatially uniform, permanent travelling waveform, and self-similar solutions. The conditions under which a particular kind of wave is initiated were investigated by a discussion of the ordinary differential equations governing any of the special solutions and by numerical integration of an initial value problem. Curiously, the treatment of Merkin and Sadiq (1996) reveals the possibility of a stable travelling wave propagating through the system, leaving behind a temporally unstable stationary state. Under these conditions, spatiotemporal chaotic behaviour is seen to develop after the passage of the wave.

3.11 Conclusions

This chapter was mainly concerned with the large time asymptotic analysis of nonlinear parabolic partial differential equations via some constructive approaches such as the balancing argument, matched asymptotic expansions, as well as some ad hoc approaches. In Section 3.2, we have derived exact travelling wave solutions of the Burgers equation satisfying specific conditions at $x = \pm\infty$. These solutions were shown to be large time asymptotics for solutions of initial value problems for the Burgers equation (3.2.1) with the step function as the initial data. This was confirmed numerically following the work of Shih (1991). In Section 3.3, we have discussed the solution of an initial boundary value problem for the Burgers equation describing vertical nonhysteretic flow of water in nonswelling soil. This problem was solved exactly via an application of the Cole–Hopf transformation (3.3.8). The large time behaviour of the solution thus obtained has a simple form, referred to as the profile at infinity. Experimental and analytical results for this problem were found to be in good agreement (Clothier et al. 1981). We have also analysed a more general initial boundary value problem (3.3.23)–(3.3.25) following Joseph and Sachdev (1993). The large time asymptotic behaviour of the solution of this problem was shown to contain the profile at infinity solution of Clothier et al. (1981) as a special case. In Section 3.4, we have presented an asymptotic analysis of an initial boundary value problem describing infiltration of water into a homogeneous soil, following Vanaja and Sachdev (1992). First, the existence of travelling wave solutions with constant end conditions at $\pm\infty$ was discussed. Travelling wave solutions given by (3.4.9) and (3.4.10) describe the large time behaviour of solutions of (3.4.7) and (3.4.8). It was confirmed numerically that the solution of (3.4.7) satisfying (3.4.8), for particular choices of $D(u)$ and $K(u)$, approaches the travelling wave solution as time becomes large. Section 3.5 is concerned with a stable profile describing cross-field diffusion in a toroidal multiple plasma. Following Berryman and Holland (1982), we have discussed the asymptotic behaviour of the solution of (3.5.9) and (3.5.10) with initial data satisfying the condition $n(x,0) \geq n_0$ (see below (3.5.10)). It was shown that the large time behaviour is given by $\ln(n/n_0) \sim A \exp\left(-\pi^2 t\right) \phi_1(x)$, $\phi_1(x) = 2^{1/2} \sin(\pi x)$. This was achieved via the derivation of several rigourous inequalities involving the integrals of the dependent variable and its derivatives. In Section 3.6, we have discussed in detail the large time solutions of the radially symmetric version (3.6.5) for the N-dimensional nonlinear diffusion equation (3.6.1) subject to the conditions (3.6.6)–(3.6.8) for the parametric range $n = 2/N$, $N > 2$. Equation (3.6.1) appears in many applications: spreading of microscopic droplets, diffusion of impurities in silicon, and so on. We may recall that equation (3.5.9) discussed in Section 3.5 is a special case of (3.6.1) with $n = 1$, $N = 1$. However, in Section 3.5, we discussed the large time behaviour of solutions of (3.5.1) with fixed boundaries. The large time behaviour of solutions of (3.6.5)–(3.6.8) has an 'unusual' self-similar form when $n = 2/N$, $N > 2$. Furthermore, solutions of (3.6.5)–(3.6.8) preserve mass for this parametric range. The asymptotic analysis here closely followed the work of King (1993). In Section 3.7, we have discussed large time asymptotic behaviour of the periodic solutions of some generalised Burgers equations, following the work of Sachdev and his collaborators (Sachdev et al. 2003, 2005). It does not seem possible to exactly linearise the generalised Burgers equations via a Cole–Hopf-like transformation. Thus, we must deal with these partial differential equations directly. We resorted to a certain perturbative approach, where we started with the solution of the linearised form of the original nonlinear partial differential equation, satisfying appropriate boundary conditions. This solution was improvised by including nonlinear effects. We may mention that this perturbative approach is reminiscent of the method frequently used by Bender and Orszag (1978) for nonlinear ordinary differential equations. We have presented in detail the large time asymptotic solution for the nonplanar Burgers equation and summarised the results for some other generalised Burgers equations. The problems discussed here satisfy periodic initial conditions. In Section 3.8, we discussed the asymptotic behaviour of solutions of some generalised Burgers equations. Here, we first introduced time and a similarity variable as the new independent variables and transformed the given partial differential equation to the one which, in a sense, embeds all possible similarity solutions using the so-called balancing argument. Then, we looked for the relative importance of the terms in the transformed partial differential equation for large time. This led to different sets of parametric ranges and reduced partial differential equations relevant to those parametric ranges. We have presented the asymptotic analysis of solutions of the nonplanar Burgers equation (3.8.1), following Grundy et al. (1994) in some detail. The balancing argument resulted in three parametric ranges: $\alpha > 1/(j+1), \alpha = 1/(j+1), \alpha < 1/(j+1)$. It was shown that in the parametric range $\alpha > 1/(j+1)$, the diffusion dominates convection whereas for $\alpha < 1/(j+1)$, the convection dominates diffusion. For $\alpha = 1/(j+1)$, all the terms are equally important and the ordinary differential equation governing self-similar solution of (3.8.1) describes the large time behaviour of solutions with integrable initial data. We have also summarised the results of Grundy (1988) for a more general partial differential equation (3.8.86) with integrable initial data. Later in this section we have discussed N-wave solutions of some generalised Burgers equations. The construction of asymptotic N-wave solutions for these equations mimics the structure of the corresponding solution of the plane Burgers equation. In Section 3.9, we have treated an initial boundary value problem involving generalised Fisher's equation following Leach and Needham (2001). A complete asymptotic structure of solutions of an initial boundary value problem as $t \to \infty$ was presented. It was shown that the permanent travelling wave (PTW) with minimum speed $v = v^* > 2$ evolves from the initial profile only when the function $F(u)$ in (3.9.1) satisfies the inequality $F(u) \nleq u$ for all $u \in [0,1]$. For a general Fisher's equation, the minimum speed of propagation v^* for travelling wave solutions was also determined. In Section 3.10, we have presented periodic travelling wave solutions of the oscillatory reaction–diffusion system (3.10.1) and (3.10.2) in one space dimension, following Sherratt (2003). Sherratt's numerical study of (3.10.1) and (3.10.2) in one space dimension with Dirichlet condition $u = v = 0$ at $x = 0$ and a 'random initial condition' showed the development of periodic travelling waves. It was also observed that these travelling waves are independent of the initial conditions used.

References

Abramowitz, M., Stegun, I. A. (1972) *Handbook of Mathematical Functions*, Dover, New York.

Bear, J. (1972) *Dynamics of Fluids in Porous Media*, Elsevier, New York.

Bender, C. M., Orszag, S. A. (1978) *Advanced Mathematical Methods for Scientists and Engineers*, McGraw-Hill, New York.

Berryman, J. G. (1977) Evolution of a stable profile for a class of nonlinear diffusion equations with fixed boundaries, *J. Math. Phys.* 18, 2108–2115.

Berryman, J. G., Holland, C. J. (1978a) Nonlinear diffusion problem arising in plasma physics, *Phys. Rev. Lett.* 40, 1720–1722.

Berryman, J. G., Holland, C. J. (1978b) Evolution of a stable profile for a class of nonlinear diffusion equations II, *J. Math. Phys.* 19, 2476–2480.

Berryman, J. G., Holland, C. J. (1980) Stability of the separable solution for fast diffusion, *Arch. Rational Mech. Anal.* 74, 379–388.

Berryman, J. G., Holland, C. J. (1982) Asymptotic behavior of the nonlinear diffusion equation $n_t = (n^{-1}n_x)_x$, *J. Math. Phys.* 23, 983–987.

Bowen, M., King, J. R. (2001) Asymptotic behaviour of the thin film equation in bounded domains, *Euro. J. Appl. Math.* 12, 135–157.

Bramson, M. D. (1978) Maximal displacement of branching Brownian motion, *Comm. Pure Appl. Math.* 31, 531–581.

Brezis, H., Peletier, L. A., Terman, D. (1986) A very singular solution of the heat equation with absorption, *Arch. Rat. Mech. Anal.* 95, 185–209.

Broadbridge, P., Knight, J. K., Rogers, C. (1988) Constant rate rain fall infiltration in a bounded profile: solution of a nonlinear model, *Soil Sci. Soc. Am. J.* 52, 1526–1533.

Broadbridge, P., Rogers, C. (1990) Exact solution for vertical drainage and redistribution in soils, *J. Engrg. Math.* 24, 25–43.

Calogero, F., Lillo, S. De. (1991) The Burgers equation on the semiline with general boundary conditions at the origin, *J. Math. Phys.* 32, 99–105.

Carslaw, H. S., Jaeger, J. C. (1959) *Conduction of Heat in Solids*, Clarendon Press, Oxford.

Cazenave, T., Escobedo, M. (1994) A two-parameter shooting problem for a second-order differential equation, *J. Differential Eq.* 113, 418–451.

Clothier, B. E., Knight, J. H., White, I. (1981) Burgers' equation: Application to field constant-flux infiltration, *Soil Sci.* 132, 255–261.

Dawson, C. N., Van Duijn, C. J., Grundy, R. E. (1996) Large time asymptotics in contaminant transport in porous media, *SIAM J. Appl. Math.* 56, 965–993.

Dix, D. B. (2002) Large-time behaviour of solutions of Burgers' equation, *Proc. Roy. Soc. Edinburgh Sect. A* 132, 843–878.

Drake, J. R. (1973) Plasma losses to an octupole hoop, *Phys. Fluids* 16, 1554–1555.

Drake, J. R., Greenwood, J. R., Navratil, G. A., Post, R. S. (1977) Diffusion coefficient scaling in the Wisconsin levitated octupole, *Phys. Fluids* 20, 148–155.

Escobedo, M., Zuazua, E. (1991) Large time behaviour for convection-diffusion equations in $\mathbf{R}^n$, *J. Funct. Anal.* 100, 119–161.

Escobedo, M., Vázquez, J. L., Zuazua, E. (1993) Asymptotic behaviour and source-type solutions for a diffusion-convection equation, *Arch. Rational Mech. Anal.* 124, 43–65.

Esteban, J. R., Rodriguez, A., Vázquez, J. L. (1988) A nonlinear heat equation with singular diffusivity, *Comm. Partial Differential Eq.* 13, 985-1039.

Fife, P. C. (1979) *Mathematical Aspects of Reacting and Diffusing Systems*, Lecture Notes in Biomathematics 28, Springer-Verlag, New York.

Friedman, A., Kamin, S. (1980) The asymptotic behaviour of gas in an *n*-dimensional porous medium, *Trans. Amer. Math. Soc.* 262, 551-563.

Galaktionov, V. A., Peletier, L. A., Vázquez J. L. (2000) Asymptotics of the fast diffusion equation with critical exponent, *SIAM J. Math. Anal.* 31, 1157–1174.

Greenwood, J. R. (1975) Diffusion in the levitated toroidal octupole, Ph. D. Thesis, University of Wisconsin, Madison.

Grundy, R. E. (1988) Large time solution of the Cauchy problem for the generalized Burgers equation, Preprint, University of St. Andrews, UK.

Grundy, R. E., Sachdev, P. L., Dawson, C. N. (1994) Large time solution of an initial value problem for a generalized Burgers equation, in *Nonlinear Diffusion Phenomenon*, P. L. Sachdev and R. E. Grundy (Eds.), 68–83, Narosa, New Delhi.

Grundy, R. E., Van Duijn, C. J., Dawson, C. N. (1994a) Asymptotic profiles with finite mass in one-dimensional contaminant transport through porous media: The fast reaction case, *Quart. J. Mech. Appl. Math.* 47, 69–106.

Hadeler, K. P., Rothe, F. (1975) Travelling fronts in nonlinear diffusion equations, *J. Math. Biol.* 2, 251–263.

Harris, S. E. (1996) Sonic shocks governed by the modified Burgers equation, *Euro. J. Appl. Math.* 7, 201–222.

Hopf, E. (1950) The partial differential equation $u_t + uu_x = \mu u_{xx}$, *Comm. Pure Appl. Math.* 3, 201–230.

Joseph, K. T., Sachdev, P. L. (1993) Exact analysis of Burgers equation on semiline with flux condition at the origin, *Int. J. Non-Linear Mech.*, 28, 627–639.

Khusnytdinova, N. V. (1967) The limiting moisture profile during infiltration in to a homogeneous soil, *Prikl. Mat. Mekh.* 31, 770–776.

King, J. R. (1993) Self-similar behaviour for the equation of fast nonlinear diffusion, *Phil. Trans. Roy. Soc. London Ser. A* 343, 337–375.

King, J. R., McCabe, P. M. (2003) On the Fisher–KPP equation with fast nonlinear diffusion, *Proc. Roy. Soc Lond. A* 459, 2529–2546.

Kolmogorov, A. N., Petrovskii, I. G., Piskunov, N. S. (1937) Study of the diffusion equation with growth of the quantity of matter and its application to a biological problem, *Mos. Uni. Bull. Math.* 1–26; also in *Applied Mathematics of Physical Phenomena* F. Oliveira-Pinto and B. W. Conolly (Eds.), Horwood, Chichester (1982).

Kopell, N., Howard, L. N. (1973) Plane wave solutions to reaction-diffusion equations, *Stud. Appl. Math.* 52, 291–328.

Krzyżański, M. (1959) Certaines inégalités relatives aux solutions de l'équation parabolique linéare normale, *Bull. Acad. Polon. Sci., Ser. Math. Astr. Phys.* 7, 131–135.

Larson, D. A. (1978) Transient bounds and time-asymptotic behavior of solutions to nonlinear equations of Fisher type, *SIAM J. Appl. Math.* 34, 93–103.

Laurencot, Ph., Simondon, F. (1998) Long-time behaviour for porous medium equations with convection, *Proc. Roy. Soc. Edin.* 128A, 315–336.

Leach, J. A., Needham, D. J. (2001) The evolution of travelling waves in generalized Fisher equations via matched asymptotic expansions: Algebraic corrections, *Quart. J. Mech. Appl. Math.* 54, 157–175.

Lee-Bapty, I. P., Crighton, D. G. (1987) Nonlinear wave motion governed by the modified Burgers equation, *Phil. Trans. Roy. Soc. Lond. A* 323, 173–209.

Lees, M. (1966) A linear three-level difference scheme for quasilinear parabolic equations, *Math. Comp.* 20, 516–522.

Mayil Vaganan, B. (1994) Exact analytic solutions for some classes of partial differential equations, Ph. D. Thesis, Indian Institute of Science, Bangalore.

McCabe, P. M., Leach, J. A., Needham, D. J. (2002) A note on the nonexistence of travelling waves in a class of singular reaction diffusion problems, *Dynam. Sys. Int. J.* 17, 131–135.

McKean, H. P. (1975) Application of Brownian motion to the equation of Kolmogorov-Petrovskii-Piskunov, *Comm. Pure Appl. Math.* 28, 323–331.

Merkin, J. H., Needham, D. J. (1989) Propagating reaction-diffusion waves in a simple isothermal quadratic chemical system, *J. Engrg. Math.* 23, 343–356.

Merkin, J. H., Sadiq, M. A. (1996) The propagation of travelling waves in an open cubic autocatalytic chemical system, *IMA J. Appl. Math.* 57, 273–309.

Needham, D. J. (1992) A formal theory concerning the generation and propagation of travelling wave-fronts in reaction-diffusion equations, *Quart. J. Mech. Appl. Math.* 45, 469–498.

Parker, D. F. (1981) An approximation for nonlinear acoustics of moderate amplitude, *Acoustics Lett.* 4, 239–244.

Peletier, L. A. (1970) Asymptotic behaviour of temperature profiles of a class of non-linear heat conduction problems, *Quart. J. Mech. Appl. Math.* 23, 441–447.

Philip, J. R. (1969) Theory of infiltration, *Adv. Hydrosci.* 5, 215–296.

Philip, J. R. (1970) Flow in porous media, *Ann. Rev. Fluid Mech.* 2, 177–204.

Sachdev, P. L. (1987) *Nonlinear Diffusive Waves*, Cambridge University Press, Cambridge, UK.

Sachdev, P. L. (2000) *Self-Similarity and Beyond. Exact Solutions of Nonlinear Problems*, Chapman & Hall/CRC Press, New York.

Sachdev, P. L., Joseph, K. T. (1994) Exact representations of N-wave solutions of generalized Burgers equations, in *Nonlinear Diffusion Phenomenon*, P. L. Sachdev and R. E. Grundy (Eds.), 197–219, Narosa, New Delhi.

Sachdev, P. L., Nair, K. R. C. (1987) Generalized Burgers equations and Euler-Painlevé transcendents II, *J. Math. Phys.* 28, 997–1004.

Sachdev, P. L., Srinivasa Rao, Ch. (2000) N-wave solution of modified Burgers equation, *Appl. Math. Lett.* 13, 1–6.

Sachdev, P. L., Enflo, B. O., Srinivasa Rao, Ch., Mayil Vaganan, B., Poonam Goyal (2003) Large-time asymptotics for periodic solutions of some generalized Burgers equations, *Stud. Appl. Math.* 110, 181–204.

Sachdev, P. L., Joseph, K. T., Nair, K. R. C. (1994) Exact N-wave solutions for the nonplanar Burgers equation, *Proc. Roy. Soc. London Ser. A* 445, 501–517.

Sachdev, P. L., Joseph, K. T., Mayil Vaganan, B. (1996) Exact N-wave solutions of generalized Burgers equations, *Stud. Appl. Math.* 97, 349–367.

Sachdev, P. L., Nair, K. R. C., Tikekar, V. G. (1986) Generalized Burgers equations and Euler-Painlevé transcendents I, *J. Math. Phys.* 27, 1506–1522.

Sachdev, P. L., Srinivasa Rao, Ch., Enflo, B. O. (2005) Large-time asymptotics for periodic solutions of the modified Burgers equation, *Stud. Appl. Math.* 114, 307–323.

Sachdev, P. L., Srinivasa Rao, Ch., Joseph, K. T. (1999) Analytic and numerical study of N-waves governed by the nonplanar Burgers equation, *Stud. Appl. Math.* 103, 89–120.

Shenker, Y., Roseman, J. J. (1995) On the exponential temporal decay of solutions and their derivatives for quasilinear parabolic equations, *Z. angew. Math. Phys.* 46, 198–223.

Sherratt, J. A. (2003) Periodic travelling wave selection by Dirichlet boundary conditions in oscillatory reaction–diffusion systems, *SIAM J. Appl. Math.* 63, 1520-1538.

Shih, Shagi-Di (1991) Shock layer structure of Burgers equation, in *Conference on Nonlinear Analysis*, Fon-Che Liu and Tai-Ping Liu (Eds.), 237–257, World Scientific, River Edge, NJ.

Smoller, J. (1989) *Shock Waves and Reaction-Diffusion Equations*, Springer Verlag, Berlin.

Sneyd, J., Dale, P. D., Duffy, A. (1998) Travelling waves in buffered systems: Applications to calcium waves, *SIAM J. Appl. Math.* 58, 1178–1192.

Srinivasa Rao, Ch., Satyanarayana, E. (2008a) Large time asymptotics for periodic solutions of a generalized Burgers equation, *Int. J. Nonlinear Sci.* 5, 237–245.

Srinivasa Rao, Ch., Satyanarayana, E. (2008b) Asymptotic N-wave solutions of the nonplanar Burgers equation, *Stud. Appl. Math.* 121, 191–221.

Srinivasa Rao, Ch., Sachdev, P. L., Mythily, R. (2002) Analysis of the self-similar solutions of the nonplanar Burgers equation, *Nonlinear Anal.* 51, 1447–1472.

Van Duijn, C. J., Grundy, R. E., Dawson, C. N. (1997) Large time profiles in reactive solute transport, *Transp. Porous Media* 27, 57–84.

Van Dyke, M. (1975) *Perturbation Methods in Fluid Mechanics*, Parabolic Press, Stanford, CA.

Vanaja, V., Sachdev, P. L. (1992) Asymptotic solutions of a generalized Burgers equation, *Quart. Appl. Math.* 50, 627–640.

Weidman, P. D. (1976) On the spin-up and spin-down of a rotating fluid Part 2. Measurements and stability, *J. Fluid Mech.* 77, 709–735.

Whitham, G. B. (1974) *Linear and Nonlinear Waves*, John Wiley & Sons, New York.

Zel'dovich, Ya. B., Barenblatt, G. I. (1958) The asymptotic properties of self-modelling solutions of the non-stationary gas filtration equations, *Sov. Phys. Dokl.* 3, 44–47.

Chapter 4
Self-similar Solutions as Large Time Asymptotics for Some Nonlinear Parabolic Equations

4.1 Introduction

Nonlinear partial differential equations, scalar or systems, are extremely hard to analyse in an exact manner. For given initial/boundary conditions, it is rare to find an explicit exact solution of a physical problem. Thus, a resort to numerical solution is inevitable but it is important to have some approximate or asymptotic solution which may be used to provide some support or verification of the numerical solution. It is here that the so-called similarity solutions (which include product solutions as special cases) come in handy. For linear problems, these special solutions may be superposed and hence certain classes of initial/boundary value problems can be explicitly solved in a series form. For nonlinear problems for which no such superposition is possible, one looks for invariant properties of the governing partial differential equations (see Bluman and Kumei 1989, Olver 1986). The symmetries of the partial differential equations are exploited to find the so-called similarity solutions. Finite and infinitesimal transformations or the so-called direct approach (Clarkson and Kruskal 1989) help change a PDE or a system of PDEs to corresponding ODEs. These impose some conditions on the parameters appearing in the problem. The solutions of these (nonlinear) ODEs may correspond as such to some generalised functions as initial conditions for the PDEs. One must first analyse when these ODEs admit solutions which relate sensibly to those pertaining to PDEs. To that end, a careful existence analysis of the ODEs, subject to appropriate boundary conditions, is clearly important. This analysis yields a certain set of parameters for which the similarity solution with physical boundary conditions makes sense. Thus, before the reduced nonlinear ODEs are used to investigate the asymptotic behaviour of the original nonlinear PDEs, one must carefully study their qualitative nature. Therefore, one must first prove existence and uniqueness of the solution of the ODEs resulting from the original PDEs, subject to appropriate initial/boundary conditions. The second major step is to show, both qualitatively and quantitatively, when these similarity solutions constitute the so-called intermediate asymptotics to

P.L. Sachdev, Ch. Srinivasa Rao, *Large Time Asymptotics for Solutions of Nonlinear Partial Differential Equations*, Springer Monographs in Mathematics, DOI 10.1007/978-0-387-87809-6_4, © Springer Science+Business Media, LLC 2010

which a large class of solutions of initial/boundary value problems of the original system of PDEs tend as time becomes large.

Section 4.2 is illustrative; here we discuss in some detail the above aspects of the asymptotic analysis with reference to the heat equation. The heat equation, a linear PDE, has played an important role in the understanding of nonlinear PDEs. For example, the Cole–Hopf transformation changes the (nonlinear) Burgers equation to the heat equation, hence much analysis of the former becomes possible (Sachdev 1987). The heat equation has been studied in much detail (Widder 1975) and enjoys considerable symmetry properties (Bluman and Kumei 1989, Olver 1986). Here, following the work of Kloosterziel (1990), we show how superposition of the similarity solutions for this linear PDE helps to solve an initial value problem over an infinite domain and hence brings out the asymptotic character of the solution, essentially via the first nonzero term of the expansion.

In a remarkable paper, Kamenomostskaya (1973), called Kamin in her subsequent publications, showed that, just as the fundamental solution of the heat equation describes asymptotic behaviour of its solution with compactly supported initial data, a certain similarity solution of the unsteady filtration equation

$$u_t = (u^{\lambda+1})_{xx} \tag{4.1.1}$$

with initial condition

$$W_E(x,0) = E\delta(x) \tag{4.1.2}$$

enjoys the same property; here E is an arbitrary constant and $\lambda > 0$ is a parameter. More specifically, equation (4.1.1) is invariant under the group of transformations

$$u' = cu, \quad x' = l^{-1}x, \quad t' = l^{-2}c^{-\lambda}t \tag{4.1.3}$$

and, therefore, possesses the self-similar solution

$$W_E(x,t) = E^{2/(\lambda+2)}t^{-1/(\lambda+2)}\Phi\left(xE^{-\lambda(\lambda+2)}t^{-1/(\lambda+2)}\right), \tag{4.1.4}$$

where

$$\Phi(\xi) = \begin{cases} a(\lambda)(\xi_0^2 - \xi^2)^{1/\lambda}, & \xi \le \xi_0, \\ 0, & \xi > \xi_0, \end{cases}$$

where $\xi_0 = \xi_0(\lambda)$. The interesting feature of the proof of Kamin (1973) is that it also uses a group of transformations of the type (4.1.3) to prove the asymptotic result. She studied the Cauchy problem for equation (4.1.1) on the half plane

$$S = \{(x,t) : x \in \mathbb{R}, 0 \le t < \infty\}$$

with the initial condition

$$u(x,0) = u_0(x), \tag{4.1.5}$$

where $u_0(x)$ is a continuous nonnegative function with compact support, and, for simplicity, assumed that the function $[u_0(x)]^{\lambda+1}$ satisfied the Lipschitz condition. She then proved the following theorem.

Theorem 4.1.1 *Suppose that $u(x,t)$ is a generalised solution of the Cauchy problem (4.1.1) and (4.1.5) and*

$$\int_{-\infty}^{\infty} u_0(x)dx = E_0.$$

Then

$$t^{1/(\lambda+2)}|u(x,t) - W_{E_0}(x,t)| \to 0 \qquad (4.1.6)$$

as $t \to \infty$, uniformly with respect to $x \in \mathbb{R}$.

Here, the Cauchy problem (4.1.1) and (4.1.5) possesses a generalised solution. A generalised solution of (4.1.1) with (4.1.5) satisfies the following conditions.

(i) $u(x,t)$ is continuous and nonnegative.
(ii) There exists a generalised bounded derivative $(\partial u^{\lambda+1})/(\partial x)$.
(iii) For any continuously differentiable function $f(x,t)$ with compact support,

$$\int \int_S \left(u\frac{\partial f}{\partial t} - \frac{\partial u^{\lambda+1}}{\partial x}\frac{\partial f}{\partial x} \right) dxdt + \int_{-\infty}^{\infty} u_0(x)f(x,0)dx = 0. \qquad (4.1.7)$$

The existence and uniqueness of the generalised solution of (4.1.1) was proved earlier by Oleinik et al. (1958) and Aronson (1969, 1970a, b).

In Section 4.3, we present a recent study of Barenblatt et al. (2000) which demonstrates in a vivid manner the intermediate asymptotic character of a self-similar solution of a degenerate parabolic equation. The latter describes the groundwater flow in a water-absorbing fissurised porous rock, hence the filtration–absorption equation. Numerical solution of the problem, subject to a class of initial conditions, confirms the large time asymptotic character of the self-similar solution. In Section 4.4, we discuss a class of weak similarity solutions of the porous media equation following Gilding and Peletier (1976). A detailed existence analysis of the ODE governing the similarity solution, subject to appropriate boundary conditions, is presented using a shooting argument. This analysis exemplifies how the existence of the ODEs subject to relevant initial/boundary conditions in the present context may be rigorously proved. The role of these special solutions as intermediate asymptotics is brought out in Section 4.6. In Sections 4.5–4.8, we treat several nonlinear parabolic equations – nonlinear heat conduction equation, porous media equation, heat equation with absorption, and a very singular diffusion equation – and determine sets of parameters occuring in the problem for which they possess similarity solutions. We then rigorously show how the latter may describe large time asymptotic behaviour of the solutions of original PDEs, subject to relevant initial/boundary conditions.

4.2 Self-similar solutions as large time asymptotics for linear heat equation

In this section we demonstrate with the help of analytic solution of initial/boundary value problems for the heat equation how self-similar solutions describe large time behaviour. This is in contrast to other solutions, even exact ones, which do not throw much light on the asymptotic form. We follow here a very interesting work by Kloosterziel (1990) where this idea is brought out with the help of several examples. A standard problem in heat conduction which is discussed in elementary books on PDEs relates to a slab of conducting material between $x = 0$ and $x = l$, assuming that both the ends are kept at zero temperature. The mathematical problem reduces to solving

$$C_t = DC_{xx}, \quad 0 < x < l, \, t > 0; \tag{4.2.1}$$

D is a constant, subject to the initial condition expressed as a sine series

$$C^0(x) = \sum_{n=1}^{\infty} a_n \sin(\lambda_n x), \quad a_n = \frac{2}{l} \int_0^l C^0(x) \sin(\lambda_n x) dx \tag{4.2.2}$$

with $\lambda_n = n\pi/l$, and the boundary conditions

$$C(0,t) = C(l,t) = 0. \tag{4.2.3}$$

The solution of this problem is well known (see, for example, Carslaw and Jaeger (1959)) and is given by

$$C(x,t) = \sum_{n=1}^{\infty} a_n e^{-\lambda_n^2 Dt} \sin(\lambda_n x), \quad t > 0 \tag{4.2.4}$$

or, equivalently,

$$C(x,t) = \sum_{n=1}^{\infty} A_n(t) \sin(\lambda_n x), \tag{4.2.5}$$

where

$$A_n(t) = a_n e^{-\lambda_n^2 Dt}. \tag{4.2.6}$$

The solution (4.2.5) may be interpreted as a superposition of an infinite system of product solutions of (4.2.1). For each n, the amplitude $A_n(t)$ decays exponentially; it forms an ordered point-spectrum decaying exponentially: for each $m > n$, $\lim_{t\to\infty}(A_m/A_n) = 0, A_n(0) \neq 0$. However, it is easy to see a clear asymptotic form of the solution from (4.2.5). The solution gets closer to the mth term in (4.2.4) as time increases:

$$C(x,t) \approx a_m e^{-\lambda_m^2 Dt} \sin(\lambda_m x) \tag{4.2.7}$$

and has asymptotically vanishing amplitude, where m is the smallest integer with $a_m \neq 0$. The spectrum in (4.2.4) describes a diffusive phenomenon where the small-scale irregularities in the initial data are ironed out with increasing time.

We now show that, when the initial data are slightly restricted, the similarity solutions of the heat equation give a very quick and precise description of the large time asymptotics. If the heat equation (4.2.1) is transformed to an ODE via a similarity transformation (see Sachdev 2000), that is, if we write

$$C(x,t) = t^{-m/2}F(xt^{-1/2}) \qquad (4.2.8)$$

in (4.2.1), we may solve the resulting ODE and hence write the solution in terms of a known function, which is readily interpreted. An example of this class of solutions of (4.2.1) is the so-called point source solution

$$C(x,t) = \frac{1}{2\sqrt{\pi Dt}}\,\exp(-x^2/4Dt) \qquad (4.2.9)$$

which, for all $t > 0$, satisfies

$$\int_{\mathbb{R}} C(x,t)dx = 1 \qquad (4.2.10)$$

and

$$\lim_{t \to 0+} C(x,t) = 0 \qquad (4.2.11)$$

for $x \neq 0$. It also solves an inhomogeneous form of (4.2.1) with $\delta(x)\delta(t)$ on the RHS and, therefore, serves as the Green's function. However, here we wish to study more general similarity solutions of (4.2.1) than (4.2.9) which may throw light on the asymptotic behaviour of solutions of a class of IVPs or IBVPs.

Now we seek similarity solutions of (4.2.1) for $t > 0$ in the form

$$C(x,t) = \frac{1}{a(t)}\hat{C}\left(\frac{x}{b(t)}\right), \qquad (4.2.12)$$

where we assume that $a(0) = b(0) = 1$. The solution (4.2.12) satisfies the initial condition

$$C(x,t=0) = \hat{C}(x). \qquad (4.2.13)$$

The superposed solutions $e^{-\lambda_n^2 Dt}\sin(\lambda_n x)$ in (4.2.4) are product solutions which are special cases of (4.2.12) with $b(t) \equiv 1$. The constant D in (4.2.1) can be scaled out by a simple transformation. Hence, now onwards, we assume that $D = 1$ without loss of generality. Substituting (4.2.12) into (4.2.1), we get

$$\frac{d^2\hat{C}}{ds^2} + b\frac{db}{dt}s\frac{d\hat{C}}{ds} + \frac{b^2}{a}\frac{da}{dt}\hat{C} = 0, \qquad (4.2.14)$$

where $s = x/b(t)$ is the similarity variable. Equation (4.2.14) becomes an ODE in s only if

$$\frac{b^2}{a}\frac{da}{dt} = \alpha, \quad b\frac{db}{dt} = \beta, \qquad (4.2.15)$$

where α and β are constants. Thus, (4.2.14) reduces to an ODE for $\hat{C} = \hat{C}(s)$. Three cases arise from (4.2.15): (i) $\alpha \neq 0$, $\beta = 0$; (ii) $\alpha = 0$, $\beta \neq 0$; (iii) $\alpha \neq 0$, $\beta \neq 0$. For case (i) with $\alpha > 0$ or $\alpha < 0$, we get exponentially decaying or growing solutions with trigonometric functions for $\hat{C}(s)$. For case (ii) with $a(t) =$ constant, we get explicit solutions in terms of the error function: $\hat{C}(s) = \int^s e^{-\beta u^2/2} du$. Case (iii) is more general and allows time-dependent amplitude $a(t)$ and a more general solution for $\hat{C}(s)$ than trigonometric functions. The solution of the system (4.2.15) with $a(0) = b(0) = 1$ may be found to be

$$a(t) = b^{\alpha/\beta}(t), \quad b(t) = (2\beta t + 1)^{1/2}. \tag{4.2.16}$$

Because $b(t)$ becomes zero for $\beta < 0$ at $t = |\beta|/2$ (see (4.2.12)), we discuss here only the case $\beta > 0$. We may, without loss of generality, choose $\beta = 1$. Thus, equation (4.2.14) becomes

$$\frac{d^2\hat{C}}{ds^2} + s\frac{d\hat{C}}{ds} + \alpha\hat{C} = 0 \tag{4.2.17}$$

with the solution $\hat{C}(s) = \hat{C}_\alpha(s)$, say. The similarity form of the solution (4.2.12) may now be written as

$$C(x,t) = \frac{1}{b^\alpha(t)}\hat{C}_\alpha\left(\frac{x}{b(t)}\right). \tag{4.2.18}$$

Clearly, $\hat{C}_\alpha(x)$ satisfies the differential equation

$$\frac{d^2\hat{C}}{dx^2} + x\frac{d\hat{C}}{dx} + \alpha\hat{C} = 0. \tag{4.2.19}$$

The solutions (4.2.18) cannot be used to solve arbitrary initial value problems on a finite interval. We may solve either a pure initial value problem on $\mathbb{R}$ or IBVP on the semiline $\mathbb{R}^+$ or $\mathbb{R}^-$ with appropriate boundary condition at the point $x = 0$ which corresponds to the similarity variable $s = 0$; of course, appropriate boundary conditions must be imposed as $x \to +\infty$, $-\infty$, or $\pm\infty$. Writing (4.2.19) in the Sturm–Liouville form

$$\frac{d}{dx}\left(e^{x^2/2}\frac{d\hat{C}}{dx}\right) + \alpha e^{x^2/2}\hat{C} = 0, \tag{4.2.20}$$

we observe that if $\hat{C}_\alpha$s are integrable with respect to the weight function $e^{x^2/2}$, we may derive the orthogonality relation

$$\int_{-\infty}^{\infty} \hat{C}_\nu(x)\hat{C}_\mu(x)w(x)dx = 0 \quad (\nu \neq \mu). \tag{4.2.21}$$

The solution of (4.2.20) may easily be found in terms of Hermite polynomials if $\alpha > 0$ is an integer. Writing

$$\hat{C}(x) = e^{-x^2/2}H(x) \tag{4.2.22}$$

and introducing the independent variable $y = x/\sqrt{2}$, we transform (4.2.20) to the Hermite equation

$$\frac{d^2H}{dy^2} - 2y\frac{dH}{dy} + 2(\alpha - 1)H = 0. \tag{4.2.23}$$

Now choosing $\alpha - 1 = n$, a positive integer or zero, we may write the solutions of (4.2.23) as Hermite poynomials

$$H_n(y) = (-1)^n e^{y^2} \frac{d^n}{dy^n} e^{-y^2}. \tag{4.2.24}$$

The first few Hermite polynomials are $H_0 = 1$, $H_1 = 2y$, $H_2 = 4y^2 - 2$, and so on. Thus, for $\alpha = 1 + n$, a positive integer, we get a particular set of similarity functions of the heat equation as

$$\hat{C}_\alpha(x) = e^{-x^2/2} H_n(x/\sqrt{2}), \quad n = 0, 1, 2, \ldots \tag{4.2.25}$$

whereas the solutions of (4.2.1) themselves are given by (4.2.18) and (4.2.25) with $b(t) = (1 + 2t)^{1/2}$. For general α, the similarity solutions may be expressed in terms of parabolic cylinder functions, which reduce to Hermite functions when α is a positive integer. Now we show how these solutions describe large time asymptotic behaviour of solutions of a large class of initial value problems for the heat equation on an infinite domain.

It is well known (see Higgins 1977) that the functions

$$\phi_n(x) = \frac{H_n(x)e^{-x^2/2}}{\sqrt{2^n n! \sqrt{\pi}}}, \quad n = 0, 1, 2, \ldots \tag{4.2.26}$$

are orthogonal on $\mathbb{R}$ and form a complete set in $L^2(\mathbb{R})$. This result in turn shows that the set of functions

$$\Omega_n(x) = \frac{H_n(x/\sqrt{2})e^{-x^2/2}}{\sqrt{2^n n! \sqrt{2\pi}}} \tag{4.2.27}$$

is orthonormal with respect to the weight function $w(x) = e^{x^2/2}$ on $\mathbb{R}$ and is also complete in $L^2(\mathbb{R}, w)$. Suppose that the initial function $C^0(x) = C(x, t = 0)$ is square integrable with respect to the weight function $w(x) = e^{x^2/2}$; that is,

$$\int_{-\infty}^{\infty} |C^0(x)|^2 e^{x^2/2} dx < \infty; \tag{4.2.28}$$

then,

$$C^0(x) = \sum_{n=0}^{\infty} a_n \Omega_n(x), \tag{4.2.29}$$

where

$$a_n = \int_{-\infty}^{\infty} C^0(x)\Omega_n(x)w(x)dx, \quad n = 0,1,2,\dots. \tag{4.2.30}$$

Thus, it follows from (4.2.18) and (4.2.29) that

$$C(x,t) = \sum_{n=0}^{\infty} \frac{a_n}{b(t)^{n+1}} \Omega_n\left(\frac{x}{b(t)}\right) \tag{4.2.31}$$

is the solution of the heat equation (4.2.1), subject to the initial condition (4.2.29). The first nonzero term in (4.2.31) determines its large time asymptotic behaviour.

In this section, we have shown how the series expansion in terms of the similarity solutions of the heat equation clearly brings out the large time asymptotic behaviour of the solution of IVP on an infinite domain. Because the similarity solutions considered here form a basis for $L^2(\mathbb{R}, e^{|x|^2/2})$, only those initial conditions that are square integrable with respect to the weight function $e^{|x|^2/2}$ may be expanded in terms of similarity functions. This vividly brings out the role of the first nonzero term in the series expansion as the descriptor of the large time asymptotic behaviour of the solution of (4.2.1).

4.3 Self-similar solutions as intermediate asymptotics for filtration–absorption model

An interesting parabolic equation describing groundwater flow in a (water) absorbing fissurised porous rock was first derived by Barenblatt et al. (2000). It also occurs in some biological models. We consider here only one-dimensional and axisymmetric cases. This equation in nondimensional form may be written as

$$u_t = uu_{xx} - (c-1)u_x^2. \tag{4.3.1}$$

Here, u is the (nondimensional) groundwater level, t is the time, x is the horizontal space coordinate along the impermeable bed, and $c > 1$ is the absorption coefficient. For a (purely) porous medium, this coefficient is less than one. For a fissurised rock, c can be much greater than one. Equation (4.3.1) is solved subject to an initial distribution

$$u(x,0) = u_0(x), \tag{4.3.2}$$

where $u_0(x)$ is a bounded continuous nonnegative function on $\mathbb{R}$ with compact support. Barenblatt et al. (2000) considered a self-similar solution of (4.3.1) with $c > 3/2$ in the form

$$u = B^2 \mu(t_0 - t)^{2\mu - 1} F\left(\frac{x - x_0}{x_f}\right), \tag{4.3.3}$$

$$x_f = B(t_0 - t)^\mu, \tag{4.3.4}$$

where x_0 is the point where the solution collapses at $t = t_0$, and x_f denotes the contracting half width of the support: the groundwater dome. The positive dome of water is symmetric about $x = x_0$. Thus, $F(\xi)$, $\xi = (x - x_0)/x_f$, is an even function of ξ in the interval $-1 \leq \xi \leq 1$; it vanishes for $|\xi| > 1$. B and μ in (4.3.3) and (4.3.4) are constants. It was shown by Barenblatt et al. (2000) that the self-similar solution (4.3.3) belongs to the so-called second kind (see Barenblatt 1996): conservation laws and dimensional considerations alone do not suffice to determine the similarity exponent μ; one must solve a nonlinear eigenvalue problem. The constants B, t_0, and x_0 depend on the initial condition and must be determined by matching the self-similar solution with the numerical solution of the Cauchy problem at the earlier non-self-similar stage.

It becomes possible to find the above self-similar solution explicitly. Substituting (4.3.3) into (4.3.1), one may obtain the nonlinear ODE

$$FF'' - (c-1)F'^2 - \xi F' + \frac{2\mu - 1}{\mu}F = 0, \quad \xi = \frac{x - x_0}{B(t_0 - t)^\mu}. \tag{4.3.5}$$

Barenblatt et al. (2000) sought solutions of (4.3.5) which have a maximum at $\xi = 0$ and a front at $\xi = 1$ where $F(1) = 0$. $F'(\xi)$ is also required to be continuous at $\xi = 1$. Thus, we have the following boundary conditions for (4.3.5):

$$F'(0) = 0, \quad F(1) = 0, \quad F'(1) = -\frac{1}{c-1}. \tag{4.3.6}$$

In view of (4.3.6), we seek a solution of (4.3.5) in the form

$$F(\xi) = \sum_{n=1}^{\infty} a_n (1 - \xi^2)^n. \tag{4.3.7}$$

One may easily check that $a_1 = 1/2(c-1)$, $a_n = 0$, $n \geq 2$. Thus, we have the explicit solution of (4.3.5) and (4.3.6) as

$$F(\xi) = \frac{1 - \xi^2}{2(c-1)}, \quad \xi = \frac{x - x_0}{B(t_0 - t)^\mu}, \quad \mu = \frac{c-1}{2c-3}. \tag{4.3.8}$$

The self-similar solution may, therefore, be written as

$$u_s = \frac{1}{2(2c-3)}B^2(t_0 - t)^{1/(2c-3)}\left[1 - \frac{(x - x_0)^2}{B^2(t_0 - t)^{2(c-1)/(2c-3)}}\right]_+,$$

$$x_f = B(t_0 - t)^{(c-1)/(2c-3)}. \tag{4.3.9}$$

For nongrowing solutions we must have $c > 3/2$; the time of collapse $t = t_0$ is assumed to be finite. The solution (4.3.9) is a weak solution of (4.3.1) which is positive in the interval $[-x_f, x_f]$ and vanishes outside. The form (4.3.9) must change for other values of c. For $1 < c < 3/2$, $\mu = (c-1)/(2c-3)$ becomes negative and the compact support contracts but the collapse time is infinite. In this case, if we replace μ and t_0 by $-\mu$ and $-t_0$, respectively, we may express the solution (4.3.9) in the form

$$u_s = \frac{1}{2(3-2c)} B^2 (t_0 + t)^{-1/(3-2c)} \left[1 - \frac{(x-x_0)^2}{B^2 (t_0 + t)^{-2(c-1)/(3-2c)}} \right]_+ ,$$

$$x_f = B(t_0 + t)^{-(c-1)/(3-2c)}. \tag{4.3.10}$$

Now t_0 is merely an additive constant; the solution (4.3.10) for $1 < c < 3/2$ vanishes in an infinite time.

It is interesting to consider the limiting behaviour of the solution (4.3.9) as $c \to (3/2)+$. Writing $c = (3/2) + \varepsilon$, and, therefore, $\mu = (1/2) + (1/4\varepsilon)$, where $\varepsilon > 0$ is a small parameter, we have

$$(t_0 - t)^\mu = t_0^{(1/4\varepsilon)+(1/2)} \left(1 - \frac{t}{t_0} \right)^{(1/4\varepsilon)+(1/2)} , \quad (t_0 - t)^{2\mu-1} = t_0^{1/2\varepsilon} \left(1 - \frac{t}{t_0} \right)^{1/2\varepsilon} \tag{4.3.11}$$

and, therefore,

$$x_f = B t_0^{(1/4\varepsilon)+(1/2)} \left(1 - \frac{t}{t_0} \right)^{(1/4\varepsilon)+(1/2)} . \tag{4.3.12}$$

Thus, as $\varepsilon \to 0$ such that $4\varepsilon t_0$ tends to a certain constant θ and $B^2 t_0^{1/2\varepsilon+1}$ tends to another constant, $C^2\theta$, say, the solution (4.3.9) assumes the form

$$u = C^2 e^{-2t/\theta} \left[1 - \frac{(x-x_0)^2}{C^2\theta e^{-2t/\theta}} \right], \quad x_f = C\theta^{1/2} e^{-t/\theta}. \tag{4.3.13}$$

Barenblatt et al. (2000) pointed out that, for $0 < c < 1$, the case of weak absorption, the compact support grows instead of contracting, though at a slower rate than for the special case of (4.3.1) with $c = 0$. For the latter, the solution (4.3.9) reduces to a self-similar solution of the first kind (see Barenblatt 1996). The degenerate case $c = 1$ was considered in detail by King (1993).

Chertock (2002) interpreted the intermediate asymptotic character of the self-similar solutions (4.3.3) and (4.3.4) in the sense of their stability. A self-similar solution is called stable if the solution of any perturbed problem (with a sufficiently small perturbation) can be represented as sum of a self-similar solution (4.3.3) with a constant B' (can be different from B) and a perturbation term which, relative to the first term, tends to zero as $t \to \infty$. Restricting our discussion to the case $c > 3/2$, we first write (4.3.1) in terms of $w_s = u_s^2(x,t)$ (see (4.3.9)):

$$2\sqrt{w_s} (w_s)_t = 2w_s (w_s)_{xx} - c((w_s)_x)^2. \tag{4.3.14}$$

This new variable is convenient because the derivative $\partial u_s / \partial x$ is discontinuous at the boundary $\xi = 1$; the derivative $(w_s)_x$, however, is continuous at $\xi = 1$. The function $w_s = w_s(x,t)$, in view of (4.3.3), can be written as

$$w_s = B^4\mu^2(t_0-t)^{4\mu-2}F^2(\xi), \quad \xi = \frac{x-x_0}{B(t_0-t)^\mu} \tag{4.3.15}$$

for $|\xi| \leq 1$ and $w_s \equiv 0$ for $|\xi| \geq 1$. The function $F(\xi)$ and the constant μ are given by (4.3.8). Now write the perturbed solution as

$$w_s(\xi,t) = B^4\mu^2(t_0-t)^{4\mu-2}[F^2(\xi)+\delta^2\phi(\xi,\tau)], \tag{4.3.16}$$

where δ is a small parameter. A simple argument shows that $\tau = -\mu\log(t_0-t)$ is a more convenient variable for the perturbation term than t itself. It is clear that $|\xi|=1$ is not the boundary of the solution. It is now displaced and may be defined by $-1-\beta_2(\tau) \leq \xi \leq 1+\beta_1(\tau)$. The perturbed boundaries are, in general, not symmetrically placed so that $\beta_1(\tau) \neq \beta_2(\tau)$. The displacement of the boundaries is proportional to the small parameter δ. This may be seen by writing $\xi = 1+\beta_1(\tau)$ in (4.3.16), linearising it, and recalling that $F^2(1) = (F^2)'(1) = 0$ and $w = 0$ at $\xi = 1+\beta_1(\tau)$. Thus, we have

$$\frac{\beta_1^2(\tau)}{2}(F^2)''(1)+\delta^2\phi(1,\tau) = 0. \tag{4.3.17}$$

It, therefore, follows that β_1 is proportional to δ. A similar argument gives the equation for $\beta_2(\tau)$. Substituting (4.3.16) into (4.3.14) and linearising, we get the following equation for the perturbation ϕ.

$$\phi_\tau = L_\xi\phi \equiv \frac{1}{2(c-1)}\left[(1-\xi^2)\phi_{\xi\xi}+2(c+1)\xi\phi_\xi+\frac{4(c+1)}{1-\xi^2}\phi-2(2c+1)\phi\right]. \tag{4.3.18}$$

The functions ϕ and ϕ_ξ must be continuous at the boundaries $\xi = \pm 1$; that is, $\phi = \phi_\xi = 0$ there. Equation (4.3.18) has the series solution

$$\phi(\xi,\tau) = \sum_{n=0}^{\infty} a_n e^{-\lambda_n\tau}\Phi_n(\xi) = \sum_{n=1}^{\infty} a_n(t_0-t)^{\mu\lambda_n}\Phi_n(\xi), \tag{4.3.19}$$

where $\Phi_n(\xi)$ is the eigenfunction with λ_n as the corresponding eigenvalue for the operator L_ξ. For the self-similar solutions to be stable we must show that all the eigenvalues λ_n are nonnegative and the corresponding set of eigenfunctions is complete. It turns out that it is possible to solve the eigenvalue problem for $\Phi_n(\xi)$ in terms of Jacobi polynomials:

$$\Phi_n(\xi) = (1-\xi^2)^{c+1}P_n^{(c,c)}(\xi), \quad \lambda_n = \frac{(n+1)(n+2c)}{2(c-1)}, \tag{4.3.20}$$

where $P_n^{(c,c)}(\xi)$ is the Jacobi polynomial of degree n defined by

$$P_n^{(c,c)}(\xi) = \frac{n+c}{n(n+2c)}\left[(2n+2c-1)\xi P_{n-1}^{(c,c)}(\xi)-(n+c-1)P_{n-2}^{(c,c)}(\xi)\right],$$

$$P_0^{(c,c)}(\xi) = 1, \quad P_1^{(c,c)}(\xi) = (c+1)\xi. \tag{4.3.21}$$

The set $\{\Phi_n(\xi)\}$ is complete and the eigenvalues λ_n for $c > 3/2$ are nonnegative. Hence the stability of the self-similar solution for $c > 3/2$ is proved.

We may remark that (4.3.5) possesses other solutions if we do not impose the condition that $F(\xi)$ has a maximum at $\xi = 0$. For example, if $F(\xi)$ has local extrema in $[-1, 1]$, a local minimum at $\xi = 0$ with $F(0) = 0$, say, then one may derive the solution

$$
u = \frac{B^2(t_0 + t)^{1/(c-2)}}{2(2c - 3)} \left[\frac{(x - x_0)^2}{B^2(t_0 + t)^{(c-1)/(c-2)}} \right.
$$

$$
\left. - \left(\frac{x - x_0}{B(t_0 + t)^{(c-1)/2(c-2)}} \right)^{1/(2-c)} \right]_+ , \tag{4.3.22}
$$

$$
x_f = B(t_0 + t)^{(c-1)/2(c-2)}, \quad -x_f(t) \le x \le x_f(t), \tag{4.3.23}
$$

and $u(x, t) \equiv 0$ outside the interval $[-x_f(t), x_f(t)]$. Here, $t_0 > 0$ is an additive constant and $3/2 < c < 2$. A similar form of the solution exists for $1 < c < 3/2$. The time of collapse here is again infinite.

For $c = 7/4$, we have a very simple solution of the BVP (4.3.5)–(4.3.6), namely,

$$
F(\xi) = \frac{4}{3}\xi^3(1 - \xi), \quad \mu = -1, \tag{4.3.24}
$$

leading to

$$
u = \frac{4}{3B}(x - x_0)^3 \left(1 - \frac{x - x_0}{t_0 + t} \right)_+, \quad x_f = B(t_0 + t)^{-1}. \tag{4.3.25}
$$

Before we turn to the numerical validation of the asymptotically stable nature of the solutions discussed above, we briefly study the axisymmetric case corresponding to (4.3.1). Here, $u = u(r, t)$ is governed by

$$
u_t = u \left(u_{rr} + \frac{u_r}{r} \right) - (c - 1)u_r^2. \tag{4.3.26}
$$

The self-similar solution of (4.3.26) is sought in the form

$$
u(r, t) = B^2 \mu (t_0 - t)^{2\mu - 1} F(\xi), \tag{4.3.27}
$$

where the similarity variable

$$
\xi = \frac{r}{B(t_0 - t)^\mu} \tag{4.3.28}
$$

is defined over $[0, 1]$. In this case (4.3.26) becomes

$$
FF'' + \frac{F}{\xi}F' - (c - 1)F'^2 - \xi F' + \frac{2\mu - 1}{\mu}F = 0. \tag{4.3.29}
$$

The function F satisfies the same conditions as before:

$$F'(0) = 0, \quad F(1) = 0, \quad F'(1) = -\frac{1}{c-1}. \tag{4.3.30}$$

For the case for which F possesses a maximum at $\xi = 0$, the solution may be found to be

$$F(\xi) = \frac{1-\xi^2}{2(c-1)}, \quad \mu = \frac{c-1}{2(c-2)}, \tag{4.3.31}$$

yielding

$$u = \frac{B^2}{4(c-2)}(t_0 - t)^{1/(c-2)}\left[1 - \frac{r^2}{B^2(t_0-t)^{(c-1)/(c-2)}}\right]_+, \quad r \in [0, r_f(t)], \tag{4.3.32}$$

where the radius of the front is given by

$$r_f = B(t_0 - t)^{(c-1)/2(c-2)}. \tag{4.3.33}$$

Here, we assume that $c > 2$ and $u \equiv 0$ outside the interval $[0, r_f(t)]$. For $1 < c < 2$, μ is negative (see (4.3.31)); the solution, therefore, may be sought in the form

$$u(r,t) = -B^2\mu(t_0 + t)^{2\mu - 1}F(\xi), \quad \xi = \frac{r}{B(t_0+t)^{\mu}}. \tag{4.3.34}$$

This form leads again to the eigenvalue problem (4.3.29) and (4.3.30). The solution in this case is found to be

$$u = \frac{B^2}{4(2-c)}(t_0 + t)^{1/(c-2)}\left[1 - \frac{r^2}{B^2(t_0+t)^{(c-1)/(c-2)}}\right]_+, \tag{4.3.35}$$

$$r_f = B(t_0 + t)^{(c-1)/2(c-2)}. \tag{4.3.36}$$

For the limiting case $c = 2$, $\mu \to \pm\infty$ (see (4.3.31)), (4.3.29) assumes the simple form

$$\frac{d}{d\xi}\left(\frac{\xi}{F}F' + \frac{\xi^2}{F}\right) = 0 \tag{4.3.37}$$

and, subject to the conditions (4.3.30), has the solution

$$F(\xi) = \frac{1}{\alpha - 2}(\xi^2 - \xi^{\alpha}), \quad \alpha \neq 2, \tag{4.3.38}$$

where α is either zero or a constant greater than one. In the manner of the solution of Barenblatt et al. (2000) for the planar case, detailed earlier, the limiting self-similar solution of (4.3.26) for $c \to 2+0$, $\mu \to +\infty$, may be obtained as

$$u = \frac{r^2}{D(\alpha - 2)}\left[1 - \frac{r^{\alpha - 2}}{C^{\alpha - 2}}e^{t(\alpha - 2)/D}\right], \tag{4.3.39}$$

where C is a constant; the radius of the front now is given by

$$r_f = Ce^{-t/D}. \tag{4.3.40}$$

The limiting case $c \to 2 - 0$, $\mu \to -\infty$ of the self-similar solution (4.3.27) again has the form (4.3.39) and (4.3.40).

Chertock (2002) also carried out linear stability analysis of the self-similar solution for the axisymmetric case in the manner of the planar case and concluded that this solution is stable both when $1 < c < 2$ and when $c > 2$.

Barenblatt et al. (2000) demonstrated numerically how the self-similar solution (4.3.9) of (4.3.1) attracts other solutions of (4.3.1) subject to the nonsymmetric initial conditions with compact support, namely,

$$u(x,0) = \begin{cases} u_0(x), & x_L(0) \le x \le x_R(0), \\ 0, & \text{otherwise.} \end{cases} \tag{4.3.41}$$

The constant c in (4.3.1) was chosen to be 1.75. Two distinct finite difference schemes were used to ensure accurate large time behaviour. For the numerical study, the first initial condition, $u_0(x)$, represented a smooth block, a homogeneous water level distribution smoothly going to zero at the edges (see Figure 4.1). Figure 4.1 shows $u(x,t)/u_{\max}(t)$, where $u_{\max}(t)$ is the maximum water level at each time. It is clearly seen that the solution curves with increasing time collapse to the parabola representing the self-similar solution (4.3.8). The time of collapse t_0 and the constant B in the self-similar solution were found from the numerical solution by a suitable matching; these depend on the specific initial conditions. The second example related to the nonsymmetric initial condition

$$u(x,0) = \begin{cases} -4x^2 + 4x, & 0 < x < 1/2 \\ -\frac{4}{9}x^2 + \frac{4}{9}x + \frac{8}{9}, & \frac{1}{2} < x < 2. \end{cases} \tag{4.3.42}$$

Figures 4.2 and 4.3 show the numerical solution of (4.3.1) subject to (4.3.42) at different times. It is observed that, after some time, the solution assumes a symmetric form and becomes self-similar. The arbitrary constants t_0 and B in (4.3.9) turn out to be different from those for the earlier example. A third example with a self-similar solution itself as the initial condition confirmed the veracity of the numerical solution (see Figures 4.4 and 4.5); in this case, the scaled solution collapses to a single curve, the self-similar solution for all time.

Chertock (2002) first recovered some of the numerical results of Barenblatt et al. (2000) and then proceeded to show that the axisymmetric self-similar solutions (4.3.32) describe the large time behaviour for the solutions of (4.3.26) (for $c = 2.25$) subject to axisymmetric initial condition with compact support.

It was found by Ughi (1986) and Dal Passo and Luckhaus (1987) that the solutions of (4.3.1) and (4.3.2) for $c = 1$ are not uniquely determined by the initial data $u_0(x)$. A reference may be made to the work of Bertsch et al. (1992) for a study of nonuniqueness of the solutions of the filtration–absorption equation for some other values of c.

In this section, we have studied the asymptotic behaviour of an initial value problem for a (degenerate) parabolic filtration–absorption equation. A family of

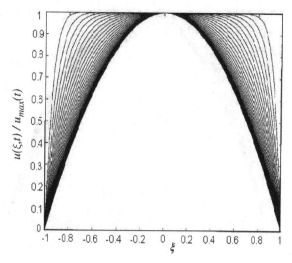

Fig. 4.1 Numerical solution of Cauchy problem for (4.3.1) with $c = 1.75$, subject to a 'smoothed block' type initial data (4.3.41) at different times in the scaled coordinates. The solution collapses to its self-similar asymptotic form (4.3.9). (Barenblatt et al. 2000. Copyright © 2000, National Academy of Sciences, USA. Reprinted with permission. All rights reserved.)

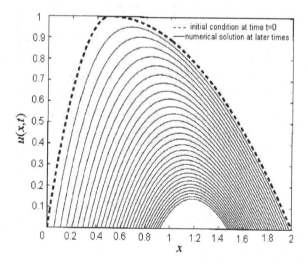

Fig. 4.2 Numerical solution of (4.3.1) with $c = 1.75$ subject to the nonsymmetric initial profile (4.3.42) at different times: initial profile (- - -), numerical solution (—). (Barenblatt et al. 2000. Copyright © 2000, National Academy of Sciences, USA. Reprinted with permission. All rights reserved.)

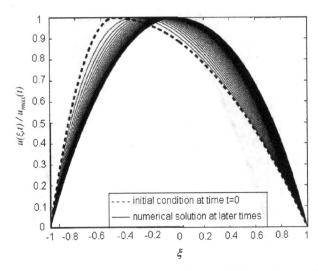

Fig. 4.3 Same as in Figure 4.2 with scaled coordinates $u(\xi,t)/u_{\max}(t)$ and ξ. (Barenblatt et al. 2000. Copyright © 2000, National Academy of Sciences, USA. Reprinted with permission. All rights reserved.)

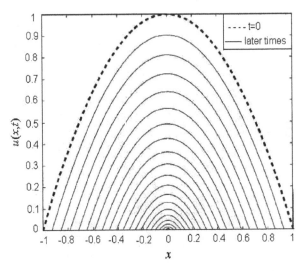

Fig. 4.4 Numerical solution of (4.3.1) with $c = 1.75$ subject to the initial profile given by the self-similar solution (4.3.9) at different times: initial profile (- - -), numerical solution (—). (Barenblatt et al. 2000. Copyright © 2000, National Academy of Sciences, USA. Reprinted with permission. All rights reserved.)

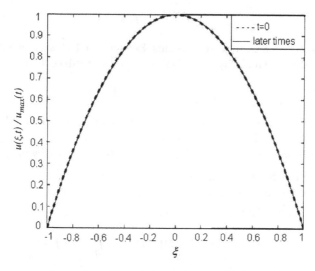

Fig. 4.5 Same as in Figure 4.4 with scaled coordinates $u(\xi,t)/u_{\max}(t)$ and ξ. (Barenblatt et al. 2000. Copyright © 2000, National Academy of Sciences, USA. Reprinted with permission. All rights reserved.)

self-similar solutions for this problem is constructed. The numerical study confirms that these special solutions form intermediate asymptotics for a wide class of initial conditions. We have also considered an axisymmetric filtration–absorption equation and constructed a family of self-similar solutions analogous to those for the one-dimensional case. These self-similar solutions attract solutions of problems with more general initial conditions.

4.4 Analysis of a class of similarity solutions of the porous media equation

In this section, we exemplify the analysis of nonlinear ODEs subject to certain boundary conditions. These boundary value problems arise from the similarity reduction of nonlinear PDEs with appropriate initial/boundary conditions. This analysis helps pick up sets of parameters for each of which the solution of the BVP for the ODE exists and is unique. One may then discover which of these solutions constitute intermediate asymptotics for the original PDEs, subject to relevant initial/boundary conditions.

In this section, we discuss the existence and uniqueness of weak solutions with compact support for the boundary value problem

$$(f^m)'' + p\eta f' = qf, \quad 0 < \eta < \infty, \tag{4.4.1}$$

$$f(0) = U, \quad f(\infty) = 0; \tag{4.4.2}$$

here p, q, and $U > 0$ are arbitrary constants. Equation (4.4.1) is a generalisation of the ODE obtained by similarity reduction of the porous media equation

$$u_t = (u^m)_{xx}, \quad m > 1 \tag{4.4.3}$$

(see Section 4.6). Here, we follow the work of Gilding and Peletier (1976), who extended an earlier work of Atkinson and Peletier (1971).

A function f is a weak solution of (4.4.1) if it satisfies the following conditions.

(i) f is bounded, continuous, and nonnegative on $[0, \infty)$.
(ii) $(f^m)(\eta)$ has a continuous derivative with respect to η on $(0, \infty)$.
(iii) f satisfies the equation

$$\int_0^\infty \phi' \{(f^m)' + p\eta f\} d\eta + (p+q) \int_0^\infty \phi f d\eta = 0$$

for all $\phi \in C_0^1(0, \infty)$.

In the sequel, we prove the following theorem.

Theorem 4.4.1 *Suppose that $U > 0$. Then the boundary value problem (4.4.1)–(4.4.2) has a weak solution with compact support if and only if $p \geq 0$ and $2p + q > 0$. Furthermore, this weak solution of (4.4.1)–(4.4.2) is unique.*

To prove this theorem, we pose the following boundary value problem for (4.4.1),

$$f(0) = U, \tag{4.4.4}$$
$$f(a) = 0, \quad (f^m)'(a) = 0, \tag{4.4.5}$$

where $a > 0$ is a real number. Using a shooting argument with $a > 0$ as the shooting parameter, we first prove the following theorem for the existence and uniqueness of classical solutions for (4.4.1) with the boundary conditions (4.4.4) and (4.4.5).

Theorem 4.4.2 *Suppose that $U > 0$. Then the boundary value problem (4.4.1), (4.4.4), and (4.4.5) has a unique solution and there exists a unique $a(U) > 0$ such that $f(\eta; a(U))$ is positive on $(0, a)$ if and only if $p \geq 0$ and $2p + q > 0$.*

Atkinson and Peletier (1971) showed that if $f(\eta_0) > 0$ for some η_0, then f is a classical solution of (4.4.1) in a neighbourhood $(\eta_0 - \varepsilon, \eta_0 + \varepsilon)$. Thus, the solution f appearing in Theorem 4.4.2 is a classical solution of (4.4.1). The proof of Theorem 4.4.2 follows three steps: (i) proof of existence and uniqueness of solution of (4.4.1) satisfying (4.4.5) in a neighbourhood of $\eta = a$ via a contraction principle; (ii) extension of this solution up to $\eta = 0$ as a positive solution (if possible); and (iii) proving the existence of $a(U)$ (when possible) such that $f(0, a(U)) = U$.

In Lemma 4.4.3, we determine necessary conditions on the parameters p and q for the existence of a nontrivial weak solution of (4.4.1) with compact support.

Lemma 4.4.3 *There exists a nontrivial weak solution of (4.4.1) with a compact support only when $p > 0$ or $p = 0$, $q > 0$.*

Proof: Suppose that $f(\eta; a)$ is a nontrivial weak solution of (4.4.1) with compact support. Then $f > 0$ in $(a - \varepsilon, a)$ and $f = 0$ in $[a, \infty)$ for some $a > 0$ and $\varepsilon > 0$. It follows that f is a classical solution of (4.4.1) on $(a - \varepsilon, a)$ and satisfies (4.4.5) at $\eta = a$; that is, $f(a) = 0$, $(f^m)'(a) = 0$. Integrating (4.4.1) from η to a, where $a - \varepsilon < \eta < a$, we get

$$- (f^m)'(\eta) = p\eta f(\eta) + (p+q) \int_{\eta}^{a} f(\xi) d\xi. \tag{4.4.6}$$

The continuity of f and $(f^m)'$ ensures the existence of $\eta_0 \in (a - \varepsilon, a)$ such that $f'(\eta_0) < 0$. This implies that the LHS of (4.4.6) is positive at $\eta = \eta_0$ and, therefore, p and $p + q$ cannot both be less than zero. Thus, $p = 0$ implies that $q > 0$. Now consider the case $p < 0$. This requires that $p + q > 0$ and hence $q > 0$. We easily check from (4.4.1) that f cannot have a maximum as long as f is positive. Therefore, f does not assume a maximum at any point in $(a - \varepsilon, a)$. Thus, $f' < 0$ on $(a - \varepsilon, a)$. It follows from (4.4.6) that

$$- m f^{m-2}(\eta) f'(\eta) - p\eta \le (p+q)(a - \eta), \tag{4.4.7}$$

where we have used the fact that $f(\xi) \le f(\eta)$ for $\xi \in (\eta, a)$, $a - \varepsilon < \eta < a$. As $\eta \to a$ in (4.4.7), LHS becomes positive, and the RHS tends to zero, a contradiction. Thus we have shown that $p = 0$, $q > 0$ or $p > 0$ are the only cases for which a nontrivial weak solution of (4.4.1) exists with a compact support.

Explicit solution of (4.4.1), (4.4.4) and (4.4.5) when $p = 0$, $q > 0$.

With $p = 0$, $q > 0$, (4.4.1) becomes

$$(f^m)'' = q(f^m)^{1/m}. \tag{4.4.8}$$

Substituting $f^m = g$ in (4.4.8) and integrating we get

$$(g')^2 = \frac{2qm}{m+1} g^{(m+1)/m}, \tag{4.4.9}$$

where we have used (4.4.5). Solving (4.4.9) for g and using (4.4.5)$_1$, we obtain

$$f^m = g = \left\{ \frac{q(m-1)^2(a-\eta)^2}{2m(m+1)} \right\}^{m/(m-1)}, \quad 0 < \eta < a.$$

Thus

$$f(\eta; a) = \left\{ \frac{q(m-1)^2(a-\eta)^2}{2m(m+1)} \right\}^{1/(m-1)}, \quad 0 < \eta < a, \tag{4.4.10}$$

is the unique solution of the problem (4.4.1) satisfying (4.4.5). We observe that

$$f(0;a) = \left\{ \frac{q(m-1)^2 a^2}{2m(m+1)} \right\}^{1/(m-1)}.$$

Because $m > 1$, $f(0;a)$ is a continuous function of a with $f(0;0) = 0$ and $f(0;\infty) = \infty$; furthermore, f is a continuous and monotonically increasing function of a. This implies that, for a given $U > 0$, there exists a unique $a(U)$ such that $f(0;a(U)) = U$. Therefore, $f(\eta;a(U))$ is the unique solution of (4.4.1) satisfying (4.4.4) and (4.4.5). An easy calculation shows that

$$a(U) = \left[\frac{2m(m+1)U^{m-1}}{q(m-1)^2} \right]^{1/2}.$$

We give below an elementary lemma for the case $p > 0$.

Lemma 4.4.4 *Suppose that $0 < b < a$ and f is a positive solution of (4.4.1) on $[b,a)$ satisfying (4.4.5). Then the following results hold.*

(i) $f'(\eta) < 0$ on $[b,a)$ provided that $p + q \geq 0$.
(ii) *Suppose that $p + q < 0$ and $f'(\eta_0) = 0$ for some $\eta_0 \in [b,a)$. Then f has a maximum at η_0 and $\eta_0 < a(p+q)/q$.*

Suppose that f is a positive solution of (4.4.1) and (4.4.5) on $[0,a)$. Then

$$f'(0) \begin{cases} > 0, & p+q < 0, \\ = 0, & p+q = 0, \\ < 0, & p+q > 0. \end{cases}$$

Proof: Integration of (4.4.1) from $\eta \in [b,a)$ to a yields

$$-(f^m)'(\eta) = p\eta f(\eta) + (p+q) \int_\eta^a f(\xi) d\xi. \tag{4.4.11}$$

Because $p > 0$, the RHS of (4.4.11) is positive when $p + q \geq 0$ and hence $(f^m)'(\eta) < 0$. This implies that $f'(\eta) < 0$ on $[b,a)$. Again, if $p + q < 0$ then $q < 0$ (because $p > 0$). By (4.4.1), $f''(\eta_0) < 0$ when $f'(\eta_0) = 0$. Thus f has a maximum at $\eta = \eta_0$ and is strictly decreasing on (η_0,a); that is, $f'(\eta) < 0$ on (η_0,a). Putting $\eta = \eta_0$ in (4.4.11), we have

$$0 = p\eta_0 f(\eta_0) + (p+q) \int_{\eta_0}^a f(\xi) d\xi$$
$$> p\eta_0 f(\eta_0) + (p+q)(a-\eta_0) f(\eta_0);$$

therefore, $p\eta_0 + (p+q)(a-\eta_0) < 0$ or $\eta_0 < a(p+q)/q$. With $\eta = 0$, (4.4.11) becomes

$$-(f^m)'(0) = (p+q) \int_0^a f(\xi) d\xi. \tag{4.4.12}$$

The result for $f'(0)$ follows immediately from (4.4.12). In particular, $f'(0) < 0$ when $p + q \geq 0$.

In the next lemma, we prove the local existence and uniqueness of a solution of (4.4.1) satisfying (4.4.5). This is accomplished by formulating an equivalent integral equation following the work of Atkinson and Peletier (1971) and Gilding and Peletier (1976).

Lemma 4.4.5 *Suppose that p is greater than zero and q is any real number. Then, for any $a > 0$, equation (4.4.1) with initial condition (4.4.5) at $\eta = a$ has a unique positive solution in a neighbourhood $(a - \varepsilon, a)$ of a; here, $\varepsilon > 0$ is a constant.*

Proof: Suppose that f is a positive solution in a left neighbourhood of $\eta = a$. By Lemma 4.4.4, $f'(\eta) < 0$ for $\eta \in (a - \varepsilon, a)$ for some $\varepsilon > 0$. Let $\eta = \sigma(f)$ where σ is the inverse of f on $(a - \varepsilon, a)$. Rewriting (4.4.11), we have

$$(f^m)'(\eta) = q\eta f(\eta) + (p + q) \int_{\eta}^{a} \xi f'(\xi) d\xi. \tag{4.4.13}$$

With $\sigma(f) = \eta$ in (4.4.13) we have

$$\frac{d\sigma}{df} = \frac{mf^{m-1}}{qf\sigma(f) - (p+q)\int_0^f \sigma(\phi)d\phi}; \tag{4.4.14}$$

equation (4.4.14) is an integrodifferential equation for $\sigma = \sigma(f)$. Integrating (4.4.14) from 0 to f, we obtain

$$\sigma(f) - a = m \int_0^f \frac{\phi^{m-1}d\phi}{q\phi\sigma(\phi) - (p+q)\int_0^\phi \sigma(\psi)d\psi}. \tag{4.4.15}$$

Let

$$\tau(f) = 1 - a^{-1}\sigma(f). \tag{4.4.16}$$

Then, equation (4.4.15) becomes

$$\tau(f) = \frac{m}{a^2} \int_0^f \frac{\phi^{m-1}d\phi}{p\phi + q\phi\tau(\phi) - (p+q)\int_0^\phi \tau(\psi)d\psi}. \tag{4.4.17}$$

By using the Banach–Cacciopoli contraction mapping principle (see Hartman 1964), we now show that equation (4.4.17) admits a unique positive solution in a right neighbourhood of $f = 0$. Let X be the set of all bounded functions $\tau(f)$ on $[0, \gamma]$, $\gamma > 0$, satisfying

$$0 \leq \tau(f) \leq \rho = \frac{p}{2(|q| + |p + q|)}. \tag{4.4.18}$$

Let $\|..\|$ be the sup norm defined on X. Then X is a complete metric space. Define

$$M(\tau)(f) = \frac{m}{a^2} \int_0^f \frac{\phi^{m-1} d\phi}{p\phi + q\phi\tau(\phi) - (p+q)\int_0^\phi \tau(\psi)d\psi}, \quad \tau(f) \in X. \quad (4.4.19)$$

First we show that M maps X into X over $[0, \gamma_0]$, $\gamma \leq \gamma_0$. Let $\tau \in X$. Clearly,

$$p\phi + q\phi\tau(\phi) - (p+q) \int_0^\phi \tau(\psi)d\psi$$
$$\geq p\phi - |q|\phi\tau(\phi) - |p+q|\|\tau\|\phi$$
$$\geq p\phi - (|q| + |p+q|)\|\tau\|\phi$$
$$\geq \frac{p\phi}{2}, \quad (4.4.20)$$

where we have used (4.4.18). Therefore, from (4.4.19), we have

$$M(\tau)(f) \leq \frac{2m}{pa^2} \int_0^f \phi^{m-2} d\phi$$
$$= \frac{2m f^{m-1}}{pa^2(m-1)}$$
$$\leq \frac{2m\gamma^{m-1}}{pa^2(m-1)}. \quad (4.4.21)$$

Thus, $M(\tau)$ is well defined on X and $M(\tau): [0, \gamma] \to \mathbb{R}$ is nonnegative and continuous. The RHS of (4.4.21) suggests that we may find γ_0, $\gamma \leq \gamma_0$ such that $\|M(\tau)\| \leq \rho$, $\tau \in X$. Thus M maps X into X for $\gamma \leq \gamma_0$.

In the next step, we show that M is a contraction map on X. Let $\tau_1, \tau_2 \in X$, and $\gamma \leq \gamma_0$. Then

$$\|M(\tau_1) - M(\tau_2)\| \leq \frac{4m}{p^2 a^2} \int_0^f \phi^{m-3} (|q||\phi|\|\tau_1 - \tau_2\| + |p+q| \int_0^\phi \|\tau_1 - \tau_2\| d\psi) d\phi$$
$$\leq \frac{4m}{(m-1)p^2 a^2} (|q| + |p+q|)\gamma^{m-1} \|\tau_1 - \tau_2\|.$$

Therefore, there exists $\gamma_1 \in (0, \gamma_0]$ such that if $\gamma \leq \gamma_1$, M is a contraction on X. By the Banach–Cacciopoli contraction principle, M has a unique fixed point in X and hence equation (4.4.17) has a unique solution. This, in turn, implies that there exists a unique positive solution of (4.4.1), (4.4.5) in an interval $(a - \varepsilon, a)$ for some $\varepsilon > 0$.

Now we enquire whether it is possible to extend $f(\eta; a)$ back to $\eta = 0$. This extension may be carried out uniquely as long as f is bounded and positive. The following three cases may arise.

(A) $f(\eta) \to \infty$ as $\eta \downarrow \eta_1$ for some $\eta_1 \in [0, a)$.
(B) $f(\eta)$ may be continued back to $\eta = 0$.
(C) $f(\eta) \to 0$ as $\eta \downarrow \eta_2$ for some $\eta_2 \in (0, a)$.

In the next lemma, we prove that a positive solution $f(\eta; a)$ of (4.4.1) and (4.4.5) cannot be unbounded. This will rule out the possibility (A) above.

Lemma 4.4.6 *Suppose that $p > 0$ and $b \in [0,a)$. Furthermore, let f be a positive solution of (4.4.1) and (4.4.5) on (b,a). Then f is bounded on (b,a) and*

$$\sup_{(b,a)} f(\eta) \leq \left[\frac{(m-1)a^2}{2m} \max\{p, 2p+q\} \right]^{1/(m-1)}.$$

Proof: We prove this lemma for the following two cases: (i) $p+q \geq 0$, (ii) $p+q < 0$.

Case (i). $p+q \geq 0$.

Because, for this case, $f' < 0$ on (b,a) by Lemma 4.4.4, $f(\eta) \geq f(\xi)$, $\xi \in (\eta, a)$. By (4.4.11),

$$-(f^m)'(\eta) \leq p\eta f(\eta) + (p+q)f(\eta)(a-\eta), \quad b \leq \eta < a$$

or

$$-mf^{m-2}f' \leq a(p+q) - q\eta, \quad b \leq \eta < a. \tag{4.4.22}$$

Integrating (4.4.22) from η to a gives

$$\frac{m}{m-1}f^{m-1}(\eta) \leq [pa + q(a-\eta)/2](a-\eta), \quad b \leq \eta \leq a. \tag{4.4.23}$$

Thus,

$$\frac{m}{m-1} \sup_{(b,a)} f^{m-1}(\eta) \leq \frac{1}{2}a^2(2p+q). \tag{4.4.24}$$

Case (ii). $p+q < 0$.

By equation (4.4.11),

$$-(f^m)'(\eta) \leq p\eta f(\eta), \quad b \leq \eta < a$$

or

$$-mf^{m-2}f' \leq p\eta, \quad b \leq \eta < a. \tag{4.4.25}$$

Integrating (4.4.25) from η to a, we have

$$\frac{m}{m-1}f^{m-1}(\eta) \leq \frac{1}{2}p(a^2 - \eta^2), \quad b \leq \eta \leq a. \tag{4.4.26}$$

This, in turn, implies that

$$\frac{m}{m-1} \sup_{(b,a)} f^{m-1}(\eta) \leq \frac{1}{2}pa^2. \tag{4.4.27}$$

Observe that the bounds in (4.4.24) and (4.4.27) are independent of b and, therefore, $f(\eta)$ cannot be unbounded as η decreases from $\eta = a$.

In the manner of the proof for Lemma 4.4.6, we arrive at the following lower bounds for the positive solution f of (4.4.1) and (4.4.5) on an interval $[b, a)$ for $p+q \geq 0$ and $p+q < 0$:

(i) $p+q \geq 0$,

$$\frac{m}{m-1} f^{m-1}(\eta) \geq \frac{1}{2} p(a^2 - \eta^2), \quad b \leq \eta \leq a. \tag{4.4.28}$$

(ii) $p+q < 0$,

$$\frac{m}{m-1} f^{m-1}(\eta) \geq \left[pa + \frac{1}{2} q(a-\eta) \right] (a-\eta),$$

$$\geq \frac{1}{2}(2p+q)(a^2 - \eta^2), \quad \max\{b, \eta_0\} \leq \eta \leq a. \tag{4.4.29}$$

Lemma 4.4.7 below and the discussion that follows clearly bring out the parametric ranges of p and q for which the cases (B) and (C) may hold (see above Lemma 4.4.6).

Lemma 4.4.7 *Suppose that f is a positive solution of (4.4.1) and (4.4.5) in a left neighbourhood of $\eta = a$, and $p > 0$. Then $f(\eta) > 0$ on $[0, a)$ when $2p+q > 0$.*

Proof: Integrating (4.4.11) from η to a, we have

$$f^m(\eta) = p\eta \int_\eta^a f(\xi)d\xi + (2p+q) \int_\eta^a (\xi - \eta)f(\xi)d\xi. \tag{4.4.30}$$

It is easy to see from (4.4.30) that, if $2p+q > 0$, then $f(\eta) > 0$ on $(0, a)$. It may also be shown that $f > 0$ on $(0, a)$ and $f(0) = 0$ when $2p+q = 0$. Moreover, f cannot remain positive when $2p+q < 0$ on $(0, a)$ (see (4.4.30)) (see Gilding and Peletier 1976 for details). In the next proposition, we find the bounds for $f(0)$. To that end, we define

$$\lambda = \frac{2p+q}{p}, \quad \mu = 1 - \left(\frac{p+q}{q}\right)^2, \quad A = \left[\frac{m-1}{2m} pa^2\right]^{1/(m-1)}.$$

Proposition 4.4.8 *Suppose that $p > 0$ and $2p+q > 0$. Then, the following bounds hold for $f(0)$.*

(i) $\lambda^{1/m} A \leq f(0) \leq \lambda^{1/(m-1)} A$ *if* $p+q \geq 0, \lambda \geq 1$.
(ii) $(\mu\lambda)^{1/(m-1)} A \leq f(0) \leq \lambda^{1/m} A$ *if* $p+q \leq 0, 0 < \lambda \leq 1$.

Proof: (i) Putting $\eta = 0$ in (4.4.24) and simplifying, we get

$$f(0) \leq \left[\left(\frac{m-1}{2m} pa^2\right) \frac{2p+q}{p} \right]^{1/(m-1)},$$

$$\leq A\lambda^{1/(m-1)}.$$

To prove the lower bound, put $\eta = 0$ in (4.4.30). Then,

$$f'''(0) = (2p+q) \int_0^a \xi f(\xi) d\xi. \tag{4.4.31}$$

Using (4.4.28) in (4.4.31) and simplifying, we have $f(0) \geq \lambda^{1/m} A$.
(ii) For $p+q \leq 0$, we find from (4.4.26) and (4.4.29) that

$$\lambda^{1/(m-1)} A[1 - (\eta/a)^2]^{1/(m-1)} \leq f(\eta) \leq A[1 - (\eta/a)^2]^{1/(m-1)}, \quad \eta_0 \leq \eta \leq a. \tag{4.4.32}$$

We recall that $\eta = \eta_0$ is the point at which f attains its maximum. Because $f(\eta) \leq f(\eta_0)$ on $[0, \eta_0]$, the second inequality in (4.4.32) holds for $0 \leq \eta \leq a$. Making use of the upper bound for f on $[0, a)$ from (4.4.32) in (4.4.31) and simplifying, we obtain $f(0) \leq \lambda^{1/m} A$. To prove the lower bound, we use (4.4.31) and write

$$f'''(0) \geq (2p+q) \int_{a^*}^a \xi f(\xi) d\xi \tag{4.4.33}$$

inasmuch as, in view of Lemma 4.4.4, $a^* = a(p+q)/q$ and $\eta_0 \leq a^*$. Now substituting the lower bound for $f(\eta)$ from (4.4.32) in (4.4.33) and simplifying we arrive at the lower bound for $f(0)$.

In the next proposition, we prove that, when $p > 0$, $2p + q \geq 0$, the solution $f(\eta)$ of (4.4.1) and (4.4.5) is monotonically increasing with respect to the parameter a (see equation (4.4.5)) for a fixed η.

Proposition 4.4.9 *Suppose that $p > 0$ and $2p + q \geq 0$. Furthermore, let $f(\eta; a_1)$ and $f(\eta; a_2)$ be solutions of (4.4.1) and (4.4.5) on $(0, a_1)$ and $(0, a_2)$, respectively, where $a_1 > a_2$. Then $f(\eta; a_1) > f(\eta; a_2)$ everywhere on $(0, a_2)$.*

Proof: For convenience, we let $f(\eta; a_i) = f_i(\eta)$, $i = 1, 2$. Suppose, for contradiction, that the conclusion in Proposition 4.4.9 is not true. Then there exists an $\overline{\eta} \in (0, a_2)$ such that $f_1(\overline{\eta}) = f_2(\overline{\eta})$ and $f_1(\eta) > f_2(\eta)$ on $(\overline{\eta}, a_2)$. From (4.4.30), we have

$$f_i''(\overline{\eta}) = p\overline{\eta} \int_{\overline{\eta}}^{a_i} f_i(\xi) d\xi + (2p+q) \int_{\overline{\eta}}^{a_i} (\xi - \overline{\eta}) f_i(\xi) d\xi, \quad i = 1, 2. \tag{4.4.34}$$

This implies that

$$p\overline{\eta} \int_{\overline{\eta}}^{a_2} (f_1 - f_2) d\xi + (2p+q) \int_{\overline{\eta}}^{a_2} (\xi - \overline{\eta})(f_1 - f_2) d\xi$$

$$+ p\overline{\eta} \int_{a_2}^{a_1} f_1(\xi) d\xi + (2p+q) \int_{a_2}^{a_1} (\xi - \overline{\eta}) f_1(\xi) d\xi = 0. \tag{4.4.35}$$

Because the second and fourth terms on the LHS of (4.4.35) are nonnegative and the other two terms are positive, we arrive at a contradiction. Hence the proposition 4.4.9.

Now we proceed to prove Theorem 4.4.2. We have already proved in Lemma 4.4.5 the local existence of a solution about $\eta = a$ for (4.4.1) and (4.4.5). This unique local solution may be extended back to $\eta = 0$ as a positive solution with $f(0) > 0$ if and only if when $2p+q > 0$ (see Lemma 4.4.7). Now if we can prove that there exists $a(U)$ such that $f(0; a(U)) = U$, then Theorem 4.4.2 is proved. To that end, we use the following result due to Barenblatt (1952). Suppose that $f(\eta; a)$ is a solution of (4.4.1) and (4.4.5) on $(0, a)$; then $\mu^{-2/(m-1)} f(\mu\eta; \mu a)$ is a solution of (4.4.1) and (4.4.5) on $(0, \mu a)$ for any $\mu > 0$. Let $\mu = a^{-1}$. Then,

$$f(0; a) = a^{2/(m-1)} f(0; 1) = U. \qquad (4.4.36)$$

Because $f(0; 1) > 0$ for $2p+q > 0$, $p > 0$, we get a unique root $a = a(U)$ of (4.4.36). Thus, $f(\eta; a(U))$ is the unique solution of (4.4.1), (4.4.4), and (4.4.5). Theorem 4.4.2 follows if we add that, for $p = 0$, we have already constructed the explicit solution (4.4.10).

We observe that

$$f(\eta) = \begin{cases} f(\eta; a), & 0 \le \eta < a \\ 0, & a \le \eta < \infty, \end{cases} \qquad (4.4.37)$$

is a weak solution of (4.4.1) and (4.4.5). Now we must show that, given $U > 0$, (4.4.37) is the only solution of (4.4.1), (4.4.4), and (4.4.5) with compact support. Suppose that $f(\eta)$ is a weak solution of the problem (4.4.1) and (4.4.2) with compact support. By Lemma 4.4.7, this is possible only if $2p+q > 0$. Moreover,

$$f(\eta) \begin{cases} > 0 & \text{on } [0, a), \\ = 0 & \text{on } [a, \infty), \ a > 0. \end{cases}$$

By Theorem 4.4.2, this is also the unique solution. Thus, we have proved Theorem 4.4.1.

In this section, we have proved that the second-order nonlinear ODE (4.4.1) governing the self-similar solution of the porous media equation subject to the boundary conditions (4.4.2) has a weak solution with compact support if and only if $p \ge 0$ and $2p+q > 0$. We prove in Section 4.6 that this subset of solutions describes large time asymptotic behaviour of the solutions of the porous medium equation subject to a certain class of initial/boundary conditions.

4.5 Similarity solutions of nonlinear heat conduction problems as large time asymptotics

This section concerns the large time asymptotics for the solutions of nonlinear heat conduction problems following the work of Peletier (1970). We consider the quasilinear PDE

$$u_t = (k(u)u_x)_x, \qquad (4.5.1)$$

which describes the conduction of heat into a semi-infinite homogeneous solid oc-
cupying the length $0 \leq x < \infty$. Here, $u = T - T_0$, T being the temperature in the
solid; T_0 is its value far from its face (initially). The temperature at the face $x = 0$
is denoted by $T_1 > T_0$. The coefficient of thermal conductivity, $k(u)$, is assumed to
be continuously differentiable and positive for all values of u and $k'(u) < 0$ for all
nonnegative values of u. This applies specifically to the situation in metals and in
certain crystals.

The initial temperature profile $u(x,0) = u_0(x)$ is assumed to be smooth; moreover, $u_0 \geq 0$ for all $x \geq 0$ and behaves as

$$u_0(x) = O\left\{\text{erfc}\left(\frac{x}{2\sqrt{k(0)}}\right)\right\} \quad \text{as } x \to \infty. \tag{4.5.2}$$

Besides, this initial function satisfies compatibility conditions at $(0,0)$:

$$u_0(0) = T_1 - T_0, \quad \frac{d}{dx}\left\{k(u_0)\frac{du_0}{dx}\right\} = 0. \tag{4.5.3}$$

These conditions guarantee smoothness of the solution up to the boundary $x = 0$.

If we look for a solution of (4.5.1) in the similarity form $u(x,t) = f(\eta)$, $\eta = x/(t+1)^{1/2}$, we check from (4.5.1) that $f(\eta)$ satisfies the ODE

$$\{k(f)f'\}' + \frac{1}{2}\eta f' = 0, \tag{4.5.4}$$

where the prime denotes the derivative with respect to η. The idea now is to con-
nect the solution of (4.5.4) with that of (4.5.1) with the initial condition $u_0(x)$ satisfying (4.5.2) and (4.5.3). As Peletier (1970) succinctly points out, the temperature profile will be given by $f(\eta)$ itself only if at $t = 0$, the initial profile happens
to be $u_0 = f(x)$. This is clearly a very special solution and, therefore, of limited
importance. The significance of $f(\eta)$ is much more than being this special solution. Peletier (1970) showed that, whatever the initial profile, subject only to certain
smoothness requirements and the compatibility conditions (4.5.3), the evolving profile will tend to the similarity profile $f(\eta)$ as $t \to \infty$. This convergence will be like
$|u(x,t) - f(\eta)| = O\left(t^{-\lambda}\right)$ as $t \to \infty$, where the constant λ lies between 0 and 1/7
and depends on $u_0(x)$ and $k(u)$, as we state more precisely later.

In the proof of the convergence theorem several properties of the solution $u(x,t)$
and the similarity solution $f(\eta)$ are made use of. Here we briefly state these results.
We recapitulate the results concerning the solution $f(\eta)$ here for ready reference.
According to a theorem of Bailey et al. (1968), a solution $f(\eta,A)$ of (4.5.4) with the
boundary conditions

$$f(0,A) = A, \quad \lim_{\eta \to \infty} f(\eta,A) = 0 \tag{4.5.5}$$

exists provided we can find functions $g(\eta,A)$ and $h(\eta,A)$ which satisfy the inequalities

$$\{k(g)g'\}' + \tfrac{1}{2}\eta g' > 0, \tag{4.5.6}$$

$$\{k(h)h'\}' + \tfrac{1}{2}\eta h' < 0, \tag{4.5.7}$$

$$\text{and } g < h, \tag{4.5.8}$$

for $\eta > 0$ and which tend to zero as $\eta \to \infty$. Furthermore, not only does the solution of this BVP for (4.5.4) exist, it also satisfies the inequalities

$$g(\eta,A) \le f(\eta,A) \le h(\eta,A). \tag{4.5.9}$$

Peletier (1970) showed that the functions $g(\eta,A)$ and $h(\eta,A)$ may be identified as

$$g(\eta,A) = -\beta \log \left[1 - \left\{ 1 - \exp\left(\frac{-A}{\beta}\right) \right\} \mathrm{erfc}\left(\frac{\eta}{2\sqrt{k(A)}}\right) \right], \tag{4.5.10}$$

$$h(\eta,A) = A\, \mathrm{erfc}\left(\frac{\eta}{2\sqrt{k(0)}}\right), \tag{4.5.11}$$

where

$$\beta = k(A) \left\{ \max_{0 \le f \le A} |k'(f)| \right\}^{-1} \tag{4.5.12}$$

(see also Shampine 1973). Observe that $g(0,A) = h(0,A) = A$.

Peletier (1970) also proved that the solution $f(\eta,A)$ satisfying the BCs (4.5.5) at $\eta = 0$ and $\eta \to \infty$ has the limiting behaviour

$$f = O\left\{ \mathrm{erfc}\left(\frac{\eta}{2\sqrt{k(0)}}\right) \right\} \quad \text{as } \eta \to \infty. \tag{4.5.13}$$

The initial boundary conditions for the solution $u(x,t)$ of (4.5.1) may be written as

$$u(0,t) = U = u_0(0), \quad u = u_0(x) \text{ at } t = 0, \tag{4.5.14}$$

where it is assumed that $u_0(x)$ is twice continuously differentiable and has bounded first and second derivatives on $[0,\infty)$. The initial function $u_0(x)$ also satisfies the compatibility conditions (4.5.3) and the large distance behaviour $u_0 = O\{\mathrm{erfc}(x/2\sqrt{k(0)})\}$ as $x \to \infty$ (see (4.5.13)). With these conditions on $u_0(x)$, a simple modification of two theorems of Oleinik and Kruzhkov (1961) guarantees that a smooth unique solution of (4.5.1) and (4.5.14) exists in the half-strip $H^+ := 0 \le t \le T$, $x \ge 0$, where T is any finite positive number. Moreover, u, u_x, u_{xx}, and u_t are all continuous and bounded on H^+.

The main theorem requires the following result. Let u and v be two solutions of (4.5.1) in H^+ and let $u \le v$ at $t = 0$ and at the boundary $x = 0$. Then $u \le v$ in H^+.

This result may be proved via the following maximum principle due to Krzyżański (1959). Let $z(x,t)$ be a bounded solution of the differential inequality

$$z_t \le a(x,t)z_{xx} + b(x,t)z_x + c(x,t)z \tag{4.5.15}$$

in H^+. Suppose that the coefficients $a(x,t), b(x,t)$, and $c(x,t)$ are bounded continuous functions of x and t and that $a(x,t) > 0$. If $z \leq 0$ on $x = 0$ and $t = 0$, then $z \leq 0$ in H^+.

To use this result we introduce the function $\bar{u} = \int_0^u k(s)ds$ and $\bar{v} = \int_0^v k(s)ds$. Then it easily follows from (4.5.1) that

$$\bar{u}_t = k(u)\bar{u}_{xx} \tag{4.5.16}$$

and

$$\bar{v}_t = k(v)\bar{v}_{xx}. \tag{4.5.17}$$

Writing $w = \bar{u} - \bar{v}$, we easily show that w satisfies the equation

$$w_t = k(u)w_{xx} + [k(u) - k(v)]\bar{v}_{xx} = k(u)w_{xx} + \beta(x,t)w, \tag{4.5.18}$$

where

$$\beta(x,t) = \left\{\frac{k'(\theta_1)}{k(\theta_2)}\right\}\bar{v}_{xx}; \tag{4.5.19}$$

here θ_1 and θ_2 are chosen suitably between u and v. Such a choice of θ_1 and θ_2 is possible due to the Cauchy mean value theorem. In view of assumptions on the functions u and v and their derivatives, $k(u)$ and $\beta(x,t)$ are bounded in H^+. Furthermore,

$$w = \int_v^u k(s)ds \tag{4.5.20}$$

is bounded and negative or zero on $x = 0$ and $t = 0$. Hence, direct application of Krzyżański's maximum principle (1959) gives $w \leq 0$; that is, $u \leq v$ in H^+.

In particular, if we choose $u = u(x,t)$ and $u = f(\eta,A)$ as two solutions of (4.5.1) for any $A > 0$ and find a constant $B > 0$ such that

$$0 \leq u_0(x) \leq f(x,B), \tag{4.5.21}$$

then

$$0 \leq u(x,t) \leq f(\eta,B) \quad \text{in } H^+. \tag{4.5.22}$$

If

$$u_0 = O\left\{\text{erfc}\left(\frac{x}{2\sqrt{k(0)}}\right)\right\} \quad \text{as } x \to \infty,$$

the number B in (4.5.21) can always be identified (see (4.5.13)). Because (4.5.22) holds for any strip $x \geq 0, 0 \leq t \leq T$, we may let T tend to infinity therein.

The main result regarding the asymptotic solution of the IBVP for (4.5.1) may now be stated (Peletier 1970). Let $u(x,t)$ be the solution of the IBVP for (4.5.1) subject to the conditions $u = U$ at $x = 0$ and $u = u_0(x)$ at $t = 0$. Let $k(s) > 0$ for all s and let $k'(s) < 0$ for $s \geq 0$. Moreover, let $u_0(x)$ be smooth and satisfy the compatibility conditions at $(0,0)$ (see (4.5.3)) and the limiting behaviour

$$u_0 = O\left\{ \operatorname{erfc}\left(\frac{x}{2\sqrt{k(0)}} \right) \right\} \quad \text{as } x \to \infty.$$

Then one may find numbers $M > 0$ and $\lambda \in (0, 1/7)$ such that

$$|u(x,t) - f(\eta, U)| \le M(t+1)^{-\lambda}, \quad x \ge 0, \, t \ge 0. \tag{4.5.23}$$

To prove this theorem we define $\bar{u} = \int_0^u k(s)ds$ and $\bar{f} = \int_0^f k(s)ds$. Then, $y = \bar{u} - \bar{f}$ may be shown to satisfy the equation

$$L(y) \equiv k(u)y_{xx} + \beta(x,t)y - y_t = 0, \tag{4.5.24}$$

where β is given by (4.5.19). Because

$$\bar{f}_{xx} = \{k(f)f'\}'(t+1)^{-1} = -\frac{1}{2}\eta(t+1)^{-1}f', \tag{4.5.25}$$

we have

$$\beta(x,t) = -\frac{1}{2}\eta(t+1)^{-1}f'\left\{ \frac{k'(\theta_1)}{k(\theta_2)} \right\}. \tag{4.5.26}$$

Now we introduce the comparison function

$$z(x,t) = (t+1)^{-\lambda}w(\eta), \tag{4.5.27}$$

where $w(\eta)$ is a positive, continuous, and piecewise differentiable function, which is prescribed later. Now,

$$L(z) = \frac{z}{t+1}\left\{ k(u)\frac{w''}{w} + \gamma + \lambda + \frac{1}{2}\eta\frac{w'}{w} \right\}, \tag{4.5.28}$$

where

$$\gamma = -\frac{1}{2}\eta f'\left\{ \frac{k'(\theta_1)}{k(\theta_2)} \right\}$$

(see (4.5.24)). Because $f' < 0$ and $k' < 0$, we infer that $\gamma \le 0$. Therefore, we have

$$L(z) \le \frac{z}{t+1}\left\{ k(u)\frac{w''}{w} + \lambda + \frac{1}{2}\eta\frac{w'}{w} \right\}. \tag{4.5.29}$$

Peletier (1970), following Serrin (1967) (see Section 5.6), chose the following form of the function $w(\eta)$,

$$w(\eta) = 1 + \frac{1}{4}\left\{ 1 - \left(\frac{\eta}{N}\right)^2 \right\}, \quad 0 \le \eta < N, \tag{4.5.30}$$

$$= \exp\left\{ 1 - \left(\frac{\eta}{N}\right) \right\}, \quad \eta > N.$$

This function satisfies the conditions (see below (4.5.27)) envisaged earlier; the constant N is specified presently. It is easy to check that

$$k(u)\frac{w''}{w} + \lambda + \frac{1}{2}\eta\frac{w'}{w} = \lambda - \frac{2k(u)+\eta^2}{4N^2 w} \qquad (4.5.31)$$

for $0 \le \eta < N$ and

$$k(u)\frac{w''}{w} + \lambda + \frac{1}{2}\eta\frac{w'}{w} = \frac{k(u)}{N^2} + \lambda - \frac{\eta}{2N} \qquad (4.5.32)$$

for $\eta > N$.

Now, if we choose $N^2 = k(0)/\delta$ and $\lambda \le (1/2) - \delta, \delta \in (0, 1/2)$, the RHS of (4.5.32) is negative. For (4.5.31), we choose B in (4.5.21) so large such that $u_0(x) \le f(x, B)$; this is possible in view of the asymptotic behaviour (4.5.2) and (4.5.13) of $u_0(x)$ and $f(\eta)$, respectively. With this choice, $u \le B, k(u) \ge k(B)$ and, therefore,

$$\lambda - \frac{2k(u)+\eta^2}{4N^2 w} \le \lambda - \frac{2k(B)}{5k(0)}\delta \le 0 \qquad (4.5.33)$$

if

$$\lambda \le \lambda_0 = \frac{2}{5}\left\{\frac{k(B)}{k(0)}\right\}\delta.$$

With $N^2 = k(0)/\delta$ and $\lambda \le \min\{(1/2) - \delta, \lambda_0\} < 1/7$, it follows from (4.5.31) to (4.5.32) that z satisfies the inequality

$$L(z) \le 0, \quad \eta \ne N. \qquad (4.5.34)$$

We may observe from (4.5.30) that

$$\lim_{\varepsilon \to 0} w'(N+\varepsilon) = -N^{-1} \quad (\varepsilon > 0) \qquad (4.5.35)$$

and

$$\lim_{\varepsilon \to 0} w'(N-\varepsilon) = -(2N)^{-1} \quad (\varepsilon > 0), \qquad (4.5.36)$$

therefore z has a concave corner at $\eta = N$ (see (4.5.27)). Now we define the function

$$\Phi(x,t) = -M_1 z(x,t) + y(x,t), \qquad (4.5.37)$$

where the constant M_1 is chosen so large that Φ is not positive at $x = 0$ and $t = 0$. The asymptotic behaviour of $u_0(x)$ and $f(x, U)$ for $x \to \infty$ ensures that such an M_1 can always be found. Because $L(y) = 0$ (see (4.5.24)) and $L(z) \le 0$ for $\eta \ne N$ we have

$$L(\Phi) = -M_1 L[z] + L[y] \ge 0, \quad \eta \ne N \qquad (4.5.38)$$

and, because z has a concave corner at $\eta = N$, it follows from Krzyżański's maximum principle (1959) enunciated earlier that $\Phi \le 0$; that is, $y(x,t) \le M_1 z(x,t)$ in H^+. By considering the function $\Psi = -M_2 z - y$ in a similar manner, we may show that $-y \leq M_2 z$. Therefore, by using the definition of y and z, we find that

$$|\bar{u} - \bar{f}| \leq \frac{5}{4} \max \{M_1, M_2\}(t+1)^{-\lambda}. \tag{4.5.39}$$

We observe from the definition of $\bar{u}$ and $\bar{f}$ (see below (4.5.23)) that

$$|u - f| < \{k(B)\}^{-1} |\bar{u} - \bar{f}|. \tag{4.5.40}$$

We conclude from (4.5.39) and (4.5.40) that

$$|u - f| \leq M(t+1)^{-\lambda}, \tag{4.5.41}$$

where $M = (5/4) \{k(B)\}^{-1} \max \{M_1, M_2\}$. Because M does not depend on the choice of T, the estimate (4.5.41) holds for all $t \geq 0$.

We may observe that this analysis due to Peletier (1970) closely follows the seminal work of Serrin (1967) for Prandtl boundary layer equations (see Section 5.6). It is constructive, rigourous, and highly instructive.

In a related study, Van Duyn and Peletier (1977b) studied the large time behaviour of solutions of (4.5.1) in $E_T = (-\infty, \infty) \times (0, T]$, $T > 0$ (a constant), subject to the initial condition

$$u(x, 0) = u_0(x), \quad -\infty < x < \infty. \tag{4.5.42}$$

They assumed that the function $k(u)$ satisfies the following conditions.

(H1) $k(s)$ is defined on $\mathbb{R}$.
(H2) $k \in C^{2+\alpha}(\mathbb{R})$, $0 < \alpha < 1$.
(H3) $k(s) \geq \Delta > 0$ for all $s \in \mathbb{R}$.

Furthermore, $u_0(x)$ satisfies the following conditions:

(H4) $u_0 \in C^{2+\alpha}(\mathbb{R})$.
(H5) $u_0(x) \to A_0$ as $x \to -\infty$ and $u_0(x) \to B_0$ as $x \to \infty$, A_0 and B_0 being arbitrary constants.
(H6) $u_0(x) - A_0 = O\left(\text{erfc}\left(-\frac{x}{2(k(A_0))^{1/2}}\right)\right)$ as $x \to -\infty$,
$u_0(x) - B_0 = O\left(\text{erfc}\left(\frac{x}{2(k(B_0))^{1/2}}\right)\right)$ as $x \to \infty$.

Existence of solutions of (4.5.1) subject to (4.5.42) was proved by Oleinik and Kruzhkov (1961). Let $f(\eta; A, B)$ be the solution of (4.5.4) subject to the conditions

$$f(-\infty) = A, \quad f(\infty) = B. \tag{4.5.43}$$

Existence and uniqueness of $f(\eta : A, B)$ was proved by Van Duyn and Peletier (1977a).

Van Duyn and Peletier (1977b) clearly demonstrated the convergence of solutions of (4.5.1) subject to (4.5.42) to the similarity solution of the former governed by (4.5.4) and subject to (4.5.43). They proved the following theorem.

Theorem 4.5.1 *Suppose that $u_0(x)$ satisfies the conditions (H4)–(H6) stated above. Furthermore, let u be a solution of (4.5.1) satisfying (4.5.42) and let $f(\eta;A_0,B_0)$ satisfy (4.5.4) and (4.5.43) with $A = A_0, B = B_0$, say. Then, for any $\varepsilon > 0$, there exists a constant $G(\varepsilon)$ such that*

$$\sup_{x\in\mathbb{R}} |u(x,t) - f(\eta;A_0,B_0)| \leq G(\varepsilon)(t+1)^{-(1-\varepsilon)/2}, \quad t \geq 0.$$

4.6 Large time asymptotics for solutions of the porous media equation

In this section, we discuss large time behaviour of solutions of initial boundary value problems for the porous media equation

$$u_t = (u^m)_{xx}. \tag{4.6.1}$$

If u denotes the density of a polytropic gas flowing through a homogeneous porous medium, then u satisfies an equation like (4.6.1) for $m \geq 2$ (see Muskat (1937)). Equation (4.6.1) also appears in Prandtl's boundary layer theory. Indeed the system of PDEs describing the latter transforms to (4.6.1) with $m = 2$ if we introduce the so-called von Mises variables and assume the pressure gradient to be zero (see Schlichting 1960). Equation (4.6.1) also appears in plasma physics (see Okuda and Dawson (1973) for details). Because the diffusion coefficient on the RHS of (4.6.1) is given by $D(u) = mu^{m-1}$, it is called fast diffusion if $0 < m < 1$ and slow diffusion if $m > 1$, corresponding, respectively, to its behaviour as $u \to 0$.

In this section, we assume that $m > 1$. The transformation

$$u(x,t) = (t+1)^\alpha f(\eta), \quad \eta = x(t+1)^{-(1+(m-1)\alpha)/2} \tag{4.6.2}$$

changes (4.6.1) to

$$(f^m)'' + \frac{1}{2}\{1+(m-1)\alpha\}\eta f' = \alpha f, \quad 0 < \eta < \infty \tag{4.6.3}$$

(see Barenblatt 1952).

Atkinson and Peletier (1971) studied a more general ordinary differential equation than (4.6.3) with $\alpha = 0$. Their analysis showed that equation (4.6.3) with $\alpha = 0$, together with the boundary conditions

$$f(0) = U > 0, \quad f \to 0 \text{ as } \eta \to \infty, \tag{4.6.4}$$

has a unique weak solution $f(\eta;U)$. Furthermore, they proved that there exists $a = a(U)$ such that $f(\eta) > 0$ on $[0,a(U))$ and $f(\eta) \equiv 0$ on $[a(U),\infty)$.

Peletier (1971) studied (4.6.1) with $m > 1$ in the half-strip $S_T = (0, \infty) \times (0, T]$, $T > 0$, subject to the condition

$$u(0,t) = U, \quad 0 < t < T, \tag{4.6.5}$$
$$u(x,0) = u_0(x), \quad 0 \leq x < \infty. \tag{4.6.6}$$

Here, $U > 0$ is a constant and $u_0(x)$ satisfies the following conditions.

(i) $u_0(x)$ is continuous and nonnegative on $[0, \infty)$.
(ii) $u_0(0) = U$ and $u_0(x) = 0$ for large values of x.
(iii) $(u_0^m)'' = 0$ at $x = 0$ and $|(u_0^{m-1})'| \leq L$ on $[0, \infty)$ for some $L > 0$.

Peletier (1971) showed that similarity solutions $f(\eta; U)$ of (4.6.1) defined by (4.6.2) with $\alpha = 0$ describe the large time behaviour of (weak) solutions $u(x,t)$ of (4.6.1) subject to the boundary conditions (4.6.5) and (4.6.6). He derived two estimates: an integral estimate and a pointwise estimate.

(a) $\int_0^\infty \eta |\tilde{u}(\eta, t) - f(\eta; U)| d\eta = O\left(t^{-1}\right)$ as $t \to \infty$; here, $\tilde{u}(\eta, t) = u(x, t)$.
(b) $|u(x,t) - f(\eta; U)| = O\left(t^{-\lambda}\right)$ as $t \to \infty$; here, $\lambda = \min(1/3, 1/(2m - 1))$, $m > 1$.

In the present section, we closely follow Gilding (1979) and obtain an estimate for the large time asymptotic behaviour of the solution of (4.6.1) satisfying (4.6.6) and $u(0,t) = \psi(t)$, a more general boundary condition than (4.6.5). Gilding (1979) essentially followed Peletier (1971) for proving the large time behaviour of weak solutions of (4.6.1). For related study, reference may be made to Kamin (1973), Bertsch (1982), Kamin and Vázquez (1991), and Vázquez (2003, 2007).

We define the domain

$$S_T = (0, \infty) \times (0, T], \quad T > 0.$$

We discuss the asymptotic behaviour of solutions of the following initial boundary value problem for the porous media equation (4.6.1).

$$u_t = (u^m)_{xx}, \quad (x,t) \in S_T, \tag{4.6.7}$$
$$u(x,0) = u_0(x), \quad 0 \leq x < \infty, \tag{4.6.8}$$
$$u(0,t) = \psi(t), \quad 0 \leq t \leq T. \tag{4.6.9}$$

We assume that $m > 1$ and the functions $u_0(x)$ and $\psi(t)$ satisfy the following conditions.

(A1) u_0 is a compactly supported nonnegative function on $[0, \infty)$ and u_0^{m-1} is Lipschitz continuous on $[0, \infty)$.
(A2) ψ is a nonnegative function on $[0, \infty)$ and ψ^m is Lipschitz continuous on $[0, \infty)$.
(A3) u_0 and ψ satisfy compatibility condition $\psi(0) = u_0(0)$.

In the manner of Oleinik, Kalashnikov, and Yui-Lin (1958), a function $u(x,t)$ is said to be a weak solution of (4.6.7)–(4.6.9) on $\overline{S}_T$ if u satisfies the following conditions.

(i) u is bounded, nonnegative, and continuous in $\overline{S}_T$.

(ii) $u(0,t) = \psi(t), \ \forall t \in [0,T]$.

(iii)   A generalised derivative of u^m with respect to x exists and is bounded in $(\delta, \infty) \times (0,T]$ for every $\delta > 0$. Furthermore, $(u^m)_x$ is square integrable in every bounded subset of S_T.

(iv) u satisfies the equation

$$\int \int_{S_T} \{\phi_x(u^m)_x - \phi_t u\}dxdt = \int_0^\infty \phi(x,0)u_0(x)dx \qquad (4.6.10)$$

for all continuously differentiable functions ϕ on $\overline{S}_T$ that vanish at $x = 0$, large x, and $t = T$. Existence of a unique weak solution for (4.6.7)–(4.6.9) was assured by Oleinik et al. (1958) under the requirement that u_0 and ψ satisfy (A1)–(A3). We may state that $u(x,t)$ is a weak solution of the problem (4.6.7)–(4.6.9) in $S = (0,\infty) \times (0,\infty)$ if u is a weak solution of (4.6.7)–(4.6.9) on every S_T, $T > 0$. We quote here some important results about the weak solutions of (4.6.7)–(4.6.9) which we use in the following.

(B1)   The weak solution of (4.6.7)–(4.6.9) can be obtained as a pointwise limit of a decreasing sequence $\{u_n(x,t)\}$ of positive classical solutions of (4.6.7) in $Q_T^n = (0,n) \times (0,T], \ n = 1,2,\ldots$.

(B2)   If u is positive at a point (x_0,t_0), then u is a classical solution of (4.6.7) in a neighbourhood of that point.

(B3)   Suppose that $u_{01}(x)$, $\psi_1(t)$ and $u_{02}(x)$, $\psi_2(t)$ are functions such that $u_{01}(x) \geq u_{02}(x)$, $\forall x \in [0,\infty)$, and $\psi_1(t) \geq \psi_2(t)$, $\forall t \in [0,T]$. If u_1 and u_2 are weak solutions of (4.6.7)–(4.6.9) with data $u_{01}(x)$, $\psi_1(t)$ and $u_{02}(x)$, $\psi_2(t)$, respectively, then $u_1 \geq u_2$ everywhere in S_T.

(B4)   If $u_0(x)$ has compact support in $[0,\infty)$, then the corresponding weak solution u of (4.6.7)–(4.6.9) has compact support in $\overline{S}_T$.

(B5)   The generalised derivative $(u^m)_x$ is continuous in S_T and $u^{m-1}(x,t)$ is Lipschitz continuous in x for any $t > 0$.

For a detailed discussion of weak solutions, reference may be made to Oleinik et al. (1958) and Aronson (1969).

Gilding and Peletier (1976) discussed the existence of weak solutions of the ordinary differential equation

$$(f^m)'' + p\eta f' = qf, \quad 0 < \eta < \infty, \qquad (4.6.11)$$

where p and q are real constants, subject to the boundary conditions (4.6.4); (4.6.3) is a special case of (4.6.11) with $p = (1 + (m-1)\alpha)/2$ and $q = \alpha$. Their study revealed that the BVP (4.6.11) and (4.6.4) has a weak solution with compact support if and only if $p \geq 0$ and $2p + q > 0$. Furthermore, this solution is unique. Thus, a weak solution of (4.6.3) with boundary conditions (4.6.4) exists if and only if $\alpha > -1/m$. Moreover, this solution is unique. Let this solution be denoted by $f(\eta;U)$. Gilding and Peletier (1976) showed that the solution $f(\eta;U)$ possesses the following features.

(C1) $f(\eta;U)$ is a positive classical solution of (4.6.3) on the interval $[0,a(U))$
and $f(\eta;U) \equiv 0$ on $[a(U),\infty)$ for some $a(U)$.
(C2) $f(\eta;U)$ is monotonic with respect to U; that is, $f(\eta;U_1) \geq f(\eta;U_2)$ for
$U_1 > U_2$ on $[0,\infty)$.
(C3)   If $U \to \infty$, then $f(\eta;U) \to \infty$ uniformly on compact subsets of $[0,\infty)$; furthermore, $a(U) \to \infty$ as $U \to \infty$.
(C4)   If $U \to 0$, then $f(\eta;U) \to 0$ uniformly on $[0,\infty)$; furthermore, $a(U) \to 0$ as
$U \to 0$.

We have discussed this problem in considerable detail in Section 4.4. Here we have
summarised the results for ready reference. To show that the similarity solution
(4.6.2) of (4.6.7) describes large time behaviour of the weak solution of (4.6.7)–
(4.6.9), Gilding (1979) proved the following important theorem.

Theorem 4.6.1 *Suppose that u_{01}, ψ_1 and u_{02}, ψ_2 satisfy the assumptions (A1)–(A3)
stated below equation (4.6.9). Further assume that, for $\alpha > -(1/m)$, ψ_1 and ψ_2
satisfy the inequalities*

$$A(t+1)^\alpha \leq \psi_1(t), \quad \psi_2(t) \leq B(t+1)^\alpha, \quad t \geq 0 \tag{4.6.12}$$

*for some positive constants A and B. If u_1 and u_2 are the weak solutions of (4.6.7)–
(4.6.9) with data u_{01}, ψ_1 and u_{02}, ψ_2, respectively, then, for $(x,t) \in S$,*

$$|u_1(x,t) - u_2(x,t)| \leq C(t+1)^\alpha \left\{ (t+1)^{-(m\alpha+1)} \left(1 + \int_0^t |\psi_1^m(s) - \psi_2^m(s)|ds \right) \right\}^\lambda;$$

*here, $\lambda = \min(1/3, 1/(2m-1))$ and C is a constant which depends on m, α, A, B,
u_{01}, and u_{02}.*

This theorem is proved via the derivation of an integral identity and Hölder continuity of weak solutions of (4.6.7)–(4.6.9). This process requires a modified result of
Peletier (1971).

Lemma 4.6.2 *Suppose that u is a weak solution of (4.6.7)–(4.6.9) in S_T for some
$T > 0$. Then the following equality holds for any $t_0 \in (0,T]$.*

$$\int_0^\infty xu(x,t_0)dx = \int_0^\infty xu_0(x)dx + \int_0^{t_0} \psi^m(t)dt. \tag{4.6.13}$$

Proof: By assumption, $u_0(x)$ has compact support. By Property (B4) (see above
(4.6.11)), the weak solution $u(x,t)$ of (4.6.7)–(4.6.9) has compact support in $\overline{S}_T$.
Thus $u(x,t) = 0$ if $x \geq \rho$, $t \in [0,T]$, for some $\rho > 0$. By property (B5) (see above
(4.6.11)), $(u^m)_x \in C(S_T)$. This, in turn, implies that $(u^m)_x = 0$ on $[\rho,\infty) \times (0,T]$.
Because u and $(u^m)_x$ are identically zero for x large (in fact, for $x \geq \rho$), equation
(4.6.10) is valid for all continuously differentiable functions on $\overline{S}_T$, which vanish at
$x = 0$ and $t = T$. We first prove (4.6.13) for $t_0 < T$; the result for $t_0 = T$ then follows
from the continuity in t. First, choose $t_0 < T$. Define

$$k(s) = \begin{cases} \exp\left[-\frac{1}{(1-s^2)}\right], & |s| < 1, \\ 0, & |s| \geq 1, \end{cases}$$

and

$$B_n(t) = \left(\int_{n(t-t_0)}^{n(T-t_0)} k(s)ds\right)\left(\int_{-\infty}^{\infty} k(s)ds\right)^{-1}, \quad n \geq 1.$$

Because $\phi(x,t) = xB_n(t)$ is continuously differentiable and vanishes for $x = 0, t = T$, it is an admissible test function for u. Putting $\phi(x,t) = xB_n(t)$ in (4.6.10), we have

$$\int\int_{S_T} \{B_n(t)(u^m)_x - xB'_n(t)u\}dxdt = \int_0^\infty xB_n(0)u_0(x)dx.$$

A simple manipulation and use of the boundary condition $u(0,t) = \psi(t)$ lead to

$$-\int_0^T B_n(t)\psi^m(t)dt - \int\int_{S_T} xB'_n(t)u(x,t)dxdt = \int_0^\infty xB_n(0)u_0(x)dx. \quad (4.6.14)$$

Letting $n \to \infty$ and using the dominated convergence theorem, we arrive at (4.6.13) for $t_0 < T$. The result (4.6.13) for $t_0 = T$ follows from continuity with respect to t_0.

In the next lemma, we obtain an integral estimate for the difference of two weak solutions of (4.6.7)–(4.6.9).

Lemma 4.6.3 *Suppose that $u_1(x,t)$ and $u_2(x,t)$ are two weak solutions of (4.6.7)– (4.6.9) in S with initial and boundary data u_{01}, ψ_1 and u_{02}, ψ_2, respectively. Then the inequality*

$$\int_0^\infty x|u_1(x,t_0) - u_2(x,t_0)|dx \leq \int_0^\infty x|u_{01}(x) - u_{02}(x)|dx + \int_0^{t_0} |\psi_1^m(t) - \psi_2^m(t)|dt \quad (4.6.15)$$

holds for any $t_0 \in (0,\infty)$.

Proof: Define

$$u_0^+(x) = \max\{u_{01}(x), u_{02}(x)\},$$
$$u_0^-(x) = \min\{u_{01}(x), u_{02}(x)\},$$
$$\psi^+(t) = \max\{\psi_1(t), \psi_2(t)\}, \quad \text{and}$$
$$\psi^-(t) = \min\{\psi_1(t), \psi_2(t)\}.$$

Because u_{01}, ψ_1 and u_{02}, ψ_2 satisfy conditions (A1)–(A3) (see below (4.6.9)), the functions u_0^+, ψ^+ and u_0^-, ψ^- also satisfy the same conditions. Let $u^+(x,t)$ and $u^-(x,t)$ be the weak solutions of (4.6.7) in S with initial and boundary data $u_0^+(x)$, $\psi^+(t)$ and $u_0^-(x)$, $\psi^-(t)$, respectively. Then, by Lemma 4.6.2, for any $t_0 \in (0,\infty)$, we have

$$\int_0^\infty x\{u^+(x,t_0) - u^-(x,t_0)\}dx = \int_0^\infty x\{u_0^+(x) - u_0^-(x)\}dx$$

$$+ \int_0^{t_0} \{(\psi^+)^m(t) - (\psi^-)^m(t)\}dt$$

$$= \int_0^\infty x|u_{01}(x) - u_{02}(x)|dx + \int_0^{t_0} |\psi_1^m(t) - \psi_2^m(t)|dt.$$

$$(4.6.16)$$

By a maximum principle (see property (B3) above (4.6.11)),

$$u^-(x,t) \le u_1(x,t), \ u_2(x,t) \le u^+(x,t), \quad (x,t) \in S.$$

This implies that

$$|u_1(x,t) - u_2(x,t)| \le u^+(x,t) - u^-(x,t), \quad (x,t) \in S. \tag{4.6.17}$$

Equation (4.6.15) follows from (4.6.16) and (4.6.17).

The following lemma helps us to derive pointwise estimates from the integral estimates (4.6.15). This can be proved following Peletier (1971).

Lemma 4.6.4 *Suppose that* $\theta(x)$ *is a nonnegative function defined on* $[0,\infty)$. *Furthermore, let* $\theta(x)$ *satisfy the following conditions.*

(i) θ *is uniformly Hölder continuous on* $[0,\infty)$ *with exponent* $\tilde{\gamma} \in (0,1]$ *and coefficient K.*

(ii) $\int_0^\infty x\theta(x)dx \le L < \infty.$

(iii) $\theta(x_0) = 0$ *for some* $x_0 \in [0,\infty).$

Then, $\theta(x)$ *satisfies the inequality*

$$\theta(x) \le C_0 K^{2/(\tilde{\gamma}+2)} L^{\tilde{\gamma}/(\tilde{\gamma}+2)}, \quad x \in [0,\infty);$$

here,

$$C_0 = \left\{ \frac{2(\tilde{\gamma}+2)}{\tilde{\gamma}} \right\}^{\tilde{\gamma}/(\tilde{\gamma}+2)}.$$

Our interest, now, is to obtain an estimate of Hölder continuity for classical solutions of (4.6.7). Then the result B1 (see below (4.6.10)) is used to derive an estimate for Hölder continuity of weak solutions of (4.6.7). The lateral boundary condition (4.6.9) at $x = 0$ suggests the use of similarity transformation (4.6.2) for (4.6.7) (see (4.6.18) below). Lemmas 4.6.5 and 4.6.6 deal with positive classical solutions of (4.6.7) and Lemma 4.6.7 deals with the Hölder continuity of the weak solutions of (4.6.7)–(4.6.9). Define

$$R = (0,H) \times (0,T], \quad H > 0, \ T > 0,$$
$$R^* = (0,H/2) \times (0,T].$$

Lemma 4.6.5 *Let* $v \in C^{2,1}(\overline{R})$ *be a positive solution of the equation*

$$v_\tau = (v^m)_{\eta\eta} + p\eta v_\eta - qv \tag{4.6.18}$$

in $\overline{R}$ such that $v_\eta \in C^{2,1}(R)$; here, p and q are two constants satisfying the inequalities $p \geq 0$ and $2p+q > 0$. Further assume that there exist positive constants A, M, and C_1 such that

$$v(\eta,\tau) \leq M, \quad (\eta,\tau) \in \overline{R}, \tag{4.6.19}$$
$$v(0,\tau) \geq A, \quad \tau \in [0,T], \tag{4.6.20}$$
$$|(v^{m-1})_\eta(\eta,0)| \leq C_1, \quad \eta \in [0,H]. \tag{4.6.21}$$

Then, the inequality

$$|v(\eta_1,\tau) - v(\eta_2,\tau)| \leq K|\eta_1 - \eta_2|^\gamma \tag{4.6.22}$$

holds for all (η_1,τ), $(\eta_2,\tau) \in \overline{R}^*$; here $\gamma = \min\{1, 1/(m-1)\}$. K appearing on the RHS of (4.6.22) is a positive constant depending on the constants m, p, q, A, M, C_1, and H.

The proof of this lemma requires considerable details, therefore we refer the reader to the original work of Gilding (1979), Oleinik and Kruzhkov (1961), Aronson (1969), and the references therein. The following lemma proves the Hölder continuity for the classical solutions of (4.6.7).

Lemma 4.6.6 *Suppose that* $\alpha > -(1/m)$. *Now, define*

$$D = \left\{(x,t): 0 < x < X(t+1)^{(1+(m-1)\alpha)/2}, 0 < t \leq T\right\},$$
$$D^* = \left\{(x,t): 0 < x < \frac{1}{2}X(t+1)^{(1+(m-1)\alpha)/2}, 0 < t \leq T\right\}.$$

Further suppose that $u(x,t) \in C^{2,1}(\overline{D})$ *is a positive classical solution of (4.6.7) in* $\overline{D}$ *with* $u_x \in C^{2,1}(D)$. *We assume that this solution u satisfies the following conditions.*

(i) $u(x,t) \leq M(t+1)^\alpha$, $(x,t) \in \overline{D}$;
(ii) $u(0,t) \geq A(t+1)^\alpha$, $t \in [0,T]$;
(iii) $|(u^{m-1})_x(x,0)| \leq C_1$, $x \in [0,X]$;

here M, A, *and* C_1 *are some positive constants. Then*

$$|u(x_1,t) - u(x_2,t)| \leq K(t+1)^{\alpha - (\gamma/2)[1+(m-1)\alpha]}|x_1 - x_2|^\gamma, \quad \forall (x_1,t), (x_2,t) \in \overline{D^*}. \tag{4.6.23}$$

The constant K *depends only on* m, α, A, M, C_1, *and* X.

Proof: Let $u(x,t) = (t+1)^\alpha v(\eta)$ where $\eta = x(t+1)^{-[1+(m-1)\alpha]/2}$, $\tau = \log(t+1)$. Then v satisfies (4.6.18) and all the conditions of Lemma 4.6.5 with $p = (1/2)[1+(m-1)\alpha]$, $q = \alpha$ for some $\alpha > -(1/m)$. Hence we obtain (4.6.23).

The following lemma discusses the Hölder continuity of weak solutions of (4.6.7) on $S = (0,\infty) \times (0,\infty)$.

Lemma 4.6.7 *Suppose that* $u(x,t)$ *is a weak solution of (4.6.7)–(4.6.9) in S and suppose that, for some* $\alpha > -(1/m)$, *there exist constants* $B \geq A > 0$, *such that*

$$A(t+1)^\alpha \le \psi(t) \le B(t+1)^\alpha, \quad t \ge 0. \tag{4.6.24}$$

Then

$$|u(x_1,t) - u(x_2,t)| \le K(t+1)^{\alpha - (\gamma/2)[1+(m-1)\alpha]}|x_1 - x_2|^\gamma, \quad \forall\, (x_1,t),\, (x_2,t) \in \overline{S};$$
$$\tag{4.6.25}$$

here $\gamma = \min\{1, 1/(m-1)\}$ and K is a constant depending on m, α, A, B, and u_0.

Proof: By Property (C3) (see above (4.6.12)), there exists a $U \ge B$ such that $f(\eta;U) \ge u_0(\eta)$ on $[0,\infty)$; $f(\eta;U)$ is the weak solution of (4.6.3) with compact support satisfying the boundary conditions (4.6.4). Let $u(x,t;U)$ be the corresponding similarity solution of (4.6.7), given by(4.6.2). Then,

$$u(0,t;U) = U(t+1)^\alpha \ge B(t+1)^\alpha \ge \psi(t), \quad t \ge 0, \tag{4.6.26}$$

and

$$u(x,0;U) = f(x;U) \ge u_0(x), \quad x \ge 0. \tag{4.6.27}$$

By the maximum principle (see B3 above (4.6.11)), $u(x,t;U) \ge u(x,t)$ in S. This implies that

$$u(x,t) \le M_0(t+1)^\alpha, \quad \forall\, (x,t) \in S,$$

$$u(x,t) \equiv 0 \quad \text{in } E^*,$$

where $E^* = \left\{(x,t) \in S: x \ge a(U)(t+1)^{[1+(m-1)\alpha]/2}\right\}$; M_0 is some positive constant and $a(U) = \sup\{\eta : f(\eta;U) > 0\}$. Define

$$X = 2a(U) \quad \text{and} \quad D = \left\{(x,t): 0 < x < X(t+1)^{(1+(m-1)\alpha)/2}, \, 0 < t \le T\right\}.$$

Suppose that $\{u_n(x,t)\}$ is a decreasing sequence of positive classical solutions of (4.6.7) in Q_T^n (see B1 above (4.6.11)). Furthermore, $u(x,t)$ may be obtained from u_n as a pointwise limit. Assume that (i) $u_n \in C^{2,1}(\overline{Q_T^n})$, (ii) $(u_n)_x \in C^{2,1}(Q_T^n)$, and (iii) $|(u_n^{m-1})_x(x,0)| \le C_1 \ \forall\, x \in (0,n)$, where C_1 is a constant independent of n. Then, by Dini's theorem, $u_n \to u$ uniformly on $\overline{D}$ as $n \to \infty$. This implies that, given $M > M_0$, there exists N large enough such that u_n is defined on $\overline{D}$, and, (iv) $u_n(x,t) \le M(t+1)^\alpha, \ \forall\, (x,t) \in \overline{D} \ \forall n \ge N$. Observe that u_n, $n \ge N$ satisfies the hypotheses of Lemma 4.6.6 (see (i)–(iv) here and (4.6.24)). Therefore, by Lemma 4.6.6, we have for $n \ge N$,

$$|u_n(x_1,t) - u_n(x_2,t)|$$

$$\le K(t+1)^{\alpha - (\gamma/2)[1+(m-1)\alpha]}|x_1 - x_2|^\gamma, \quad \forall\, (x_1,t),\, (x_2,t) \in \overline{S_T \setminus E^*};\tag{4.6.28}$$

here K depends on the constants m, α, A, M, C_1, and X but not on n. Because $u_n \to u$ as $n \to \infty$ pointwise in S_T and $u \equiv 0$ in E^* and because $T > 0$ is arbitrary, we have (4.6.25).

Proof of Theorem 4.6.1: We recall that if $y_1(x)$ and $y_2(x)$ are Hölder continuous with exponent γ and coefficients K_1 and K_2, respectively, in a set Ω, then $|y_1(x) - y_2(x)|$ is Hölder continuous with exponent γ and coefficient $(K_1 + K_2)$ in Ω. Because u_1 and u_2 are two weak solutions of (4.6.7)–(4.6.9), they are Hölder continuous on $[0, \infty)$ for a fixed t. This, in turn, implies that $|u_1(x,t) - u_2(x,t)|$ is Hölder continuous with respect to x on $[0, \infty)$ for a fixed t. Let the exponent and the coefficient in the Hölder's inequality be γ and K, respectively. By Lemma 4.6.3, we have

$$\int_0^\infty x|u_1(x,t) - u_2(x,t)|dx \le C_1 \left[1 + \int_0^t |\psi_1^m(s) - \psi_2^m(s)|ds\right] \equiv L(t).$$

Because u_1 and u_2 are weak solutions of (4.6.7)–(4.6.9) with compact support, there exists an x_0 in $(0, X)$ such that $|u_1(x_0,t) - u_2(x_0,t)| = 0$. Then, by Lemma 4.6.4 with $\theta = |u_1(x,t) - u_2(x,t)|$, we have

$$|u_1(x,t) - u_2(x,t)| \le C_0 K(t)^{2/(\gamma+2)} L(t)^{\gamma/(\gamma+2)}, \quad \forall x \in [0, \infty), \qquad (4.6.29)$$

where

$$C_0 = \left\{\frac{2(\gamma+2)}{\gamma}\right\}^{\gamma/(\gamma+2)}.$$

Note that, here $K(t) = (K_1 + K_2)(t+1)^{\alpha - (\gamma/2)[1+(m-1)\alpha]}$; K_1 and K_2 are constants corresponding to u_1 and u_2 in the Hölder's inequality (4.6.25). Note that $\gamma/(\gamma+2) = \min(1/3, 1/(2m-1)) = \lambda$, (say). Therefore, by (4.6.29),

$$|u_1(x,t) - u_2(x,t)| \le C(t+1)^{2\{\alpha - (\gamma/2)[1+(m-1)\alpha]\}/(\gamma+2)}$$
$$\times \left(1 + \int_0^t |\psi_1^m(s) - \psi_2^m(s)|ds\right)^\lambda. \qquad (4.6.30)$$

Because $2\{\alpha - (\gamma/2)[1 + (m-1)\alpha]\}/(\gamma+2) = \alpha - \lambda(m\alpha + 1)$,

$$|u_1(x,t) - u_2(x,t)| \le C(t+1)^\alpha \left[(t+1)^{-(m\alpha+1)}\left(1 + \int_0^t |\psi_1^m(s) - \psi_2^m(s)|ds\right)\right]^\lambda.$$

Thus, the proof of Theorem 4.6.1 is complete. Following closely the proof of Theorem 4.6.1, we may prove the following theorem (Gilding (1979)).

Theorem 4.6.8 *Suppose that u_1 and u_2 are two weak solutions of (4.6.7)–(4.6.9) in S with initial and boundary data u_{01}, ψ_1 and u_{02}, ψ_2, respectively, satisfying the assumptions (A1)–(A3) (see below (4.6.9)). Suppose further that*

$$A\exp(\alpha t) \le \psi_1(t), \quad \psi_2(t) \le B\exp(\alpha t), \quad t \ge 0;$$

here A and B are some positive constants and $\alpha > 0$. Then,

$$|u_1(x,t) - u_2(x,t)| \le C\exp(\alpha t)\left[\exp(-m\alpha t)\left(1 + \int_0^t |\psi_1^m(s) - \psi_2^m(s)|ds\right)\right]^\lambda,$$

$$(x,t) \in S,$$

where $\lambda = \min(1/3, 1/(2m-1))$ and C is a constant which depends only on the constants m, α, A, B, u_{01}, u_{02}.

Using Theorem 4.6.1, we can see that $u(x,t)$ converges, as $t \to \infty$, to the similarity solution (4.6.2) uniformly with respect to x. To prove this assertion, suppose that $u_1(x,t)$ is the weak solution of (4.6.7)–(4.6.9) with boundary data $\psi_1(t) \sim U(t+1)^\alpha$ as $t \to \infty$. Further suppose that $u_2(x,t)$ is the similarity solution of (4.6.7) given by (4.6.2) with $f(0) = U$. This, in turn, implies that $u_2(0,t) = U(t+1)^\alpha$. Then, it is easy to see that

$$\int_0^t |\psi(s)^m - U^m(s+1)^{\alpha m}| ds = o\left(t^{m\alpha+1}\right) \quad \text{as } t \to \infty.$$

Then, by Theorem 4.6.1, $(t+1)^{-\alpha}|u_1(x,t) - u_2(x,t)| \to 0$ as $t \to \infty$. Hence the result.

In this section, we have obtained a pointwise estimate for the weak solutions $u(x,t)$ of the initial boundary value problem (4.6.7)–(4.6.9) as $t \to \infty$; this estimate involves the boundary condition $u(0,t) = \psi(t)$. This, in turn, shows that the solution $u(x,t)$ of the IBVP (4.6.7)–(4.6.9) converges to the similarity solution of (4.6.7) as $t \to \infty$. Thus, following the work of Gilding (1979), we have proved that similarity solutions of (4.6.1) subject to relevant initial/boundary conditions constitute intermediate asymptotics for the problem (4.6.7)–(4.6.9).

4.7 Large time behaviour of solutions of a dissipative semilinear heat equation

In this section we consider the initial value problem for the semilinear parabolic equation

$$u_t - \triangle u + u^p = 0 \quad \text{for } x \in \mathbb{R}^N,\ t > 0, \tag{4.7.1}$$

$$u(x,0) = u_0(x) \quad \text{for } x \in \mathbb{R}^N, \tag{4.7.2}$$

subject to the condition

$$u_0(x) \sim A|x|^{-\alpha} \quad \text{as } |x| \to \infty, \tag{4.7.3}$$

where $u_0(x)$ is a continuous, nonnegative, and bounded function and the parameters p, N, A, and α satisfy the conditions $p > 1$, $N \geq 1$, $A > 0$, and $\alpha > 0$.

We follow closely the work of Herraiz (1999). First, we give a formal derivation of the asymptotic behaviour of the solutions of (4.7.1)–(4.7.3) for the parametric range $\alpha < 2/(p-1)$. This formal approach resembles that of Grundy (1988), and Grundy, Van Duijn, and Dawson (1994). Afterwards, we prove that, for $N > \alpha =$

$2/(p-1)$, self-similar solutions of (4.7.1) form the large time asymptotics. This is achieved via a construction of suitable supersolution and subsolution of an equation related to (4.7.1).

Kamin and Peletier (1985) studied (4.7.1) subject to the initial condition

$$u(x,0) = \phi(x), \quad x \in \mathbb{R}^N, \tag{4.7.4}$$

where $N \geq 1$ and $p > (N+2)/N$. Here $\phi \geq 0$ and $\phi \in L^\infty(\mathbb{R}^N)$. Furthermore, ϕ satisfies

$$\lim_{|x| \to \infty} |x|^\alpha \phi(x) = A > 0. \tag{4.7.5}$$

It was assumed that $0 < \alpha < N$. They proved that the large time behaviour of non-negative solutions of (4.7.1) satisfying (4.7.4) and (4.7.5) is described by a self-similar solution of the heat equation when $2/(p-1) < \alpha < N$. They also showed that self-similar solutions of (4.7.1) are large time asymptotics for the nonnegative solutions of (4.7.1) satisfying (4.7.4) and (4.7.5) when $\alpha = 2/(p-1)$. The case $\alpha < 2/(p-1)$ was discussed by Gmira and Veron (1984).

Escobedo and Kavian (1987) studied (4.7.1) with the intial data $u(x,0) = u_0(x)$ satisfying $u_0(x) \neq 0$, $0 \leq u_0(x) \leq A\exp(-a|x|^2)$, $x \in \mathbb{R}^N$. Here A and a are positive constants and $1 < p < 1+2/N$. Their analysis showed that, for large time, solutions of (4.7.1) behave as do the positive self-similar solutions of equation (4.7.1).

A reference may be made to Galaktionov et al. (1986), Herraiz (1998), Kwak (1998), and Cazenave et al. (2001) for a related study. Now we present below a study of solutions of (4.7.1)–(4.7.3) following Herraiz (1999).

The existence of a unique, classical, and global solution of (4.7.1)–(4.7.3) is well established (Friedman 1964). If we drop the Laplacian in (4.7.1) and integrate with respect to t, a natural upper bound immediately results:

$$u(x,t) \leq ((p-1)t)^{-1/(p-1)}. \tag{4.7.6}$$

Herraiz (1999) introduced the form

$$u(x,t) = t^{-1/(p-1)}\Phi(y,\tau), \quad y = xt^{-1/2}, \quad \tau = \log t, \tag{4.7.7}$$

where $y = xt^{-1/2}$ is the standard similarity variable. Equation (4.7.1), in view of (4.7.7), becomes

$$\Phi_\tau = \triangle\Phi + \frac{1}{2}y.\nabla\Phi + \frac{1}{p-1}\Phi - \Phi^p. \tag{4.7.8}$$

It seems 'reasonable' to assume that, as $\tau \to \infty$, the solution of (4.7.8) tends to a global, bounded, and nonnegative stationary solution of the same equation. Excluding for the moment the nonconstant self-similar asymptotic form, we may infer from (4.7.8) that either

$$\Phi(y,\tau) \to c_* \equiv (p-1)^{-1/(p-1)} \text{ as } \tau \to \infty \tag{4.7.9}$$

or

$$\Phi(y, \tau) \to 0 \text{ as } \tau \to \infty. \tag{4.7.10}$$

These forms of asymptotics do not hold uniformly on $\mathbb{R}^N$. These have to be compatible with the requirement that, for $|y| = |x|t^{-1/2} \gg 1$, the solution remains close to the initial values. Thus

$$u(x,t) \sim A|x|^{-\alpha} \text{ as } t \to \infty \text{ for } |y| = |x|t^{-1/2} \gg 1. \tag{4.7.11}$$

In fact, we may verify that either (4.7.9) and (4.7.11) or (4.7.10) and (4.7.11) hold as $t \to \infty$. The first case comes about only if

$$\alpha < \frac{2}{p-1}. \tag{4.7.12}$$

To show that we write

$$\Phi(y, \tau) = c_* + \Psi(y, \tau) \tag{4.7.13}$$

so that (4.7.8) becomes

$$\Psi_\tau = \triangle\Psi + \frac{1}{2}y.\nabla\Psi - \Psi + F(\Psi), \tag{4.7.14}$$

where $F(\Psi) = O\left(\Psi^2\right)$ as $\Psi \to 0$. Ignoring the nonlinear term (in (4.7.14)) and seeking solutions for the radial case, we may write

$$\Psi(y, \tau) = ke^{-\lambda\tau}\zeta(|y|) \tag{4.7.15}$$

and obtain

$$\zeta''(s) + \left(\frac{N-1}{s} + \frac{s}{2}\right)\zeta'(s) = (1-\lambda)\zeta(s) \text{ for } s = |y| > 0. \tag{4.7.16}$$

We seek solution of (4.7.16) subject to the conditions

$$\zeta(s) \text{ is bounded at } s = 0, \quad \zeta(s) \sim s^{2(1-\lambda)} \text{ as } s \to \infty; \tag{4.7.17}$$

the second condition easily follows from (4.7.16) as $s \to \infty$. Equation (4.7.16) subject to the conditions (4.7.17) has its solution provided that $0 < \lambda < 1$. In view of (4.7.13), (4.7.15), and (4.7.17), we may seek an expansion for Φ of the following form,

$$\Phi(y, \tau) \sim c_* + ke^{-\lambda\tau}|y|^{2(1-\lambda)} \text{ as } |y| \to \infty, \tag{4.7.18}$$

where k and λ may be easily determined; here $0 < \lambda < 1$. The expansion (4.7.18) breaks down when $ke^{-\lambda\tau}|y|^{2(1-\lambda)} = O(1)$. Thus, we introduce the new variable $\xi = ye^{-\lambda\tau/2(1-\lambda)}$ in (4.7.8) and obtain

$$\Phi_\tau = \frac{\xi.\nabla\Phi}{2(1-\lambda)} + \frac{1}{p-1}\Phi - \Phi^p + e^{-\Gamma\tau}\triangle\Phi, \tag{4.7.19}$$

where $\Gamma = \lambda/(1-\lambda) > 0$ is some constant. The operators ∇ and $\triangle$ now apply with respect to the new variable ξ. Restricting ourselves again to radial solutions and noting that $\Phi_\tau = o(1)$ and $e^{-\Gamma\tau}\triangle\Phi = o(1)$ as $\tau \to \infty$, (4.7.19) reduces in this limit to the first-order ODE of Bernoulli type,

$$\frac{|\xi|\Phi'(|\xi|)}{2(1-\lambda)} + \frac{1}{p-1}\Phi - \Phi^p = 0, \tag{4.7.20}$$

with the general solution

$$\Phi(|\xi|) = \left((p-1) + C|\xi|^{2(1-\lambda)}\right)^{-1/(p-1)}, \tag{4.7.21}$$

where C is an arbitrary constant. Thus, with (4.7.3), (4.7.18), and (4.7.21) in view, we arrive at the following asymptotic behaviour of the solution $u(x,t)$ for $t \gg 1$:

$$u(x,t) \sim t^{-1/(p-1)}\left(c_* + ke^{-\lambda\tau}\zeta(|y|)\right) \text{ for } |\xi| \ll 1,$$

$$\sim t^{-1/(p-1)}\left((p-1) + C|\xi|^{2(1-\lambda)}\right)^{-1/(p-1)} \text{ for } |\xi| = O(1),$$

$$\sim A|x|^{-\alpha} \text{ for } |\xi| \gg 1. \tag{4.7.22}$$

The constants k, λ, and C in (4.7.22) must be found by matching solution forms in different domains. For example, we may require that

$$t^{-1/(p-1)}\left((p-1) + C|\xi|^{2(1-\lambda)}\right)^{-1/(p-1)} \sim A|x|^{-\alpha} \text{ for } |\xi| \gg 1, \tag{4.7.23}$$

yielding the relations

$$C = A^{-(p-1)}, \quad \lambda = 1 - \frac{1}{2\beta} \text{ with } \beta = \frac{1}{\alpha(p-1)}. \tag{4.7.24}$$

Because $0 < \lambda < 1$, we must have $\alpha < 2/(p-1)$. Taking these values of the constants into account we may match the first two relations in (4.7.22) in different regions under the assumptions, $|y| \gg 1$ and $|\xi| = O(1)$, and obtain

$$t^{-1/(p-1)}\left(c_* + ke^{-\lambda\tau}\zeta(|y|)\right) \sim t^{-1/(p-1)}\left((p-1) + C|\xi|^{2(1-\lambda)}\right)^{-1/(p-1)}. \tag{4.7.25}$$

This relation holds provided that

$$k = -C(p-1)^{-1/(p-1)-2}. \tag{4.7.26}$$

Thus, we arrive at the following result. Let $u(x,t)$ be the solution of (4.7.1)–(4.7.3) such that $\alpha < 2/(p-1)$. Then, for sufficiently large time t, we have

$$u(x,t) = t^{-1/(p-1)} \left(c_* - kt^{-\lambda} \zeta \left(\frac{|x|}{\sqrt{t}} \right) \right) (1+o(1)) \quad \text{for } |x| \leq f(t),$$

$$= \left((p-1)t + A^{1-p}|x|^{\alpha(p-1)} \right)^{-1/(p-1)} (1+o(1)) \quad \text{for } f(t) \leq |x| \leq g(t),$$

$$= A|x|^{-\alpha}(1+o(1)) \quad \text{for } g(t) \leq |x|, \tag{4.7.27}$$

where $f(t)$ and $g(t)$ are arbitrary functions such that, for $t \gg 1$, we have

$$t^{1/2} \ll f(t) \ll t^{\beta} \quad \text{with } \beta = \frac{1}{\alpha(p-1)}, \quad t^{\beta} \ll g(t),$$

and $\lambda = 1 - 1/2\beta$; $\zeta(r)$ is the solution of (4.7.16) and (4.7.17) (where s is replaced by r).

Thus Herraiz (1999) showed the existence of three asymptotic regions: an external region, an internal region, and a transition region. In the external region, the solution of (4.7.1)–(4.7.3) behaves as does the initial profile itself. The effect of nonlinearity is observed in the internal region and the nonlinear effects die away gradually in the transition zone. We may point out that the asymptotic results known previously were mostly given in the regions of the form $|x| \leq ct^{1/2}$. However, Herraiz (1999) gave global asymptotic expansions (see (4.7.27)).

Now we turn to the similarity solution of (4.7.1) and its asymptotic nature. For this purpose, write

$$u(x,t) = (t+1)^{-1/(p-1)} \Phi(y,\tau), \quad y = x(t+1)^{-1/2}, \quad \tau = \log(t+1) \tag{4.7.28}$$

(see (4.7.7)). Then $\Phi(y,\tau)$ satisfies (4.7.8). If we substitute $\Phi(y,\tau) = g(|y|)$, then (4.7.8) transforms to

$$g''(s) + \left(\frac{N-1}{s} + \frac{s}{2} \right) g'(s) + \frac{1}{p-1}g(s) - g(s)^p = 0 \quad \text{for } s > 0; \tag{4.7.29}$$

here $s = |y| = |x|/\sqrt{t+1}$. Now we impose the special conditions

$$g(0) = \mu, \quad g'(0) = 0. \tag{4.7.30}$$

Existence and uniqueness of the solution of the IVP (4.7.29) and (4.7.30) were discussed in detail by Brezis et al. (1986) and Escobedo et al. (1995).

Herraiz (1999) discussed the asymptotic nature of the similarity solution (4.7.28) of (4.7.1) in radial symmetry, governed by (4.7.29) and (4.7.30). He proved the following theorem.

Theorem 4.7.1 *Suppose that $N > \alpha = 2/(p-1)$. Furthermore, let $u(x,t)$ be the solution of (4.7.1)–(4.7.3) and $w_\mu(x,t) = (t+1)^{-1/(p-1)}g_\mu(|x|/\sqrt{t+1})$, where $g_\mu(|x|/\sqrt{t+1})$ is the solution of (4.7.29)–(4.7.30). Then there exists a unique $\mu(A)$ such that*

$$u(x,t) = w_\mu(x,t)(1+o(1)) \quad \text{as } t \to \infty \text{ uniformly for } x \in \mathbf{R}^N.$$

To prove this theorem, Herraiz (1999) constructed a supersolution and a subsolution of (4.7.8) in such a way that these supersolutions and subsolutions behave as do the self-similar solutions of (4.7.1) for large time. This, in turn, results in the large time behaviour of solutions of (4.7.1)–(4.7.3).

Let $\varepsilon > 0$ be sufficiently small such that $1/(p-1) + \varepsilon < N/2$. Define

$$\Phi^+(y, \tau) = (1+\varepsilon)g(|y|) + M_1 \exp(-\varepsilon\tau)G(|y|),$$
$$\Phi^-(y, \tau) = ((1-\varepsilon)g(|y|) - M_2 \exp(-\varepsilon\tau)G(|y|))_+ .$$

Here $G(s) \equiv G_\varepsilon(s)$ is a positive bounded solution of

$$G''(s) + \left(\frac{N-1}{s} + \frac{s}{2}\right)G'(s) + dG(s) = 0, \quad d = \frac{1}{p-1} + \varepsilon, \quad s > 0$$

satisfying $G(s) \sim s^{-2((1/(p-1))+\varepsilon)}$ as $s \to \infty$. Furthermore, $g(s)$ is the unique positive solution of (4.7.29)–(4.7.30) satisfying $g(s) \sim As^{-2/(p-1)}$ as $s \longrightarrow \infty$. Let

$$L_0(\Phi) \equiv \Phi_{ss} + \left(\frac{N-1}{s} + \frac{s}{2}\right)\Phi_s + \frac{1}{p-1}\Phi, \tag{4.7.31}$$

where $s = |y|$. It is easy to see that

$$L_0((1+\varepsilon)g) = (1+\varepsilon)g(s)^p \leq (1+\varepsilon)^p g(s)^p,$$
$$L_0(M_1 G(s)) = -M_1 \varepsilon G(s) < 0$$

(see the definitions of g and G). Then

$$\Phi_\tau^+ - L_0\left(\Phi^+\right) + \Phi^{+p} \geq \Phi_\tau^+ - ((1+\varepsilon)g(s))^p + M_1\varepsilon G(s)e^{-\varepsilon z} + \Phi^{+p}$$
$$= -((1+\varepsilon)g(s))^p + ((1+\varepsilon)g(s) + M_1 e^{-\varepsilon\tau}G(s))^p \geq 0$$

for ε small. Thus Φ^+ is a supersolution of (4.7.8).

Again

$$L_0((1-\varepsilon)g(s)) = (1-\varepsilon)g(s)^p,$$
$$L_0\left(M_2 \exp(-\varepsilon\tau)G(s)\right) = -\varepsilon M_2 \exp(-\varepsilon\tau)G(s).$$

Then, for $(1-\varepsilon)g(s) > M_2 e^{-\varepsilon z}G(s)$,

$$L_0((1-\varepsilon)g(s) - M_2 e^{-\varepsilon\tau}G(s)) = (1-\varepsilon)g(s)^p + M_2\varepsilon\exp(-\varepsilon\tau)G(s)$$

This implies that

$$\Phi_\tau^- - L_0\left(\Phi^-\right) + \Phi^{-p} \leq -(1-\varepsilon)^p g(s)^p + ((1-\varepsilon)g(s) - M_2\exp(-\varepsilon\tau)G(s))^p \leq 0.$$

Thus Φ^- is a subsolution of (4.7.8). Observe that the jump of Φ_{ss}^- has the right sign at $(1-\varepsilon)g(s) = M_2\exp(-\varepsilon\tau)G(s)$. Now we are ready to prove that the self-similar

solution (4.7.28) describes the large time behaviour of solutions of (4.7.1)–(4.7.3).
At first, we show that

$$\Phi^-(y,0) \leq \Phi(y,0) \leq \Phi^+(y,0) \quad \text{for all } y. \tag{4.7.32}$$

This inequality follows easily from (4.7.3) for $|y| \geq M$; M is sufficiently large. The
boundedness of $u(y,0) = \Phi(y,0)$ gives that $\Phi(y,0) \leq k$ for some $k > 0$ and $|y| \leq M$.
Choose M_1 and M_2 in such a way that $k \leq M_1 G(M)$ and $M_2 = (1-\varepsilon)g(0)/G(M)$.
This implies that

$$(1-\varepsilon)g(y) \leq (1-\varepsilon)g(0) = G(M)M_2 \leq G(y)M_2 \quad \text{for } |y| \leq M$$

and hence $\Phi^-(y,0) = 0$ on $|y| \leq M$. Furthermore,

$$\Phi(y,0) \leq k \leq M_1 G(M) \leq M_1 G(y) + (1+\varepsilon)g(y) \equiv \Phi^+(y,0).$$

This proves (4.7.32). By the maximum principle,

$$\Phi^-(y,\tau) \leq \Phi(y,\tau) \leq \Phi^+(y,\tau), \quad y \in \mathbb{R}^N.$$

Because $g(|y|) \gg \exp(-\varepsilon\tau)G(|y|)$ as $\tau \to \infty$ for any $y \in \mathbb{R}^n$ and ε arbitrarily small,
we have Theorem 4.7.1.

It may be pointed out that Theorem 4.7.1 is also true for the parametric ranges
$N \leq \alpha = 2/(p-1)$. Thus the space dimension N plays no role for the parametric
ranges $\alpha = 2/(p-1)$ and $\alpha < 2/(p-1)$. We refer the reader to the original work
of Herraiz (1999) for other cases. He made use of a blend of matched asymptotic
techniques, integral results, and comparison methods to arrive at the results.

In this section we have studied, using formal asymptotic methods, the asymptotic
behaviour of solutions of (4.7.1) in N dimensions for $\alpha < 2/(p-1)$ when the initial
profile (4.7.2) is continuous, bounded, and nonnegative and has algebraic decay as
$|x| \to \infty$ (see (4.7.2) and (4.7.3)). We have also proved that for the case $N > \alpha = 2/(p-1)$ the self-similar solution (4.7.28) describes large time behaviour of the
solutions of (4.7.1) and (4.7.2) subject to (4.7.3).

4.8 Large time asymptotics for the solutions of a very fast diffusion equation

In this section, we study the solutions of the singular diffusion equation

$$u_t = \Delta(\log u); \tag{4.8.1}$$

this equation is termed singular because the diffusivity coefficient $1/u$ tends to zero
or infinity depending on whether u tends to infinity or zero. It appears in many applications; for example, it describes limiting density distribution of gases obeying

the Boltzmann equation (Kurtz 1973). Equation (4.8.1) is also the central limit approximation to Carleman's model of the Boltzmann equation (see Carleman 1957).

We discuss large time asymptotics of the 'maximal' solutions of the nonlinear parabolic equation

$$u_t = \Delta(\log u), \quad (x,t) \in \mathbb{R}^N \times (0,\infty), \qquad (4.8.2)$$

subject to the Cauchy data

$$u(x,0) = u_0(x), \quad x \in \mathbb{R}^N, N \geq 2; \qquad (4.8.3)$$

$u_0 \geq 0$ is a locally integrable function in $\mathbb{R}^N$. We follow here the work of Guo (1996) (see also Hsu 2005). Equation (4.8.2) may also be viewed as a special case of the more general equation

$$u_t = \nabla.(u^{m-1}\nabla u) \qquad (4.8.4)$$

with $m = 0$. Esteban et al. (1988) discussed the existence and uniqueness of 'maximal' solutions of (4.8.4) for the parametric range $-1 < m \leq 0, N = 1$. They proved the following theorems.

Theorem 4.8.1 *Suppose that $u_0 \in L^1_{loc}(\mathbb{R})$, $u_0 \geq 0$, and $u_0 \not\equiv 0$. Then there exists a function $u > 0$ in $C^\infty(\mathbb{R} \times (0,\infty)) \cap C([0,\infty);L^1_{loc}(\mathbb{R}))$ satisfying the following.*

(i)   u solves (4.8.4) in $\mathbb{R} \times (0,\infty)$.
(ii) $u \to u_0$ as $t \to 0$ in $L^1_{loc}(\mathbb{R})$.

Theorem 4.8.2 *For every $u_0 \in L^1_{loc}(\mathbb{R})$, there is a unique maximal solution of (4.8.4) with initial condition (4.8.3). This maximal solution is characterized by the following decay conditions.*

$$u^m(x,t) = O(|x|) \text{ as } |x| \to \infty, \text{ when } -1 < m < 0;$$
$$-\ln(u) \leq O(|x|) \text{ as } |x| \to \infty, \text{ when } m = 0$$

uniformly in (τ, T), $\tau > 0$.

Analogous results were also proved for $u_0 \in L^1(\mathbb{R})$ when $-1 < m \leq 0$. They also derived some asymptotic properties of the maximal solution of (4.8.4). Vázquez (1992) proved the nonexistence of solutions of (4.8.4) with finite mass for the parametric ranges : (i) $m \leq 0, N > 3$; (ii) $m < 0, N = 2$; (iii) $m \leq -1, N = 1$ (see also Herrero (1989, 1991) for $m \leq -1$ in one dimension).

A discussion on the existence of integrable solutions for (4.8.4), when $m > 0$, may be found in Bénilan and Crandall (1981).

Zhang (1993) studied the large time behaviour of (4.8.4), with $-1 < m \leq 0, N = 1$, subject to the initial condition $u(x,0) = u_0(x)$. Assuming that $u_0 \geq 0$, $u_0 - H \in L^1(\mathbb{R})$, Zhang proved that the maximal solution of (4.8.4) with $u_0(x)$ as the initial data converges to a similarity solution of (4.8.4) as $t \to \infty$. For precise conditions on $u_0(x)$, see Zhang (1993) and $H(x)$ is the Heaviside function here. A study of relevant similarity solutions of (4.8.4) may be found in Van Duijn et al. (1988).

In this section, we include both the analysis of the ODE governing the similarity solution and asymptotic nature of the latter (see Section 4.6 for a discussion on the aymptotic behaviour of the solutions of $u_t = \Delta(u^m)$ for $m > 1$).

$u(x,t)$ is a solution of (4.8.2)–(4.8.3) when

(i) $u \in C^\infty(\mathbb{R}^N \times (0,\infty)) \cap C^0([0,\infty); L^1_{loc}(\mathbb{R}^N))$.
(ii) u is positive for $t > 0$.
(iii) u satisfies (4.8.2) in the classical sense.
(iv) $u(.,t) \to u_0$ in $L^1_{loc}(\mathbb{R}^N)$ as $t \to 0$.

We discuss the large time asymptotics for (4.8.2) and (4.8.3) in three steps, following Guo (1996). In the first step, we prove the existence of a family of self-similar solutions $S_{A,p}(x,t)$ of (4.8.2) satisfying

$$S_{A,p}(x,t)|x|^{-p} \to A > 0 \text{ as } |x| \to \infty.$$

In the next step, we quote the results concerning the existence of the maximal solution of (4.8.2) and (4.8.3) with continuous, nonnegative initial data $u_0(x)$ satisfying

$$\liminf_{|x|\to\infty} u_0(x)|x|^{-p} > 0, \quad p > -2.$$

The maximal solution $u(x,t)$ of (4.8.2) and (4.8.3) satisfies the decay condition

$$u(x,t) \geq c|x|^p, \quad |x| \text{ sufficiently large}, \tag{4.8.5}$$

uniformly in t in any compact subset of $(0,\infty)$ for some $c > 0$ and $p > -2$. Finally, we prove that the maximal solution of (4.8.2) and (4.8.3) behaves as does the self-similar solution $S_{A,p}(x,t)$ for large time.

We write

$$u(x,t) = t^{-\alpha}\phi(|x|/t^\sigma), \tag{4.8.6}$$

where $\sigma > 0$ is a constant and $\alpha = 2\sigma - 1$. Substituting (4.8.6) into (4.8.2), we get

$$(\phi^{-1}\phi')' + \frac{N-1}{r}\phi^{-1}\phi' + \sigma r\phi' + \alpha\phi = 0, \quad r > 0, \tag{4.8.7}$$

where $r = |x|/t^\sigma$. The conditions required on ϕ are $\phi'(0) = 0$, $\phi > 0$ on $[0,\infty)$.

In the manner of Brezis et al. (1986) and Guo (1995, 1996) posed an initial value problem for (4.8.7) with

$$\phi(0) = \eta > 0, \quad \phi'(0) = 0, \tag{4.8.8}$$

where η is a constant. He assumed that $\sigma \neq 1/2$ is positive. The case $\sigma = 1/2$ gives only a constant solution for the IVP (4.8.7) and (4.8.8). Using fixed point arguments, one may prove local existence and uniqueness of solutions of the IVP (4.8.7) and (4.8.8) (see, for example, Section 4.4). The transformation

$$w(r) = \ln\phi(r) \text{ or } \phi(r) = \exp(w(r)) \tag{4.8.9}$$

transforms (4.8.7) to

$$w'' + \frac{N-1}{r}w' + \sigma r \exp(w)w' + \alpha \exp(w) = 0, \quad r > 0. \tag{4.8.10}$$

The condition $(4.8.8)_2$ becomes $w'(0) = 0$. Let

$$\rho(r) = \exp\left(\sigma \int_0^r s \exp(w(s))ds\right). \tag{4.8.11}$$

Equation (4.8.10) now assumes the form

$$(r^{N-1}\rho(r)w')' = -\alpha r^{N-1}\rho \exp(w). \tag{4.8.12}$$

Integration of (4.8.12) from 0 to r gives

$$w'(r) = -\frac{\alpha}{r\rho(r)} \int_0^r \left(\frac{s}{r}\right)^{N-2} s \exp(w(s))\rho(s)ds. \tag{4.8.13}$$

Because $\alpha = 2\sigma - 1$, w is monotonically increasing for $\sigma < 1/2$ and monotonically decreasing for $\sigma > 1/2$. From (4.8.11), we have,

$$\rho'(s) = \sigma s \exp(w(s))\rho(s). \tag{4.8.14}$$

Because $\alpha < 0$ for $\sigma < 1/2$, we have from (4.8.13) the inequality

$$w'(r) \leq -\frac{\alpha}{r\rho(r)} \int_0^r s \exp(w(s))\rho(s)ds$$

$$= -\frac{\alpha}{r\rho(r)} \int_0^r \frac{\rho'(s)}{\sigma}ds$$

$$\leq -\frac{\alpha}{\sigma r};$$

see (4.8.14). Thus,

$$0 < w'(r) \leq -\frac{\alpha}{\sigma r} \quad \text{for } \sigma < 1/2. \tag{4.8.15}$$

Similarly, it may be shown that

$$-\frac{\alpha}{\sigma r} \leq w'(r) < 0 \quad \text{if } \sigma > 1/2. \tag{4.8.16}$$

Equations (4.8.15) and (4.8.16) imply that the solutions of (4.8.7) and (4.8.8) are global. It follows from (4.8.15) to (4.8.16) that $w'(r) \to 0$ as $r \to \infty$. Now, we prove that $w(r) \to \infty$ as $r \to \infty$ for $\sigma < 1/2$. Recall that w is monotone increasing for $\sigma < 1/2$ (see (4.8.15)). Assume, for contradiction, that $w(r) \to w_0$ as $r \to \infty$ where w_0 is a fixed positive constant. This, in turn, implies that

$$w(1) \leq w(r) \leq w_0, \quad r \geq 1. \tag{4.8.17}$$

Dividing (4.8.10) by r and hence integrating from 1 to r, we obtain

$$\frac{w'(r)}{r} - w'(1) + \sigma \left[e^{w(r)} - e^{w(1)} \right] + N \int_1^r \frac{w'(s)}{s^2} ds = -\int_1^r \frac{\alpha}{s} e^{w(s)} ds. \quad (4.8.18)$$

Dividing (4.8.15) by r^2 and integrating from 1 to ∞, we have

$$0 < \int_1^\infty \frac{w'}{r^2} dr \leq -\frac{\alpha}{2\sigma}. \quad (4.8.19)$$

Using (4.8.17), we check that

$$-\alpha \int_1^r \frac{e^{w(s)}}{s} ds \geq -\alpha e^{w(1)} \ln r. \quad (4.8.20)$$

We also observe that the LHS of (4.8.18) is a finite quantity as $r \to \infty$ (see (4.8.19)) whereas its RHS tends to ∞ (see (4.8.20)). Thus, we arrive at a contradiction. Therefore, $w(r) \to \infty$ as $r \to \infty$. Using similar arguments, we may show that $w(r) \to -\infty$ as $r \to \infty$ when $\sigma > 1/2$. From (4.8.11) we observe that $\rho(r) \to \infty$ exponentially as $r \to \infty$ for $\sigma < 1/2$. Now we show that $\rho(r)$ also tends to ∞ exponentially as $r \to \infty$ when $\sigma > 1/2$. An integration of the first inequality in (4.8.16) from 1 to r shows that

$$\exp(w(r)) \geq \exp(w(1)) r^{-\alpha/\sigma}, \quad r \geq 1. \quad (4.8.21)$$

Recalling that $\alpha = 2\sigma - 1$, we have $-(\alpha/\sigma) + 1 = -1 + (1/\sigma) > -1$; it follows from (4.8.11) that $\rho(r) \to \infty$ exponentially as $r \to \infty$. Now we find the asymptotic behaviour of the similarity function ϕ, governed by (4.8.7), as $r \to \infty$.

Theorem 4.8.3 *Assume that $N \geq 2$ and $\sigma > 0$. Then*

$$\lim_{r \to \infty} r^{\alpha/\sigma} \phi(r) = A > 0; \quad (4.8.22)$$

here, $\alpha = 2\sigma - 1$.

To prove (4.8.22), we first show that

$$\lim_{r \to \infty} \frac{r\phi'(r)}{\phi(r)} = -\frac{\alpha}{\sigma}. \quad (4.8.23)$$

Later, we find that

$$\lim_{r \to \infty} r^\lambda \left[\frac{r\phi'(r)}{\phi(r)} + \frac{\alpha}{\sigma} \right] = 0; \quad (4.8.24)$$

here (the parameter) $\lambda > 0$ is such that $\lambda < 2$ if $\sigma < 1/2$ and $\lambda < 1/\sigma$ if $\sigma > 1/2$. An integration of (4.8.24) would lead to (4.8.22). We integrate (4.8.12) from 0 to r to obtain

$$w'(r) = -\frac{\alpha}{r^{N-1} \rho(r)} \int_0^r s^{N-1} \rho(s) e^{w(s)} ds. \quad (4.8.25)$$

Substituting (4.8.9) in (4.8.25) and rewriting, we obtain

$$\frac{r\phi'(r)}{\phi(r)} = -\frac{\alpha}{r^{N-2}\rho(r)} \int_0^r s^{N-1}\phi(s)\rho(s)ds. \qquad (4.8.26)$$

Letting $r \to \infty$ on both sides of (4.8.26) and using L'Hospital's rule, we obtain

$$\lim_{r \to \infty} \frac{r\phi'(r)}{\phi(r)} = -\lim_{r \to \infty} \frac{\alpha}{(N-2)r^{-2}\phi^{-1}(r) + \sigma}, \qquad (4.8.27)$$

provided that the limit on the RHS of (4.8.27) exists. To prove (4.8.23), it suffices to show that $r^2\phi(r) \to \infty$ as $r \to \infty$. Because $\phi(r) = \exp(w(r))$ and $w(r) \to \infty$ as $r \to \infty$ for $\sigma < 1/2$, it follows that $r^2\phi(r) \to \infty$ as $r \to \infty$. For $\sigma > 1/2$,

$$r^2\phi(r) = r^2\exp(w(r)) \geq \exp(w(1))r^{2-\alpha/\sigma} = \exp(w(1))r^{1/\sigma} \to \infty,$$

as $r \to \infty$ (see (4.8.21)). Therefore, the limit (4.8.23) follows immediately from (4.8.27).

Again,

$$\lim_{r \to \infty} r^\lambda \left[\frac{r\phi'(r)}{\phi(r)} + \frac{\alpha}{\sigma} \right] = \frac{\alpha}{\sigma} \lim_{r \to \infty} \frac{r^{N-2}\rho - \sigma \int_0^r s^{N-1}\phi(s)\rho(s)ds}{r^{N-\lambda-2}\rho}$$

$$= \frac{\alpha}{\sigma} \lim_{r \to \infty} \frac{(N-2)r^\lambda}{(N-\lambda-2) + \sigma r^2\phi}, \qquad (4.8.28)$$

after using L'Hospital's rule and (4.8.14). Thus, we have

$$\lim_{r \to \infty} r^\lambda \left[\frac{r\phi'(r)}{\phi(r)} + \frac{\alpha}{\sigma} \right] = \frac{\alpha}{\sigma} \lim_{r \to \infty} \frac{(N-2)r^{\lambda-2}}{(N-\lambda-2)r^{-2} + \sigma\phi(r)}. \qquad (4.8.29)$$

For $\sigma < 1/2$ and $\lambda < 2$, the RHS of (4.8.29) tends to zero as $r \to \infty$ (recall that for $\sigma < 1/2$, $\phi \to \infty$ as $r \to \infty$; see below (4.8.20)).

Now let $\sigma > 1/2$. Rewriting (4.8.28) as

$$\lim_{r \to \infty} r^\lambda \left[\frac{r\phi'(r)}{\phi(r)} + \frac{\alpha}{\sigma} \right] = \frac{\alpha}{\sigma} \lim_{r \to \infty} \frac{(N-2)r^{\lambda-1/\sigma}}{(N-\lambda-2)r^{-1/\sigma} + \sigma\phi r^{2-1/\sigma}} \qquad (4.8.30)$$

and observing that

$$(N-\lambda-2)r^{-1/\sigma} + \sigma\phi r^{2-1/\sigma} \geq (N-\lambda-2)r^{-1/\sigma} + \sigma e^{w(1)}, \qquad (4.8.31)$$

where we have used (4.8.21), we find that

$$\lim_{r \to \infty} r^\lambda \left[\frac{r\phi'(r)}{\phi(r)} + \frac{\alpha}{\sigma} \right] \leq \frac{\alpha}{\sigma} \lim_{r \to \infty} \frac{(N-2)r^{\lambda-1/\sigma}}{(N-\lambda-2)r^{-1/\sigma} + \sigma e^{w(1)}}$$

$$\to 0 \text{ as } r \to \infty, \text{ for } \lambda < 1/\sigma.$$

This proves (4.8.24). An integration of (4.8.24) gives (4.8.22).

The constant $A > 0$ appearing in (4.8.22) depends on the initial condition $\phi(0) = \eta$; that is, $A = A(\eta)$. Define

$$g(r) = \frac{1}{\eta}\phi\left(\frac{r}{\sqrt{\eta}}\right). \tag{4.8.32}$$

Then, $g(r)$ satisfies (4.8.7) and $g(0) = 1$ (see (4.8.8)). By Theorem 4.8.3,

$$\lim_{r\to\infty} r^{\alpha/\sigma}\phi(r) = A(\eta).$$

Therefore,

$$\lim_{r\to\infty}(r/\sqrt{\eta})^{\alpha/\sigma}\phi(r/\sqrt{\eta}) = A(\eta)$$

or

$$\lim_{r\to\infty}\frac{r^{\alpha/\sigma}}{\eta^{-1+\alpha/2\sigma}}\left\{\frac{1}{\eta}\phi(r/\sqrt{\eta})\right\} = A(\eta). \tag{4.8.33}$$

The term in the braces in (4.8.33) is simply $g(r)$. Recalling that $g(0) = 1$, we have

$$\lim_{r\to\infty} r^{\alpha/\sigma}g(r) = A(1) = A(\eta)\eta^{-1+\alpha/2\sigma}.$$

Thus, for $\sigma > 0$, $A > 0$, and $\alpha = 2\sigma - 1$,

$$A(\eta) = A(1)\eta^{1/2\sigma}. \tag{4.8.34}$$

Because $A(\eta)$ assumes all values in $(0,\infty)$ for $\eta \in (0,\infty)$, there exists a unique positive solution of the problem (4.8.7) with the boundary condition

$$\phi'(0) = 0, \quad \lim_{r\to\infty} r^{\alpha/\sigma}\phi = A \tag{4.8.35}$$

for any given $A > 0$. Recasting (4.8.6) as

$$u(x,t) = |x|^{-\alpha/\sigma}(|x|/t^\sigma)^{\alpha/\sigma}\phi(|x|/t^\sigma) \tag{4.8.36}$$

and letting $t \to 0$, we obtain

$$u(x,0) = A|x|^{-\alpha/\sigma}$$

where we have used $(4.8.35)_2$. Thus we have shown the existence of a similarity solution of (4.8.2) with the initial condition

$$u(x,0) = A|x|^p, \quad p = -2 + 1/\sigma, \tag{4.8.37}$$

where $p > -2$ and $A > 0$. Let this similarity solution be denoted by $S_{A,p}$. By Theorem 4.8.3,

$$\lim_{|x|\to\infty} |x|^{-p}S_{A,p}(x,t) = A, \tag{4.8.38}$$

uniformly in t in any compact subset of $(0, \infty)$.

Guo (1996) proved a comparison principle and a global existence theorem for (4.8.2). We state his results without proof.

Theorem 4.8.4 *Let $Q = \mathbb{R}^N \times (0, \infty)$ and $B_R = \{x \in \mathbb{R}^N : |x| < R\}$ and let $\mathscr{L}u = u_t - \Delta(\log u)$. Suppose that the following assumptions hold.*

(i) $u, v \in C^\infty(Q)$, $u > 0$, $v > 0$ for $t > 0$.
(ii) $\mathscr{L}u \leq \mathscr{L}v$ in Q in the classical sense.
(iii) v satisfies the inequality $v(x, t) \geq c|x|^p$ for $|x|$ sufficiently large, uniformly in t in any compact subset of $(0, \infty)$; here c is a positive constant and $p > -2$.
(iv)

$$\lim_{t \downarrow 0} \int_{B_R} (u - v)_+(x, t)dx = 0$$

for any $R > 0$. Here $(u - v)_+ := \max(u - v, 0)$.

Then $u \leq v$ in Q.

Theorem 4.8.5 *Suppose that $u_0(x)$ is continuous and positive and satisfies the condition*

$$\lim_{|x| \to \infty} \inf[|x|^{-p}u_0(x)] > 0 \qquad (4.8.39)$$

for some $p > -2$. Then there exists a solution of the initial value problem (4.8.2)–(4.8.3).

Guo (1996) pointed out that the solution $u(x, t)$ of (4.8.2) and (4.8.3) satisfying (4.8.39) is maximal (see (4.8.5)); moreover, $u(x, t) \in C^0(\mathbb{R}^N \times [0, \infty))$. Guo (1996) then proved that the similarity solution $S_{A,p}$ describes large time behaviour of maximal solutions obtained in Theorem 4.8.5. More precisely, he proved the following theorem.

Theorem 4.8.6 *Let $u_0(x) > 0$ be continuous and*

$$\lim_{|x| \to \infty} |x|^{-p}u_0(x) = A, \qquad (4.8.40)$$

where $A > 0$ and $p > -2$. Further let $u(x, t)$ be the maximal solution of (4.8.2) and (4.8.3) with u_0 as the initial data. Then,

$$\lim_{t \to \infty} t^\alpha |u(x, t) - S_{A,p}(x, t)| = 0,$$

uniformly on sets $\{(x, t) : |x| \leq Ct^\sigma\}$, $\forall C > 0$; here, $\sigma = 1/(p + 2)$ and $\alpha = 2\sigma - 1$.

To prove this theorem, we use the scaling argument (see, for example, Galaktionov and Vázquez (2004), p. 26) and write

$$u_\lambda(x, t) = \lambda^{-p}u(\lambda x, \lambda^{1/\sigma}t), \qquad (4.8.41)$$

where $\lambda > 1$. Let

$$v_0(x) = \begin{cases} c(1+|x|)^p, & p > 0, \\ c|x|^p, & p \le 0, \end{cases} \tag{4.8.42}$$

where c is a constant. Then,

$$u_\lambda(x,0) \le v_0(x), \quad \forall x \in \mathbb{R}^N, \quad \lambda \ge 1,$$

for some constant c. If $v(x,t)$ is the maximal solution of (4.8.2) and (4.8.3) with initial data v_0 as in (4.8.42), then by the comparison principle (see Theorem 4.8.4), we have

$$u_\lambda(x,t) \le v(x,t) \text{ in } Q \equiv \mathbb{R}^N \times (0,\infty), \quad \forall \lambda \ge 1.$$

By the regularity theory for parabolic equations (see Ladyženskaya et al. 1968), $u_\lambda \to S_{A,p}$ uniformly in any compact subset of Q as $\lambda \to \infty$. This, in turn, implies that

$$u_\lambda(y,1) \equiv \lambda^{-p} u(\lambda y, \lambda^{1/\sigma}) \to S_{A,p}(y,1) \equiv \phi(|y|) \tag{4.8.43}$$

as $\lambda \to \infty$ uniformly on the sets $\{y : |y| \le C\}$, $\forall C > 0$. Let $\lambda = t^\sigma$ and $x = \lambda y$. Then, $\lambda \to \infty$ implies $t \to \infty$. Therefore, (4.8.43) yields

$$\lim_{t \to \infty} t^\alpha |u(x,t) - S_{A,p}(x,t)| = 0$$

uniformly on sets $\{(x,t) : |x| \le Ct^\sigma\}$ $\forall C > 0$ recalling that $S_{A,p}(x,t) = t^{-\alpha} \phi(|x|/t^\sigma)$ and $\alpha = -p\sigma$. We have thus proved that the maximal solution of the (singular) nonlinear diffusion equation (4.8.2), subject to (4.8.3), behaves as its similarity solution (4.8.6) as $t \to \infty$. A reference may be made to Guo and Man Sun (1996), Hsu (2002, 2005), and Vázquez (2006) for a related study.

In this section, we have shown that the maximal solution of IVP (4.8.2) and (4.8.3), for large time, behaves as does the similarity solution $S_{A,p}$ satisfying the condition $S_{A,p}|x|^{-p} \to A$ as $|x| \to \infty$, $p > -2$.

4.9 Conclusions

In this chapter, we have shown, with examples, that self-similar solutions describe the large time behaviour of solutions of initial/initial boundary value problems posed for some nonlinear parabolic equations. These partial differential equations include the nonlinear heat conduction equation, porous medium equation, heat equation with absorption, and a fast diffusion equation. We have also presented the study of ordinary differential equations obtained by the similarity reduction of the partial differential equations under study whenever needed. Section 4.1 has presented the introduction to this chapter. Section 4.2 has been concerned with the large time asymptotics for the linear diffusion equation on infinite domains. Following Kloosterziel (1990), we have shown that expansion of the solution of the diffusion equation on infinite/semi-infinite domains in terms of similarity solutions quickly provides the large time asymptotics. Here the initial profile is assumed to be a square-integrable function with respect to the weight function $\exp\left(x^2/2\right)$, $-\infty < x < \infty$. In Section 4.3, we have discussed the self-similar asymptotics for a filtration–absorption equation . This equation models the groundwater flow in a water-absorbing fissurised porous rock. A family of self-similar solutions for the filtration–absorption equation has been constructed and has been shown numerically to be intermediate asymptotic for a class of initial conditions with compact support following Barenblatt et al. (2000). We have also discussed the axisymmetric filtration–absorption equation in higher dimensions. We have constructed a family of self-similar solutions analogous to those in the one-dimensional case following Chertock (2002). Chertock (2002) has shown that these self-similar solutions are the intermediate asymptotics for a class of initial conditions. In Section 4.4, we have presented the existence and uniqueness of weak solutions of nonlinear ordinary differential equations subject to certain boundary conditions at $x = 0$ and $x = \infty$. These ordinary differential equations are obtained by the similarity reduction of the porous medium equation in one dimension. The shooting technique has been used to prove the existence of similarity solutions with appropriate boundary conditions following Gilding and Peletier (1976). In Section 4.6, we have shown that the similarity solutions so obtained in Section 4.4 describe the large time behaviour of the weak solutions of the porous medium equation subject to suitable initial and boundary conditions. This is achieved via the derivation of an integral identity. Section 4.5 has dealt with the large time asymptotic behaviour of the solutions of the nonlinear heat conduction equation on the semi-infinite domain. It was assumed that the coefficient of thermal conductivity depends on the temperature and is a decreasing function of temperature. At infinity, the initial temperature distribution tends to a constant T_0 and at $x = 0$, the temperature is maintained at $T_1 > T_0$. Existence of the similarity solutions of nonlinear heat conduction with appropriate boundary conditions is presented. Finally, we have shown that these similarity solutions describe the large time behaviour of the solutions of the initial boundary value problem for a wide class of initial conditions. This section followed closely the work of Peletier (1970). In Section 4.7, we have discussed the large time behaviour of the solutions of initial value problems for the heat equation with absorption in higher dimensions. The initial profile was assumed to be continuous, nonnegative, bounded, and have asymptotic behaviour $O(|x|^{-\alpha})$ as $|x| \to \infty$. We have informally discussed the asymptotic behaviour of the solutions of the initial value problem stated above for $\alpha < 2/(p-1)$. We have shown, following Herraiz (1999), that similarity solutions of the heat equation with absorption describe the large time behaviour of the solutions of the above initial value problem for $\alpha = 2/(p-1)$ via the construction of a supersolution and subsolution. In Section 4.8, we have discussed the asymptotic behaviour (as $t \to \infty$) of solutions of the initial value problem for a very fast diffusion equation in higher dimensions. This equation describes the limiting density distribution of gases obeying the Boltzman equation. The initial profile is assumed to be nonnegative and locally integrable in $\mathbb{R}^n$. Following Guo (1996), we have shown that the similarity solutions of the very fast diffusion equation describe the large time behaviour of the maximal solutions of the initial value problem posed for the very fast diffusion equation. We have also presented a detailed study of the similarity solutions of the very fast diffusion equation. Thus this chapter brought out quite clearly the importance of self-similar solutions as large time asymptotics for solutions of initial/initial boundary value problems posed for nonlinear partial differential equations.

References

Aronson, D. G. (1969) Regularity properties of flows through porous media, *SIAM J. Appl. Math.* 17, 461–467.

Aronson, D. G. (1970a) Regularity properties of flows through porous media: A counter example, *SIAM J. Appl. Math.* 19, 299–307.

Aronson, D. G. (1970b) Regularity properties of flows through porous media: The interface, *Arch. Rat. Mech. Anal.* 37, 1–10.

Atkinson, F. V., Peletier, L. A. (1971) Similarity profiles of flows through porous media, *Arch. Rat. Mech. Anal.* 42, 369–379.

Bailey, P. B., Shampine, L. F., Waltman, P. E. (1968) *Nonlinear Two Point Boundary Value Problems*, Academic Press, New York.

Barenblatt, G. I. (1952) On some unsteady motions of a liquid and a gas in a porous medium, *Prikl. Math. Mech.* 16, 67–78.

Barenblatt, G. I. (1996) *Scaling, Self-similarity, and Intermediate Asymptotics*, Cambridge University Press, Cambridge, UK.

Barenblatt, G. I., Bertsch, M., Chertock, A. E., Prostokishin, V. M. (2000) Self-similar intermediate asymptotics for a degenerate parabolic filtration-absorption equation, *PNAS* 97, 9844–9848.

Bénilan, Ph., Crandall, M. G. (1981) The continuous dependence on ϕ of solutions of $u_t - \Delta\phi(u) = 0$, *Indiana Univ. Math. J.* 30, 161–177.

Bertsch, M. (1982) Asymptotic behaviour of solutions of a nonlinear diffusion equation, *SIAM J. Appl. Math.* 42, 66–76.

Bertsch, M., Dal Passo, R., Ughi, M. (1992) Nonuniqueness of solutions of a degenerate parabolic equation, *Ann. Mat. Pura Appl.* 161, 57–81.

Bluman, G. W., Kumei, S. (1989) *Symmetries and Differential Equations*, Springer-Verlag, New York.

Brezis, H., Peletier, L. A., Terman, D. (1986) A very singular solution of the heat equation with absorption, *Arch. Rat. Mech. Anal.* 95, 185–209.

Carleman, T. (1957) Problèms Mathématiques dans la Théorie Cinétique des Gaz, Almquists-Wiksells, Uppsala.

Carslaw, H. S., Jaeger, J. C. (1959) *Conduction of Heat in Solids*, Clarendon Press, Oxford.

Cazenave, T., Dickstein, F., Escobedo, M., Weissler, F. B. (2001) Self-similar solutions of a nonlinear heat equation, *J. Math. Sci. Univ. Tokyo* 8, 501–540.

Chertock, A. (2002) On the stability of a class of self-similar solutions to the filtration-absorption equation, *Euro. J. Appl. Math.* 13, 179–194.

Clarkson, P. A., Kruskal, M. D. (1989) New similarity reductions of the Boussinesq equation, *J. Math. Phys.* 30, 2201–2213.

Dal Passo, R., Luckhaus, S. (1987) A degenerate diffusion problem not in divergence form, *J. Differential Eq.* 69, 1–14.

Escobedo, M., Kavian, O. (1987) Asymptotic behaviour of positive solutions of a non-linear heat equation, *Houston J. Math.* 13, 39–50.

Escobedo, M., Kavian, O., Matano, H. (1995) Large time behaviour of solutions of a dissipative semilinear heat equation, *Comm. Part. Differential Eq.* 20,1427–1452.

Esteban, J. R., Rodriguez, A., Vázquez, J. L. (1988) A nonlinear heat equation with singular diffusivity, *Comm. Part. Differential Eq.* 13, 985–1039.

Friedman, A. (1964) *Partial Differential Equations of Parabolic Type*, Prentice-Hall, Englewood Cliffs, NJ.

Galaktionov, V. A., Kurdyumov, S. P., Samarskii, A. A. (1986) On the asymptotic "eigenfunctions" of the Cauchy problem for a nonlinear parabolic equation, *Math. USSR Sb.* 54, 421–455.

Galaktionov, V. A., Vázquez, J. L. (2004) *A Stability Technique for Evolution Partial Differential Equations – A Dynamical Systems Approach*, Birkhäuser, Boston.

Gilding, B. H. (1979) Stabilization of flows through porous media, *SIAM J. Math. Anal.* 10, 237–246.

Gilding, B. H., Peletier, L. A. (1976) On a class of similarity solutions of the porous media equation, *J. Math. Anal. Appl.* 55, 351–364.

Gmira, A., Veron, L. (1984) Large time behaviour of the solutions of a semilinear parabolic equation in $\mathbf{R}^N$, *J. Differential Eq.* 53, 258–276.

Grundy, R. E. (1988) Large time solution of the Cauchy problem for the generalized Burgers equation, Preprint, University of St. Andrews, UK.

Grundy, R. E., Van Duijn, C. J., Dawson, C. N. (1994) Asymptotic profiles with finite mass in one-dimensional contaminant transport through porous media: The fast reaction case, *Quart. J. Mech. Appl. Math.* 47, 69–106.

Guo, J. S. (1995) Similarity solutions for a quasilinear parabolic equation, *J. Austral. Math. Soc. Ser. B* 37, 253–266.

Guo, J. S. (1996) On the Cauchy problem for a very fast diffusion equation, *Comm. Part. Differential Eq.* 21, 1349–1365.

Guo, J. S., Man Sun, I. (1996) Remarks on a singular diffusion equation, *Nonlinear Anal. Theo. Meth. Appl.* 27, 1109–1115.

Hartman, P. (1964) *Ordinary Differential Equations*, John Wiley & Sons, New York.

Herraiz, L. A. (1998) A nonlinear parabolic problem in an exterior domain, *J. Differential Eq.* 142, 371–412.

Herraiz, L. A. (1999) Asymptotic behaviour of solutions of some semilinear parabolic problems, *Ann. Inst. Henri Poincaré* 16, 49–105.

Herrero, M. A. (1989) A limit case in nonlinear diffusion, *Nonlinear Anal. Theo. Meth. Appl.* 13, 611–628.

Herrero, M. A. (1991) Singular diffusion on the line, *Preprint*.

Higgins, J. R. (1977) *Completeness and Basic Properties of Sets of Special Functions*, Cambridge University Press, Cambridge, UK.

Hsu, S. Y. (2002) Asymptotic profile of solutions of a singular diffusion equation as $t \to \infty$, *Nonlinear Anal.* 48, 781–790.

Hsu, S. Y. (2005) Large time behaviour of solutions of a singular diffusion equation in $\mathbf{R}^n$, *Nonlinear Anal.* 62, 195–206.

Kamenomostskaya, S. (Kamin) (1973) The aymptotic behaviour of the solution of the filtration equation, *Israel J. Math.* 14, 76–87.

Kamin, S., Peletier, L. A. (1985) Large time behaviour of solutions of the heat equation with absorption, *Ann. Scuola Norm. Sup. Pisa Cl. Sci.* 12, 393–408.

Kamin, S., Vázquez, J. L. (1991) Asymptotic behaviour of solutions of the porous medium equation with changing sign, *SIAM J. Math. Anal.* 22, 34–45.

King, J. R. (1993) "Instantaneous source" solutions to a singular nonlinear diffusion equation, *J. Engrg. Math.* 27, 31–72.

Kloosterziel, R. C. (1990) On the large-time asymptotics of the diffusion equation on infinite domains, *J. Engrg. Math.* 24, 213–236.

Krzyżański, M. (1959) Certaines inégalités relatives aux solutions de l'équation parabolique linéaire normale, *Bull. Acad. Pol. Sci. Sér. Math. Astr. Phys.* 7, 131–135.

Kurtz, T. G. (1973) Convergence of sequences of semigroups of nonlinear operators with an application to gas kinetics, *Trans. Amer. Math. Soc.* 186, 259–272.

Kwak, M. (1998) A semilinear heat equation with singular initial data, *Proc. Roy. Soc. Edin.* 128A, 745–758.

Ladyženskaya, O. A., Solonnikov, V. A., Ural'ceva, N. N. (1968) *Linear and Quasi-linear Equations of Parabolic Type*, Amer. Math. Soc., Providence, RI.

Muskat, M. (1937) *The Flow of Homogeneous Fluids through Porous Media*, Mc-Graw–Hill, New York.

Okuda, H., Dawson, J. M. (1973) Theory and numerical simulation on plasma diffusion across a magnetic field, *Phys. Fluids* 16, 408–426.

Oleinik, O. A., Kalashnikov, A. S., Yui-Lin, C. (1958) The Cauchy problem and boundary problems for equations of the type of unsteady filtration, *Izv. Akad. Nauk SSSR Ser. Mat.* 22, 667–704.

Oleinik, O. A., Kruzhkov, S. N. (1961) Quasi-linear second-order parabolic equations with many independent variables, *Russian Math. Surveys* 16, 105–146.

Olver, P. J. (1986) *Applications of Lie Groups to Differential Equations*, Springer-Verlag, New York.

Peletier, L. A. (1970) Asymptotic behaviour of temperature profiles of a class of nonlinear heat conduction problems, *Quart. J. Mech. Appl. Math.* 23, 441–447.

Peletier, L. A. (1971) Asymptotic behaviour of solutions of the porous media equation, *SIAM J. Appl. Math.* 21, 542–551.

Sachdev, P. L. (1987) *Nonlinear Diffusive Waves*, Cambridge University Press, Cambridge, UK.

Sachdev, P. L. (2000) *Self-Similarity and Beyond. Exact Solutions of Nonlinear Problems*, Chapman & Hall/CRC Press, New York.

Schlichting, H. (1960) *Boundary Layer Theory*, McGraw-Hill, New York.

Serrin, J. (1967) Asymptotic behaviour of velocity profiles in the Prandtl boundary layer theory, *Proc. Roy. Soc. London Ser. A* 299, 491–507.

Shampine, L. F. (1973) Concentration dependent diffusion, *Quart. Appl. Math.* 30, 441–452.

Ughi, M. (1986) A degenerate parabolic equation modeling the spread of epidemic, *Ann. Mat. Pura. Appl.* 143, 385–400.

Van Duijn, C. J., Gomes, S. M., Zhang, H. (1988) On a class of similarity solutions of the equation $u_t = (|u|^{m-1}u_x)_x$ with $m > -1$, *IMA J. Appl. Math.* 41, 147–163.

Van Duyn, C. J., Peletier, L. A. (1977a) A class of similarity solutions of the nonlinear diffusion equation, *Nonlinear Anal. Theor. Meth. Appl.*, 1, 223–233.

Van Duyn, C. J., Peletier, L. A. (1977b) Asymptotic behaviour of solutions of a nonlinear diffusion equation, *Arch. Rat. Mech. Anal.*, 65, 363–377.

Vázquez, J. L. (1992) Nonexistence of solutions for nonlinear heat equations of fast-diffusion-type, *J. Math. Pures. Appl.* 71, 503–526.

Vázquez, J. L. (2003) Asymptotic behaviour for the porous medium equation posed in the whole space, *J. Evol. Eq.* 3, 67–118.

Vázquez, J. L. (2006) *Smoothing and Decay Estimates for Nonlinear Diffusion Equations*, Oxford University Press, Oxford.

Vázquez, J. L. (2007) *The Porous Medium Equation Mathematical Theory*, Clarendon Press, Oxford.

Widder, D. V. (1975) *The Heat Equation*, Academic Press, New York.

Zhang, H. (1993) Large time behaviour of the maximal solution of $u_t = (u^{m-1}u_x)_x$ with $-1 < m \leq 0$, *Differential and Integr. Eq.* 6, 613–626.

Chapter 5
Asymptotics in Fluid Mechanics

5.1 Introduction

As we pointed out in some detail in Chapter 1, the concept of intermediate asymptotics first arose in two physical contexts: propagation in a certain spatial region of a gene, whose carriers have an advantage in the struggle for existence, and fluid mechanics, in particular, explosion and implosion phenomena involving shocks. In Chapters 3 and 4, we studied the asymptotic behaviour of solutions of nonlinear PDEs of parabolic type, which include those describing gene propagation. In the present chapter, we discuss some physical problems which arise from fluid mechanics and which are governed by hyperbolic or parabolic systems of equations. These systems admit similarity solutions of the first or second kind. The latter, in general, enjoy intermediate asymptotic character in some parametric regimes. We recall that the self-similar solutions of the first kind are fully determined by the dimensional considerations and require the solution of the resulting nonlinear ODEs with appropriate conditions at the shock and the centre of the explosion in this context, say, whereas those of the second kind involve solution of an eigenvalue problem in the reduced phase plane, the implosion problem, for example.

In Section 5.2, we study the propagation of a strong shock produced by a large explosion into a medium for which the density varies according to the power law $\rho_0(r) = kr^{-\omega}$, where r is the distance measured from the centre of the explosion and $k > 0$ and $\omega > 0$ are constants. It is first shown that the self-similar solutions describing this phenomenon change their character from one of the first kind for $\omega < 3$ to that of the second kind for $\omega > 3$. The intermediate asymptotic character of the solutions for $\omega_g(\gamma) < \omega < \omega_c(\gamma)$ is brought out by comparison with the numerical solution of the original system of nonlinear PDEs with appropriate initial conditions; here, $\omega_g(\gamma)$ and $\omega_c(\gamma)$ depend on the ratio of specific heats, $\gamma = C_p/C_v$. Section 5.3 deals with self-similar solutions of the second kind which describe a collapsing spherical cavity. Here, we show by reference to the basic work of Hunter (1960) that the numerical solution of the governing system of nonlinear PDEs with appropriate initial/boundary conditions tends, for different sets of γ, to the relevant

P.L. Sachdev, Ch. Srinivasa Rao, *Large Time Asymptotics for Solutions of Nonlinear Partial Differential Equations*, Springer Monographs in Mathematics, DOI 10.1007/978-0-387-87809-6_5, © Springer Science+Business Media, LLC 2010

self-similar solution of the second kind as the radius of the cavity tends to zero. More recent work on this problem is also summarised. Section 5.4 concerns large time behaviour of compressible flow equations with damping. It is shown, by following the work of Liu (1996), that the solutions of this hyperbolic system of equations tend, for large time, to those of a nonlinear parabolic equation; this is brought out by referring to a special class of solutions of each of these systems. This study justifies, in a limited sense, the so-called Darcy's law which is used to describe compressible flow in a porous medium. Sections 5.5 and 5.6 deal with the systems of nonlinear PDEs holding in the Prandtl boundary layer. In the former section, we take up the study of unsteady boundary layer equations governing the flow in an incompressible medium and rigorously show, following the work of Oleinik (1966a), that, under a certain set of conditions, the solutions of an unsteady system of equations tend to those of the corresponding steady equations as time becomes large. In the latter section, we deal with the basic work of Serrin (1967) which proves that the similarity solution of the steady boundary layer equations governed by the Falkner–Skan equation, a nonlinear ODE of third order, subject to relevant boundary conditions at 0 and ∞, describes the asymptotic behaviour of solutions of (steady) boundary layer equations with appropriate initial/boundary conditions on the spatial domain $0 < x < \infty$. This is accomplished in a rigorous analytical manner. We may mention that the present problem is parabolic in character.

5.2 Strong explosion in a power law density medium – Self-similar solutions of first and second kind

One of the most important examples of self-similar solutions of the first kind in fluid mechanics relates to a point source explosion into a uniform medium which was analysed by Taylor (1950), Sedov (1946), and von Neumann (1947); see Sachdev (2004) for a detailed account. This explosion results from a sudden release of a large amount of energy in a small volume. We consider here the case of a spherical explosion; it is headed by a strong shock which propagates into a medium with uniform density and zero pressure. This problem was solved by the exact similarity reduction of the governing system of nonlinear PDEs to ODEs and the solution of the latter subject to Rankine–Hugoniot conditions at the shock and zero particle velocity at the centre of the explosion. It was dealt with in Eulerian coordinates by Taylor (1950) and Sedov (1946) and Lagrangian coordinates by von Neumann (1947).

Here we consider a more general problem: a strong shock produced by a large explosion propagates into a medium for which the density varies according to power law

$$\rho_0(r) = Kr^{-\omega}, \tag{5.2.1}$$

where r is the distance measured from the centre of the explosion and $K > 0$, $\omega > 0$ are constants. The medium ahead is assumed to be an ideal gas with zero pressure.

This is a curious example where the self-similar solution changes its character from one of the first kind for $\omega < 3$ to the (so-called) second kind for $\omega > 3$ (see Chapter 1). The case $\omega < 3$, which is a straightforward generalisation of the Taylor–Sedov solution, was first treated by Korobeinikov and Riazanov (1959). They reduced the system of nonlinear PDEs governing the inviscid gas dynamic equations in spherical, cylindrical, and plane symmetries to a system of nonlinear ODEs by assuming that pressure, density, and particle velocity behind the shock may be written in the form $v = v_2 f(\lambda), \rho = \rho_2 g(\lambda), p = p_2 h(\lambda), \lambda = r/r_2$, where the subscript '2' denotes conditions immediately behind the shock, $r_2 = r_2(t)$. The nonlinear ODEs for f, g, and h were solved subject to the conditions $f(1) = g(1) = h(1) = 1$ at the shock and the particle velocity $f(0) = 0$ at the centre; the latter arises from the spherical symmetry of the problem. Indeed the system of ODEs with the above conditions was solved in a closed form. Korobeinikov and Riazanov (1959) considered only those parameters for which the solution could be extended to the centre of symmetry. The solution, however, exhibited singularities when the density exponent ω in (5.2.1) assumes values $\omega = \omega_1, \omega_2, \omega_3$, where $\omega_1 = (7 - \gamma)/(\gamma + 1), \omega_2 = (2\gamma + 1)/\gamma$, and $\omega_3 = 3(2 - \gamma)$ for the spherically symmetric case that we consider here. For these singular cases, Korobeinikov and Riazanov (1959) found limiting behaviour of the solutions by solving the governing system of ODEs directly.

Unaware of the above work, Waxman and Shvarts (1993) considered this problem in a different fashion. They did not directly impose the symmetry condition $u(0, t) = 0$ at the centre of explosion. Instead they enquired for what values of the parameter ω the similarity solution was a straightforward generalisation of the Taylor–Sedov solution. They arrived at a very interesting conclusion, namely, that Taylor–Sedov type solutions exist only for the case $\omega < 3$. The value $\omega = 3$ is exactly the point where the singularities in the solution of Korobeinikov and Riazanov (1959) appear if we assume that $\gamma > 1$.

Waxman and Shvarts (1993) showed that self-similar solutions of the first kind fail to describe the asymptotic behaviour as $t \to \infty$ for $3 \le \omega < 5$. They discovered new solutions belonging to the so-called second kind for $3 < \omega < 5$ and for $\omega \ge 5$. These solutions are distinct from the Taylor–Sedov type because they describe flows with accelerating shocks. This is in contrast to Taylor–Sedov solutions for $\omega < 3$, which are headed by decelerating shocks. The new class of solutions was found in the manner of the solutions for the converging shocks for which the shock exponent is found, not from dimensional considerations alone, but by requiring that the solution, for a given γ, starting from the shock, passes through an appropriate singular point of the reduced ODE in the sound speed square-particle velocity plane, the so-called Guderley map. In the present case, the solution must pass through a 'new' singular point in the Guderley map (see Sachdev 2004).

Now we follow the work of Waxman and Shvarts (1993). The main purpose here is to identify self-similar solutions of the second kind for $\omega > 3$, which describe the limiting behaviour which is approached asymptotically for $t \to \infty$ by flows that are initially non-self-similar. A large class of problems, which differ in boundary and initial conditions, tends in the above limit to the same asymptotic self-similar behaviour of the second kind.

The spherically symmetric flows with shocks are governed by

$$\frac{\partial}{\partial t}\ln\rho + u\frac{\partial}{\partial r}\ln\rho + \frac{\partial u}{\partial r} + 2\frac{u}{r} = 0, \tag{5.2.2}$$

$$\frac{\partial u}{\partial t} + u\frac{\partial u}{\partial r} + \frac{c^2}{\gamma}\frac{\partial}{\partial r}\ln\rho + \frac{1}{\gamma}\frac{\partial c^2}{\partial r} = 0, \tag{5.2.3}$$

$$\frac{\partial}{\partial t}\ln(c^2\rho^{1-\gamma}) + u\frac{\partial}{\partial r}\ln(c^2\rho^{1-\gamma}) = 0, \tag{5.2.4}$$

where u, ρ, and c are particle velocity, density, and sound speed, respectively. We assume that, at later time, the solution of the system (5.2.2)–(5.2.4) does not depend upon constants with the dimension of length or time deriving from the boundary conditions; the flow here must depend on only two-dimensional parameters. The typical length scale is the radius of the shock given by

$$R(t) = At^{\alpha}, \tag{5.2.5}$$

where A and α are constants. The flow variables may now be expressed in the self-similar form

$$u(r,t) = \dot{R}\xi U(\xi), \quad c(r,t) = \dot{R}\xi C(\xi),$$
$$\rho(r,t) = Bt^{\beta}G(\xi), \quad \xi = \frac{r}{R(t)}. \tag{5.2.6}$$

Here, ξ is the similarity variable. For sufficiently late times the length scale of the flow is described by $R = R(t)$.

In the Taylor–Sedov form of the solution, the early flow is described by two-dimensional parameters, the parameter K in the density law (5.2.1) and the energy of the flow which is assumed to be equal to the explosion energy E. By simple dimensional considerations we may find the dimensional constants A and B and the parameters α and β in (5.2.5) and (5.2.6) as

$$A = \varphi(\gamma, \omega)\left(\frac{E}{K}\right)^{\alpha/2}, \quad \alpha = \frac{2}{5-\omega}$$
$$B = KA^{-\omega}, \quad \beta = -\alpha\omega, \tag{5.2.7}$$

where φ is a dimensionless function of the (dimensionless) parameters γ and ω, which is determined from the constancy of the explosion energy behind the shock. Korobeinikov and Riazanov (1959) considered the case $\omega \leq 3$.

Waxman and Shvarts (1993) showed that the self-similar solutions of Taylor–Sedov type discussed above cease to hold for large R for two reasons: (i) the energy of explosion in this case tends to infinity and therefore cannot be the second dimensional parameter, and (ii) for $\omega \geq 3$ there must exist a region around the origin with a non-self-similar behaviour. For large time, when $R(t)$ diverges, the initial length and time scales which arise from the initial conditions do not characterise the physical process; these scales characterise flows at early times.

The solution in the outer region is in fact described by a self-similar solution of the second kind, as we show presently. The substitution of (5.2.5) and (5.2.6) into (5.2.2)–(5.2.4) reduces the latter to the system of ODEs,

$$\xi(U-1)G' + \xi GU' = -\left(\frac{\beta + 3\alpha U}{\alpha}\right)G, \qquad (5.2.8)$$

$$\xi G(U-1)U' + \frac{C^2}{\gamma}\xi G' + \frac{2}{\gamma}\xi CGC' = \left[\frac{1}{\alpha} - U\right]UG - \frac{2}{\gamma}C^2 G, \quad (5.2.9)$$

$$2\alpha\xi(U-1)GC' + \alpha(1-\gamma)\xi(U-1)CG' = [\beta(\gamma-1) + 2 - 2\alpha U]CG, \qquad (5.2.10)$$

which, in the (U,C) plane, becomes

$$\frac{dU}{dC} = \frac{\Delta_1(U,C)}{\Delta_2(U,C)}; \qquad (5.2.11)$$

U and C are related to ξ by

$$\frac{d\log\xi}{dU} = \frac{\Delta(U,C)}{\Delta_1(U,C)} \qquad (5.2.12)$$

and

$$\frac{d\log\xi}{dC} = \frac{\Delta(U,C)}{\Delta_2(U,C)}. \qquad (5.2.13)$$

Here,

$$\Delta = C^2 - (1-U)^2,$$

$$\Delta_1 = U(1-U)\left(1-U-\frac{\alpha-1}{\alpha}\right) - C^2\left(3U - \frac{\omega - 2\left[\frac{\alpha-1}{\alpha}\right]}{\gamma}\right),$$

$$\Delta_2 = C\left[(1-U)\left(1-U-\frac{\alpha-1}{\alpha}\right) - \frac{\gamma-1}{2}U\left(2(1-U) + \frac{\alpha-1}{\alpha}\right)\right.$$

$$\left. -C^2 + \frac{(\gamma-1)\omega + 2\left[\frac{\alpha-1}{\alpha}\right]}{2\gamma}\frac{C^2}{1-U}\right]. \qquad (5.2.14)$$

The function G in (5.2.6) may then be obtained from a quadrature of the system (5.2.8)–(5.2.10), namely,

$$C^{-2}(1-U)^\lambda G^{\gamma-1+\lambda}\xi^{3\lambda-2} = \text{constant}, \qquad (5.2.15)$$

where

$$\lambda = \frac{(\gamma-1)\omega+2\left[\frac{(\alpha-1)}{\alpha}\right]}{3-\omega}. \qquad (5.2.16)$$

It is customary to first examine the equation (5.2.11) in the (U,C)-plane to identify the required integral and then use (5.2.12) and (5.2.13) to relate U and C to the variable ξ. The strong shock conditions

$$u = \frac{2}{\gamma+1}\dot{R}, \quad \rho = \frac{\gamma+1}{\gamma-1}\rho_0, \quad p = \frac{2}{\gamma+1}\rho_0\dot{R}^2, \qquad (5.2.17)$$

in view of (5.2.6), become

$$U(1) = \frac{2}{\gamma+1}, \quad C(1) = \frac{\sqrt{2\gamma(\gamma-1)}}{\gamma+1}, \quad G(1) = \frac{\gamma+1}{\gamma-1}. \qquad (5.2.18)$$

For the Taylor–Sedov type of solution, the energy E_1 contained in the region $\xi_1 \le \xi \le 1$ corresponding to $[\xi_1 R(t) \le r \le R(t)]$ is given by

$$E_1 = \int_{\xi_1 R}^{R} dr 4\pi r^2 \rho \left\{\frac{1}{2}u^2 + \frac{1}{\gamma(\gamma-1)}c^2\right\}$$

$$= 4\pi K R^{3-\omega}\dot{R}^2 \int_{\xi_1}^{1} d\xi \, \xi^4 G \left(\frac{1}{2}U^2 + \frac{1}{\gamma(\gamma-1)}C^2\right) \qquad (5.2.19)$$

which, in view of (5.2.5)–(5.2.7), becomes

$$E_1 = \left[4\pi\left(\frac{2}{5-\omega}\right)^2 \varphi^{5-\omega} \int_{\xi_1}^{1} d\xi \, \xi^4 G \left(\frac{1}{2}U^2 + \frac{1}{\gamma(\gamma-1)}C^2\right)\right] E. \qquad (5.2.20)$$

In this case, E_1 is independent of time. In view of this constancy, the self-similar contour in the (U,C)-plane may be obtained by the energy–work done principle, namely, the work done during the interval dt by a fluid element which is at $\xi = \xi_1$ at time t on the fluid that lies in $\xi > \xi_1$ at time t equals the energy that leaves the region $\xi_1 \le \xi \le 1$ during the same time interval via the energy flux through the surface $\xi_1 = $ constant. Thus, the work done by a fluid element that lies at ξ_1 at the time t on the fluid that occupies the region $r > \xi_1 R(t)$ at that time during the interval dt is

$$4\pi r_1^2 u(r_1,t)dt\gamma^{-1}\rho(r_1,t)c^2(r_1,t) = 4\pi\gamma^{-1}KR^{2-\omega}\dot{R}^3\xi_1^5 U(\xi_1)G(\xi_1)C^2(\xi_1)dt. \qquad (5.2.21)$$

The corresponding energy that leaves the region $\xi > \xi_1$ during the time dt due to the energy flux across the surface $\xi_1 = $ constant is given by

$$4\pi r_1^2 \left[\xi_1\dot{R} - u(r_1,t)\right] dt \left(\frac{1}{2}\rho(r_1,t)u^2(r_1,t) + \frac{1}{\gamma(\gamma-1)}c^2(r_1,t)\right)$$

$$= 4\pi\gamma^{-1} KR^{2-\omega}\dot{R}^3 \xi_1^5 \left[1 - U(\xi_1)\right] G(\xi_1) \left(\frac{\gamma}{2} U^2(\xi_1) + \frac{C^2(\xi_1)}{\gamma - 1}\right) dt. \quad (5.2.22)$$

Equating (5.2.21) and (5.2.22), we arrive at the integral

$$C^2 = \frac{\gamma(\gamma - 1)}{2} \frac{U^2(1 - U)}{\gamma U - 1}. \quad (5.2.23)$$

In view of (5.2.23), equation (5.2.12) becomes

$$\frac{d \log \xi}{dU} = (\gamma + 1) \frac{\gamma U^2 - 2U + \dfrac{2}{\gamma + 1}}{U(\gamma U - 1)\left[5 - \omega - (3\gamma - 1)U\right]} \quad (5.2.24)$$

which involves only U on the RHS.

Equation (5.2.24) was analysed by Waxman and Shvarts (1993) in the appendix to their paper. It was found that, for $\omega < (7 - \gamma)/(\gamma + 1)$, U tends to $1/\gamma$ and C tends to infinity as $\xi \to 0$. On the other hand, for $\omega > (7 - \gamma)/(\gamma + 1)$, a vacuum is formed in $0 < \xi < \xi_{in}$ where $\xi = \xi_{in} > 0$ is the outer boundary of this region. The self-similar solution holds in $\xi_{in} \le \xi \le 1$ and is separated from the evacuated region $\xi < \xi_{in}$ by a particle path. We observe that the value $(7 - \gamma)/(\gamma + 1)$ of ω coincides with the singularity $\omega = \omega_1$ of Korobeinikov and Riazanov (1959) for spherical symmetry.

The case $\omega > (7 - \gamma)/(\gamma + 1)$ includes $\omega \ge 3$ for $\gamma \ge 1$. Here, the solution curve approaches the point $(C = 0, U = 1)$ as ξ tends to ξ_{in}. A simple local analysis of (5.2.15), (5.2.23), and (5.2.24) shows that

$$U(\xi) \approx 1 - \frac{3\gamma + \omega - 6}{\gamma} \log\left(\frac{\xi}{\xi_{in}}\right),$$

$$C(\xi) \approx \left[\frac{3\gamma + \omega - 6}{2} \log\left(\frac{\xi}{\xi_{in}}\right)\right]^{1/2}, \quad (5.2.25)$$

$$G(\xi) \approx \text{constant} \times \left[\log\left(\frac{\xi}{\xi_{in}}\right)\right]^{-(\gamma\omega + \omega - 6)/(3\gamma + \omega - 6)}$$

for $\xi \gtrsim \xi_{in}$. Insertion of (5.2.25) in the expression (5.2.20) for the energy E_1 in the region $\xi_1 \le \xi \le 1$ corresponding to $\xi_1 R(t) \le r \le R(t)$ shows that, for $\omega \ge 3$, $E_1 \to \infty$ as $\xi_1 \to \xi_{in}$. The (constant) energy of the blast is therefore contained in a region bounded by some point $\xi_1 = \xi_*$ and the shock. It follows that the flow resulting from the finite energy of the blast is described by a Taylor–Sedov solution in a smaller region $\xi_* \le \xi_0 \le \xi \le 1$ and is different from the self-similar solution in the intermediate layer $\xi_{in} \le \xi \le \xi_0$. This implies that some initial length and time scales influence the flow behaviour over the region $O(R)$ as $R \to \infty$, contradicting its self-similar nature. It is this fact and not the infinite energy of the blast for $\omega > 3$ alone which makes the self-similar solution of Taylor–Sedov type untenable. In any case, for $\omega > 3$, (infinite) energy of the explosion is not the relevant second parameter.

For $\omega > 3$, Waxman and Shvarts (1993) constructed the asymptotic similarity solution as follows. They considered two flow regions: the outer region lying in $r_1(t) < r(t) < R(t)$ and the inner one lying between $r = r_1(t)$ and $r = 0$. They assumed that the flow in the outer region is independent of that in the inner region and is determined for $t > t_0$ entirely by the initial flow conditions in the former at some time $t = t_0$. They also assumed that $r_1(t)/R(t)$ tends to zero as $R \to \infty$; furthermore, they surmised that the flow behaviour in the inner region does not affect the description of the flow over the scales of order R. Thus, the flow in the large outer region is assumed to be independent of that in the much smaller inner region. A self-similar solution is constructed for the outer region, which is not affected by the initial length and time scales; the latter influence only the smaller inner region.

It was shown that the inner region is bounded by a C_+ characteristic which starts from $r = r_1, t = t_0$ (Zel'dovich and Raizer 1967). The solution in the outer region is constructed in the manner self-similar solutions of the second kind are analysed (Zel'dovich and Raizer 1967). A physical solution starting from the shock must cross the sonic line $\triangle = 0$ at a singular point of (5.2.11) where $\triangle_1 = \triangle_2 = 0$, otherwise equations (5.2.12) and (5.2.13) would imply that one of the functions $U(\xi)$ and $C(\xi)$ is not single-valued. Thus the appropriate value of α for a given γ is found by starting the integration from the shock point $(U(1), C(1))$ and continuing until the integral curve crosses the sonic line $\triangle = 0$ at a singular point of (5.2.11). It is now shown that such an integral curve exists for a certain range of ω values greater than 3.

Fortunately, it becomes possible to find an explicit self-similar solution of (5.2.12) and (5.2.13) for a particular value of $\omega = \omega_a \equiv 2(4\gamma - 1)/(\gamma + 1)$, satisfying the shock conditions $[U(1), C(1)]$ and passing through the singular point $U = 2/(\gamma + 1)$, $C = (\gamma - 1)/(\gamma + 1)$. Because for $\gamma > 1, \omega_a > 3$, the exponent $\alpha = \alpha_a(\gamma)$ for this case is simply $(\gamma + 1)/2$. The explicit form of the solution for this special case is found to be

$$U(\xi) = U(1) = \frac{2}{\gamma + 1}$$

$$C(\xi) = C(1)\xi^3 = \frac{\sqrt{2\gamma(\gamma - 1)}}{\gamma + 1}\xi^3, \qquad (5.2.26)$$

$$G(\xi) = \frac{\gamma + 1}{\gamma - 1}\xi^{-8}$$

and motivates the solution for other values of ω. It also provides a valuable check on the veracity of the numerical solution for large time. We may observe that the exponent $\alpha_a = (\gamma + 1)/2$ differs considerably from the corresponding Taylor–Sedov value $2(\gamma + 1)/(7 - 3\gamma)$.

Before considering the solution for other $\omega > 3$, a numerical solution of the full flow equations (5.2.2)–(5.2.4) was found by Waxman and Shvarts (1993) for $\omega = \omega_a$, using the methods of artificial viscosity (see Sachdev 2004). The initial conditions for this particular example were zero particle velocity everywhere, constant density and pressure at the time of energy release in $r < d$, zero pressure, and density proportional to $r^{-\omega}$ for $r > d$. The results for $\omega_a = 4.25$ corresponding to $\gamma = 5/3$ from the exact analytic solution, the numerical solution, and from the Taylor–Sedov form of the solution are shown in Figure 5.1. Here, $(t/\tau)^{\alpha}/(R/d)$ is plotted against R/d for large time with $\tau = (Kd^{5-\omega}/E)^{1/2}$. It is easily seen that this quantity approaches a constant value for the explicit solution (5.2.26) with $\alpha = 4/3$ (solid line). This is in contrast to the Taylor–Sedov values for $\alpha = 8/3$ (dashed line) which continue to diverge.

The solution $C = C(U)$ for $\gamma = 5/3$ and $\omega_a = 4.25$ according to (5.2.26) (solid line), numerical solution (dashed line), and the Taylor–Sedov type solution (dashed dot line) are shown in Figure 5.2. $U(\xi)$ and $C(\xi)$ were found from (5.2.5) and (5.2.6) by first computing $R(t)$ and $dR(t)/dt$ from the numerical results. It is clear from Figure 5.2 that the numerical solution for large times agrees very well with the exact asymptotic solution given by (5.2.26). The dotted line in the figure is $U + C = 1$. There is some departure of the analytic solution from the numerical solution when it is below this line (see Waxman and Shvarts 1993).

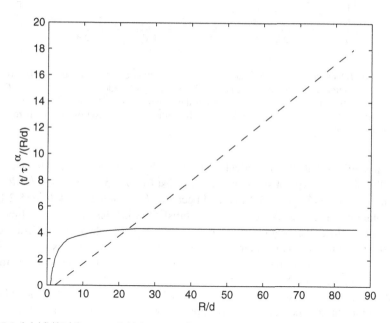

Fig. 5.1 $(t/\tau)^{\alpha}/(R/d)$ versus R/d for $\alpha = 4/3$ (—, Waxman and Shvarts 1993), and for $\alpha = 8/3$ ($---$, α_{TS}). (Waxman and Shvarts 1993. Copyright ©1993 American Institute of Physics. Reprinted with permission. All rights reserved.)

The exact solution (5.2.26) shows that it approaches the point $U = 2/(\gamma+1), C = 0$ as $\xi \to 0$. This is a singular point of (5.2.11). It is a special case of the singular point $P(U = 1/\alpha, C = 0)$ which exists for all $\omega > 0$ and α. This is the point where the self-similar solution starting from the shock (5.2.18) crosses the sonic line; it is approached in the limit $\xi \to 0$. Waxman and Shvarts (1993) also studied (C, U) curves for $\gamma = 5/3$, $\omega = 3.4$ and $\gamma = 5/3$, $\omega = 5.5$. The agreement of the self-

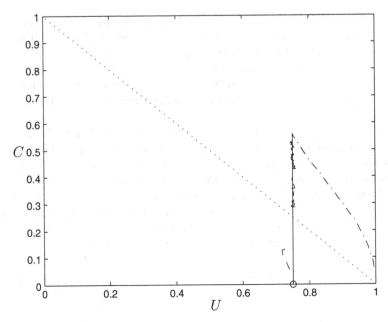

Fig. 5.2 (C,U) curves for $\gamma = 5/3$ and $\omega = 4.25$: Waxman and Shvarts self-similar solution (5.2.26) (—), numerical solution $(---)$ for $R/d = 86.6$, Taylor–Sedov solution $(-.-.-)$. *Dotted line* represents $U + C = 1$ and the circle denotes the singular point $(U = 1/\alpha, C = 0)$. (Waxman and Shvarts 1993. Copyright © 1993 American Institute of Physics. Reprinted with permission. All rights reserved.)

similar solution with the numerical solution for large times was clearly observed. The Taylor–Sedov type of solution does not exist for $\omega = 5.5$. Because $dU/dC = 0$ when $U = 1$ (see (5.2.11)–(5.2.14)) and because $U < 1$ at the shock (see (5.2.18)), the solution being considered here must satisfy the inequality $U < 1$ behind the shock. Therefore, the singular point $P(U = 1/\alpha, C = 0)$ where the integral curve crosses the sonic line is of interest only when $\alpha > 1$. Thus for this class of self-similar solutions the similarity exponent α is greater than 1; this is in contrast to the self-similar solutions of the Taylor–Sedov type for $\omega < 3$ for which $\alpha < 1$.

It is of some interest to write the present self-similar solution in the neighbourhood of the singular point P. We may approximate $\triangle$, $\triangle_1$, and $\triangle_2$ near P as

$$\triangle = -\left(1 - \frac{1}{\alpha}\right)^2,$$

$$\triangle_1 = -\frac{1}{\alpha}\left(1 - \frac{1}{\alpha}\right)\left(U - \frac{1}{\alpha}\right) + \left(\frac{\omega - 2\left[\dfrac{\alpha - 1}{\alpha}\right]}{\gamma} - \frac{3}{\alpha}\right)C^2, \quad (5.2.27)$$

$$\triangle_2 = -\frac{3}{\alpha}\left(1 - \frac{1}{\alpha}\right)\frac{\gamma - 1}{2}C.$$

We may, therefore, write an approximate solution of (5.2.11) near P as

$$U = \frac{1}{\alpha} + \begin{cases} \text{constant } C^{2/3(\gamma-1)}, & \gamma > 4/3, \\ f_1(\gamma, \omega, \alpha)C^2 \log C, & \gamma = 4/3, \\ f_2(\gamma, \omega, \alpha)C^2, & \gamma < 4/3, \end{cases} \qquad (5.2.28)$$

where

$$f_1 = \left(\frac{3}{\alpha} - \frac{\omega - 2[(\alpha-1)/\alpha]}{\gamma} \right) \frac{2\alpha}{\alpha-1},$$

$$f_2 = \left(\frac{3}{\alpha} - \frac{\omega - 2[(\alpha-1)/\alpha]}{\gamma} \right) \frac{\alpha^2}{(\alpha-1)(3\gamma-4)}. \qquad (5.2.29)$$

The approximate form of the functions $C(\xi)$ and $G(\xi)$ may hence be obtained from (5.2.13) and (5.2.15) as

$$C(\xi) = \text{constant} \times \xi^{3(\gamma-1)/2(\alpha-1)}, \qquad (5.2.30)$$

$$G(\xi) = \text{constant} \times \xi^{-(\alpha\omega-3)/(\alpha-1)}. \qquad (5.2.31)$$

A detailed numerical evaluation of the exponent $\alpha = \alpha(\omega)$ shows that α tends to unity as $\omega \downarrow \omega_g$ for some $\omega_g > 3$ ($\omega_g = 3.256$ for $\gamma = 5/3$); α tends to infinity as $\omega \uparrow \omega_c$ for some finite ω_c ($\omega_c = 7.686$ for $\gamma = 5/3$). For $3 \le \omega \le \omega_g(\gamma)$, there is no α for which the integral curve in the (U, C)-plane crosses the sonic line at the singular point P.

To summarise, the self-similar analysis of flows arising from a strong explosion into a nonuniform power law medium reveals some very fascinating features. These solutions change their character from the first kind to those of the second kind as the density exponent ω crosses the value 3. The asymptotic character of the solutions in the ranges $3 \le \omega \le \omega_g(\gamma)$ and $\omega \ge \omega_c(\gamma)$ needs further investigation.

5.3 Self-similar solutions for collapsing cavities

The generation and propagation of converging shock waves were first treated by Guderley (1942). Guderley's self-similar solutions are self-similar solutions of the second kind and are descriptors of converging cylindrical shocks close to the axis of symmetry. This work was subsequently discussed by several authors (see Zel'dovich and Raizer 1967, Whitham 1974). In particular, their analytical asymptotic behaviour in the neighbourhood of the axis was elegantly treated by Van Dyke and Guttmann (1982). We have earlier discussed converging shocks in some detail (Sachdev 2004). Here we deal with a related problem, the collapse of an empty spherical or cylindrical cavity in water and air and the asymptotic behaviour of its solution near the centre (axis) of collapse. First we discuss briefly the collapse of an empty spherical cavity in water because the analysis here is quite close to that detailed in Section 5.2. Later, we study the same problem when the medium is air; the analytical results for the latter are quite distinct from those for water.

The major work concerning cavity collapse in water is due to Hunter (1960). Here, the effects of viscosity and surface tension were neglected but compressibility of the water was allowed for by using a suitable form of the equation of state. It is envisioned that a cavity of initially infinite size has been collapsing for an infinite time. Thus the pressure in the cavity is zero and that for the liquid far from it is p_0. It is also assumed that there is a constant, finite energy E associated with the flow. Following Rayleigh's (1917) treatement of the same problem in an incompressible medium, the cavity wall was assumed to move according to the formula

$$E = 2\pi\rho' R'^3 \acute{R}'^2, \tag{5.3.1}$$

where $\acute{R} = \acute{R}(t)$ is the radius of the cavity. On integration of (5.3.1), we have

$$R'^{5/2} = \frac{5}{2}\sqrt{\left(\frac{E}{2\pi\rho'}\right)}(t_0' - t'), \tag{5.3.2}$$

where t_0' is the instant of collapse. After appropriately scaling the variables by the conditions at infinity, the problem may be reduced to solving the system

$$u_t + uu_r + \frac{1}{\gamma - 1}(c^2)_r = 0, \tag{5.3.3}$$

$$(c^2)_t + u(c^2)_r + (\gamma - 1)c^2\left(u_r + \frac{2u}{r}\right) = 0. \tag{5.3.4}$$

The boundary conditions are

$c = 1$ and $u = \dot{R}$ at the cavity wall $r = R$,

$c \to 1$ as $r \to \infty$,

$$R \sim (t_0 - t)^{2/5}, \quad u \sim -\frac{2}{5}\frac{(t_0 - t)^{1/5}}{r^2},$$

$$c^2 = 1 + \frac{2(\gamma - 1)(t_0 - t)^{-6/5}}{25}\left[\frac{(t_0 - t)^{2/5}}{r} - \frac{(t_0 - t)^{8/5}}{r^4}\right] \tag{5.3.5}$$

as $t_0 - t \to \infty$. The expression for c^2 comes from the solution of this problem assuming the medium to be incompressible. The initial conditions for the numerical solution of this problem were chosen to be (5.3.5) when $\dot{R} = -0.1$. $\gamma = c_p/c_v$ in (5.3.3) and (5.3.4) was assumed to be 7. The method of characteristics was used to solve this problem. Two significant results emerged from this numerical study: (i) $R(-\dot{R})^\tau \to$ constant as $R \to 0$ where the constant τ was found to be 1.27. (ii) For each fixed value of r/R, $-u \to \infty$ and $c \to \infty$ as $R \to 0$ in such a way that u/R and $c^2/\dot{R}^2$ become functions of r/R alone. In one of the first instances of similarity theory, Hunter (1960) was motivated by the numerical results to seek out the solution of this problem in the form

$$u = \dot{R}f(r/R), \quad c^2 = \dot{R}^2g(r/R).$$ (5.3.6)

The problem of asymptotic collapse of the cavity was studied as its radius tends to zero. Thus, any length scale for the flow, which may be derived from the boundary conditions, will be too large to be relevant as $R \to 0$. Therefore, the only suitable length scale for the flow is the cavity radius R itself. The same is true for the velocity scale because any such scale derived from the boundary conditions will be too small compared to the velocity of the cavity wall. Thus, $\dot{R}$ is the appropriate velocity scale. When (5.3.6) is substituted into (5.3.3) and (5.3.4), the latter reduce to ODEs only if $R\ddot{R}/\dot{R}^2 = \text{constant} = 1 - n^{-1}$, say. Thus, $R = A(-t)^n$ where A is a constant and $t = 0$ is the instant of the cavity collapse. The conditions $u = \dot{R}$ and $c^2 = 0$ at $r = R$ lead to the boundary conditions $f(1) = 1$ and $g(1) = 0$ at the cavity. The equation of state was assumed to be $p \propto \rho^\gamma$ where $\gamma = 7$. The reduction of the system of ODEs to a single equation is similar to that described in Section 5.2. Hunter (1960) carried out a very detailed analysis of singularities in the (Y,Z)-plane (see Section 5.2). The value of the similarity exponent n was determined by the regularity properties of the similarity solution. For $\gamma = 7$, it was found to be 0.5552. The corresponding value from the numerical solution was found to be 0.560. We may mention that τ and n are related by $\tau = n/(1-n)$ (see discussion below (5.3.5)). The similarity analysis valid only for high pressures and velocities was continued beyond the instant of the cavity collapse to describe the formation and initial propagation of the shock wave after the collapse is completed.

In a later study, Thomas et al. (1986) considered the collapse of a spherical cavity surrounded by a perfect gas initially at rest. The cavity begins to move with uniform velocity $-2c_0/(\gamma - 1)$, where c_0 is the speed of sound in the undisturbed gas. Thomas et al. (1986) showed that the cavity velocity remains practically uniform until the radius of the cavity, R, becomes a small fraction $\xi(\gamma)$ of the initial radius R_0. Then it begins to move according to the asymptotic self-similar behaviour $\dot{R} \sim R^{-\tau(\gamma)}$. However, for $1 < \gamma < \gamma_{cr}, \gamma_{cr} \approx 1.5$, the velocity $\dot{R}$ of the cavity surface remains strictly uniform for the entire period of collapse. The effect of geometry seems to be overridden by the high compressibility of the gas.

We consider first the case when the radius R of the cavity remains sufficiently close to its initial value R_0. The flow is described by a plane rarefaction wave (Courant and Friedrichs 1948) with the modulus of the velocity profile given by

$$u = \frac{2}{\gamma+1}\left(\frac{R_0 - r}{t} + c_0\right).$$ (5.3.7)

In this case, the velocity of the cavity is simply

$$\dot{R} = -\frac{2c_0}{(\gamma - 1)} = -V_i, \quad \text{say.}$$ (5.3.8)

Writing $u = dr_i/dt$ in (5.3.7) and integrating we arrive at the result

$$r_i = R_0 - \frac{2c_0 t}{\gamma - 1} \left[1 - \frac{\gamma + 1}{2} \left(\frac{r_{io} - R_0}{c_0 t} \right)^{(\gamma - 1)/(\gamma + 1)} \right], \tag{5.3.9}$$

where $r_{io} = r_i(t = 0)$. The relation (5.3.9) holds provided that $t \geq (r_{io} - R_0)/c_0$. The above analysis is valid until the geometrical compression begins to contribute significantly.

As we discussed earlier with reference to the work of Hunter (1960) for $\gamma = 7$ (see also Zel'dovich and Raizer 1967), the flow for $R \ll R_0$ is described by the self-similar solution. Here the radius of the cavity is given by

$$R = A(-t)^n, \tag{5.3.10}$$

where A is a constant which depends on the initial conditions and $n = n(\gamma)$. We may rewrite (5.3.10) as

$$\dot{R} = -CR^{-\hat{t}} \tag{5.3.11}$$

where $\hat{t} = \tau^{-1} = (1 - n)/n$ and $C = nA^{1/n}$.

As we mentioned earlier, the law (5.3.10) was motivated by Rayleigh's results for the cavity collapse in an incompressible fluid for which $\hat{t} = 3/2$ and $C = (3p_0/2\rho_0)^{1/2} R_0^{3/2}$. For an ideal gas $\hat{t}$ depends on the compressibility of the medium.

Thomas et al. (1986) carried out the numerical solution of this problem with a scheme different from Hunter's and came up with some interesting conclusions. The results we describe here refer to the asymptotic stage when $r \to R$ and $R \to 0$. Figure 5.3 shows $-\dot{R}$ as a function of the cavity radius R/R_0. Here, the undisturbed pressure p_0 and density ρ_0 were taken to be 10^6 dyn cm^{-2} and 1.6×10^{-4} g cm^{-3}, respectively. The minimum value, R_{min}/R_0, for which 'numerical saturation' took place depended on the value of γ. It was found to be $R_{min}/R_0 = 10^{-2}, 2 \times 10^{-3}, 8 \times 10^{-4}$ for $\gamma = 7, 4, 2.4$, respectively. Figure 5.3 also shows an $(R/R_0, -\dot{R})$ relation for $\gamma = \infty, 7, 4, 2.4$, and $5/3$. It may be observed that, for large γ, the cavity velocity remains constant for a short initial distance and time and then it begins to promptly conform to asymptotic self-similar behaviour. As γ decreases, the early stage of uniform velocity increases and the slope of the asymptotic line decreases until, for $\gamma \approx 5/3$, the cavity speed becomes essentially constant. For $\gamma \lesssim 5/3$, the cavity moves to the centre with almost uniform velocity $-2c_0/(\gamma - 1)$ during its entire course. In fact, there exists a value of $\gamma = \gamma_1$, $1.5 < \gamma_1 < 5/3$, below which the flow never approaches self-similar behaviour and the cavity moves inward with constant velocity right up to the point of collapse. Thomas et al. (1986) compared their conclusions regarding the asymptotic nature of the cavity collapse with those from the stability analysis of these flows by Lazarus (1982) and observed that there are no (one-dimensionally) stable asymptotic solutions for $3/2 < \gamma < 5/3$; there exist degenerate stable asymptotic solutions with $\dot{R} = $ constant for $\gamma \lesssim 3/2$. These solutions were numerically simulated by Thomas et al. (1986).

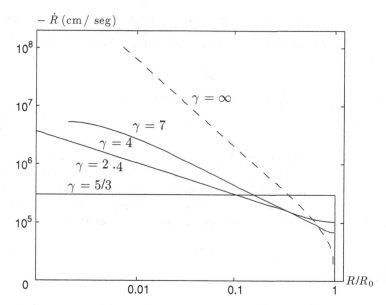

Fig. 5.3 $-\dot{R}$ versus R/R_0 with undisturbed pressure $p_0 = 10^6$ dyn cm^{-2}, density $\rho_0 = 1.6 \times 10^{-4}$ gm cm^{-3} and $\gamma = \infty, 7, 4, 2.4, 5/3$. (Thomas et al. 1986. Copyright © 1986 American Institute of Physics. Reprinted with permission. All rights reserved.)

5.4 Large time behaviour of solutions of compressible flow equations with damping

In an interesting study, Liu (1996) showed how the solutions of compressible flow equations with damping, which have a hyperbolic character, tend for large time to those of a nonlinear parabolic equation. For this purpose, he made use of a special class of solutions of each of these systems. Barenblatt (1953) had constructed a special class of solutions for the porous flow equations. Making use of this class of solutions, Liu (1996) justified, at least in a limited sense, the so-called Darcy's law which is used to describe compressible flow in a porous medium. The physical situation described by these flows is rather special and includes a vacuum front where $\rho = 0$. It is assumed that no shocks are formed in the flow.

The vector form of the isentropic compressible flow equations with damping may be written as

$$\rho_t + \nabla \cdot (\rho \bar{u}) = 0, \tag{5.4.1}$$

$$(\rho \bar{u})_t + \nabla \cdot \rho (\bar{u} \otimes \bar{u}) + \nabla p(\rho) + \alpha \rho \bar{u} = 0, \tag{5.4.2}$$

where p and ρ are related by the polytropic law

$$p(\rho) = k\rho^{\gamma}, \quad k > 0, \ \gamma > 1. \tag{5.4.3}$$

α in (5.4.2) is positive and denotes the (constant) coefficient of friction. Using Darcy's law

$$\nabla p(\rho) = -\alpha\rho\bar{u}, \tag{5.4.4}$$

equation (5.4.2) becomes the porous media equation

$$\rho_t = \alpha^{-1} \triangle p(\rho). \tag{5.4.5}$$

It was shown by Liu (1996) that a certain class of solutions of equations (5.4.1)–(5.4.3) tends, for large time, to the corresponding solutions of the porous medium equation (5.4.5). Darcy's law for the porous medium is thus shown to hold, at least in this asymptotic sense. We later summarise more general results in this context due to Hsiao and Liu (1992). Introducing the speed of the sound via

$$c^2 = p'(\rho) = k\gamma\rho^{\gamma-1} \tag{5.4.6}$$

into (5.4.1) and (5.4.2) we get

$$(c^2)_t + \nabla(c^2) \cdot \bar{u} + (\gamma - 1)c^2 \nabla \cdot \bar{u} = 0, \tag{5.4.7}$$

$$\bar{u}_t + (\bar{u} \cdot \nabla)\bar{u} + (\gamma - 1)^{-1} \nabla (c^2) = -\alpha\bar{u}. \tag{5.4.8}$$

To motivate the form of the special solution in the neighbourhood of the vacuum $\rho = c = 0$, one may observe that the trajectory of the vacuum front,

$$\Gamma \equiv \{(\bar{x},t) : \rho(\bar{x},t) \geq 0\} \cap \{(\bar{x},t) : \rho(\bar{x},t) = 0\}, \tag{5.4.9}$$

is a particle line along which

$$\frac{d\bar{x}}{dt} = \bar{u}(\bar{x}(t),t). \tag{5.4.10}$$

Equations (5.4.8) in the direction (5.4.10) may be written as

$$\frac{d\bar{u}}{dt} + \alpha\bar{u} = -(\gamma - 1)^{-1} \nabla (c^2). \tag{5.4.11}$$

Because we assume that $d\bar{u}/dt$ is finite, (5.4.11) suggests that we may write

$$c^2(\bar{x},t) = \eta(\bar{x},t)|\bar{x}(t) - \bar{x}|, \tag{5.4.12}$$

where the function $\eta(\bar{x},t)$ is differentiable right up to the vacuum front Γ (see (5.4.9)). Equation (5.4.12) implies that $c(\bar{x},t) \cong |\bar{x}(t) - \bar{x}|^{1/2}$ and therefore the characteristic speeds of the system (5.4.7) and (5.4.8) are not Lipschitz continuous near the vacuum front. Using (5.4.12) we may write

$$\rho(\bar{x},t) \cong |\bar{x}(t) - \bar{x}|^{\frac{1}{(\gamma-1)}},$$

$$p(\bar{x},t) \cong |\bar{x}(t) - \bar{x}|^{\frac{\gamma}{(\gamma-1)}}. \tag{5.4.13}$$

It turns out that the singularities of the system (5.4.7) and (5.4.8) are the same as were first observed by Barenblatt (1953) for the porous medium equation (5.4.5). Restricting himself to the plane and spherically symmetric flows with $n = 1$ and $n = 3$, respectively, Liu (1996) specialised (5.4.7) and (5.4.8) as

$$(c^2)_t + u(c^2)_x + (\gamma - 1)c^2 u_x + \frac{n-1}{x}(\gamma - 1)c^2 u = 0, \tag{5.4.14}$$

$$u_t + u u_x + \frac{1}{\gamma - 1}(c^2)_x + \alpha u = 0, \tag{5.4.15}$$

$x = \left(\sum_{i=1}^n x_i^2\right)^{1/2}$, $\bar{u} = (\bar{x}/x)u$ and sought their solutions in the form (see Sachdev 2004)

$$\rho(x,t) \equiv 0, \quad |x| > \left(\frac{e(t)}{b(t)}\right)^{1/2}, \tag{5.4.16}$$

$$c^2(x,t) = e(t) - b(t)x^2, \tag{5.4.17}$$

$$u(x,t) = a(t)x, \tag{5.4.18}$$

where the particle velocity is linear in x and the sound speed square is quadratic. Substituting (5.4.17) and (5.4.18) into (5.4.14) and (5.4.15) and equating coefficients of $x^i, i = 0, 1, 2$ to zero we obtain the following system of ODEs for the functions $e(t), b(t)$, and $a(t)$.

$$e' + n(\gamma - 1)ea = 0, \tag{5.4.19}$$

$$b' + (n\gamma - n + 2)ab = 0, \tag{5.4.20}$$

$$a' + a^2 + \alpha a - \frac{2}{\gamma - 1}b = 0. \tag{5.4.21}$$

A phase plane analysis of (5.4.20) and (5.4.21) shows that the functions $a(t)$ and $b(t)$ and hence $e(t)$ exist for all time. Thus we find that the solutions of (5.4.14) and (5.4.15) of the form (5.4.17) and (5.4.18) with given $a(0), b(0)$, and $e(0)$ exist for all $t > 0$. There also exist travelling wave solutions of (5.4.14) and (5.4.15) with $n = 1$ of the simple form

$$c^2(x,t) = D(e(t) - x), \quad x < e(t), \tag{5.4.22}$$

$$u(x,t) = a(t), \tag{5.4.23}$$

where D is a constant. Here the functions $a(t)$ and $e(t)$ are governed by

$$a' + \alpha a = \frac{D}{\gamma - 1}, \quad e' = a. \tag{5.4.24}$$

Equations (5.4.24) admit the exact solution

$$a(t) = a(0)e^{-\alpha t} + \frac{D}{\alpha(\gamma - 1)}\left(1 - e^{-\alpha t}\right), \tag{5.4.25}$$

$$e(t) = e(0) + \left(\frac{a(0)}{\alpha} - \frac{D}{\alpha^2(\gamma - 1)}\right)\left(1 - e^{-\alpha t}\right) + \frac{D}{\alpha(\gamma - 1)}t. \tag{5.4.26}$$

This class of solutions tends, as $t \to \infty$, to the simple travelling waveform

$$c^2(x,t) = (\gamma - 1)u_0\alpha(u_0 t - x),$$
$$u(x,t) = u_0, \tag{5.4.27}$$

where $x < u_0 t$ and $u_0 = D/(\alpha(\gamma - 1)) > 0$. It turns out that they are also the travelling wave solutions of the porous medium equation (5.4.5).

We may now obtain the corresponding solution of the basic system under Darcy's law. Thus, (5.4.4) in one dimension implies that

$$\frac{1}{\gamma - 1}(c^2)_x + \alpha u = 0. \tag{5.4.28}$$

The ansatz

$$c^2(x,t) = \bar{e}(t) - \bar{b}(t)x^2, \tag{5.4.29}$$

and

$$u(x,t) = \bar{a}(t)x \tag{5.4.30}$$

in (5.4.14) and (5.4.28) and so on lead to the system of ODEs

$$\bar{e}' + n(\gamma - 1)\bar{e}\bar{a} = 0, \tag{5.4.31}$$
$$\bar{b}' + (n\gamma - n + 2)\bar{a}\bar{b} = 0, \tag{5.4.32}$$
$$\alpha\bar{a} = \frac{2\bar{b}}{\gamma - 1}. \tag{5.4.33}$$

Equation (5.4.33) follows from Darcy's law (5.4.28) and is, therefore, called Darcy's line. The solution of the system (5.4.31)–(5.4.33) leads to Barenblatt's form (5.4.29) and (5.4.30) with

$$\bar{a}(t) = \frac{1}{n\gamma - n + 2}t^{-1}, \tag{5.4.34}$$

$$\bar{b}(t) = \frac{(\gamma - 1)\alpha}{2(n\gamma - n + 2)}t^{-1}, \tag{5.4.35}$$

$$\bar{e}(t) = e_0 t^{-n(\gamma - 1)/(n\gamma - n + 2)}, \tag{5.4.36}$$

where we assume that the solution passes through the point $x = 0$ at $t = 0$; thus $\bar{a}(0) = \bar{b}(0) = \infty$. The (positive) constant e_0 may be related to the total mass

$$
\begin{aligned}
m &= \Omega_{n-1} \int_0^{\sqrt{\bar{e}(t)/\bar{b}(t)}} \rho(x,t)x^{n-1}dx \\
&= \Omega_{n-1}(k\gamma)^{-1/(\gamma-1)} \int_0^{\sqrt{\bar{e}(t)/\bar{b}(t)}} (c^2(x,t))^{1/(\gamma-1)}x^{n-1}dx \\
&= \Omega_{n-1}(k\gamma)^{-1/(\gamma-1)} \int_0^{\sqrt{\bar{e}(t)/\bar{b}(t)}} (\bar{e}(t) - \bar{b}(t)x^2)^{1/(\gamma-1)}x^{n-1}dx \\
&= \Omega_{n-1}(k\gamma)^{-1/(\gamma-1)} \left(\frac{2(n\gamma-n+2)}{(\gamma-1)\alpha} \right)^{n/2} e_0^{(n\gamma-n+2)/2(\gamma-1)} \\
&\quad \times \int_0^1 (1-y^2)^{1/(\gamma-1)}y^{n-1}dy,
\end{aligned} \tag{5.4.37}
$$

where we have used (5.4.29) to determine the limit in the integral in (5.4.37) and Ω_{n-1} equals 1 and 4π for $n = 1$ and $n = 3$, respectively.

To prove formally the asymptotic nature of the Darcy solution we need to define the following trajectories;

$$
\Gamma_1 : b = \frac{\gamma-1}{2}(a^2 + \alpha a), \tag{5.4.38}
$$

$$
\Gamma_2 : b = \frac{\gamma-1}{2}\alpha a; \tag{5.4.39}
$$

see (5.4.21). It follows easily from these equations that

$$
b' < 0, \ a' > 0 \ \text{between } b\text{-axis and } \Gamma_1, \tag{5.4.40}
$$

$$
b' < 0, \ a' = 0 \ \text{on } \Gamma_1, \tag{5.4.41}
$$

$$
b' < 0, \ a' < 0 \ \text{between } a\text{-axis and } \Gamma_1. \tag{5.4.42}
$$

It follows from (5.4.20), (5.4.21), and (5.4.39) that

$$
\frac{db}{da} = \frac{1}{2}(n\gamma-n+2)(\gamma-1)\alpha \tag{5.4.43}
$$

on Γ_2. From these statements it may be checked that $a(t)$ and $b(t) \to 0$ as $t \to \infty$. We have already seen that all trajectories of Barenblatt's solution move along the line Γ_2 (see (5.4.34), (5.4.35) and (5.4.39)). One may verify from (5.4.20) and (5.4.21) that all trajectories of this system are transversal to Γ_2 at a constant angle θ given by

$$
\tan\theta = \frac{2(n\gamma-n+1)(\gamma-1)\alpha}{(n\gamma-n+2)(\gamma-1)^2\alpha^2+4}. \tag{5.4.44}
$$

Now we formally show that the trajectories of the system (5.4.20) and (5.4.21) obey the asymptotic law

$$\frac{b(t)}{a(t)} \to \frac{\gamma - 1}{2} \alpha \text{ as } t \to \infty; \tag{5.4.45}$$

that is, they tend to the Darcy line Γ_2 as $t \to \infty$. From (5.4.20) and (5.4.21) we have

$$\left(\frac{b}{a}\right)' = \frac{b'a - a'b}{a^2} = -(n\gamma - n + 1)b + \frac{\alpha b}{a} - \frac{2}{\gamma - 1}\left(\frac{b}{a}\right)^2. \tag{5.4.46}$$

Now, for a small $\varepsilon > 0$, consider the curve

$$\Gamma_\varepsilon : -(n\gamma - n + 1)ba^2 + \alpha ba - \frac{2}{\gamma - 1}b^2 = \varepsilon a^2 \tag{5.4.47}$$

(cf. (5.4.46)). The curve Γ_ε lies below Γ_2 and has the slope

$$\frac{\gamma - 1}{2}\alpha + \frac{\gamma - 1}{4}\left(\left(\alpha^2 - \frac{8}{\gamma - 1}\varepsilon\right)^{1/2} - \alpha\right) > \frac{\gamma - 1}{2}\alpha - \frac{2\varepsilon}{\alpha} \tag{5.4.48}$$

at the origin $(0,0)$. This slope tends to $((\gamma - 1)/2)\alpha$, the slope of Γ_2, as $\varepsilon \to 0$. It is also clear from (5.4.46) that, below Γ_ε, $(b/a)' > \varepsilon$. Therefore, any trajectory can remain below Γ_ε for a finite time only. Because ε is arbitrary, we may choose $\varepsilon = 0$. Thus all trajectories approach Γ_2 as $t \to \infty$.

Finally, we show that the solutions (5.4.17)–(5.4.21) of (5.4.14) and (5.4.15) with total mass m given by (5.4.37) tend to the special solutions (5.4.29) and (5.4.30) together with (5.4.34)–(5.4.36) of the porous medium equations (5.4.4) and (5.4.5):

$$(a,b,e)(t) = (\bar{a},\bar{b},\bar{e})(t) + O(1)\frac{\ln t}{t} \tag{5.4.49}$$

as $t \to \infty$. Here, the bound $O(1)$ is independent of $t \geq 1$ but varies with the trajectories of (5.4.19)–(5.4.21). Using (5.4.45) in (5.4.20) we find that

$$b' + \frac{2(n\gamma - n + 2)}{\alpha(\gamma - 1)}(1 + o(1))b^2 = 0, \tag{5.4.50}$$

where the term $o(1)$ tends to zero as $t \to \infty$. It follows from (5.4.50) that

$$b(t) = D(t)t^{-1}, \quad b'(t) = O(1)t^{-2} \tag{5.4.51}$$

for some function $D(t)$ which is positive and bounded away from zero. Introducing

$$f = a - \frac{2}{\alpha(\gamma - 1)}b, \tag{5.4.52}$$

we may write

$$f' + \alpha f + \frac{2}{\alpha(\gamma - 1)}b' = -a^2 = O(1)t^{-2}, \tag{5.4.53}$$

where we have used (5.4.45) and (5.4.51). Thus, we have

$$f(t) = a(t) - \frac{2}{\alpha(\gamma-1)}b(t) = O(1)t^{-2} = O(1)b(t)t^{-1}. \tag{5.4.54}$$

It follows that

$$|a(t)| + |b(t)| = O(1)t^{-1} \tag{5.4.55}$$

and

$$\left| a(t) - \frac{2}{\alpha(\gamma-1)}b(t) \right| = O(1)b(t)t^{-1} \tag{5.4.56}$$

as $t \to \infty$.

Using the above estimates we may compare the solution $b(t)$ of (5.4.20) and $\bar{b}(t)$ of (5.4.32):

$$b' + \frac{2(n\gamma-n+2)}{\alpha(\gamma-1)}\left(1+O(1)t^{-1}\right)b^2 = 0 \tag{5.4.57}$$

or

$$b(t) = \left[b(t_0)^{-1} + \frac{2(n\gamma-n+2)}{\alpha(\gamma-1)} \int_{t_0}^{t} \left(1+O(1)s^{-1}\right)ds \right]^{-1}$$
$$= \bar{b}(t)\left(1+O(1)\frac{\ln t}{t}\right). \tag{5.4.58}$$

Substituting (5.4.56) and (5.4.58) into (5.4.19), we get

$$e' + n(\gamma-1)e\,\bar{a}\left(1+O(1)\frac{\ln t}{t}\right) = 0 \tag{5.4.59}$$

which, on integration, gives

$$e(t) = e(t_0)e^{-\int_{t_0}^{t} n(\gamma-1)\bar{a}(s)ds}e^{-\int_{t_0}^{t} O(1)\left(\ln s/s^2\right)ds}$$
$$= A\bar{e}(t)\left(1+O(1)\int_{t_0}^{t}\frac{\ln s}{s^2}ds\right)$$
$$= A\bar{e}(t)\left(1+O(1)\frac{\ln t}{t}\right) \tag{5.4.60}$$

for some constant A. Because (5.4.1)–(5.4.3) as well as the porous medium equation (5.4.5) satisfy the same conservation of mass law

$$m = \int_{-\infty}^{\infty} \rho(x,t)dx, \tag{5.4.61}$$

it follows from (5.4.37), (5.4.58), and (5.4.60) that $A = 1$. Thus, the proof of (5.4.49) is complete.

In a related study, Hsiao and Liu (1992) considered the system (5.4.1) and (5.4.2) with $n = 1$, expressed in Lagrangian coordinates, namely,

$$v_t - u_x = 0, \tag{5.4.62}$$

$$u_t + (p(v))_x + \alpha u = 0, \quad \alpha > 0, \quad p'(v) < 0, \tag{5.4.63}$$

and approximated it according to Darcy's law. Thus, we have

$$v_t = -\frac{1}{\alpha}(p(v))_{xx}, \tag{5.4.64}$$

$$(p(v))_x = -\alpha u. \tag{5.4.65}$$

Hsiao and Liu (1992) again attempted to prove the time asymptotic equivalence of the systems (5.4.62), (5.4.63) and (5.4.64), (5.4.65) by referring to their solutions subject to a class of initial conditions with the same behaviour at $x = \pm\infty$.

Specifically, equations (5.4.62) and (5.4.63) were first solved subject to initial conditions which satisfy the conditions

$$(u,v)(x,0) \to (u_\pm, v_\pm) \quad \text{as } x \to \pm\infty. \tag{5.4.66}$$

The solution $\bar{v}(x,t)$ of (5.4.64) must also satisfy the end conditions

$$\bar{v}(\pm\infty, t) = v_\pm. \tag{5.4.67}$$

The corresponding values for $\bar{u}(\pm\infty, t)$ may be obtained from $-(1/\alpha)(p(\bar{v}))_x$ (see (5.4.65)). Hsiao and Liu (1992) made use of the self-similar solution of (5.4.64) of the form

$$v^*(x,t) = \phi\left(\frac{x}{\sqrt{t}}\right) \equiv \phi(\xi), \quad -\infty < \xi < \infty, \tag{5.4.68}$$

which must also satisfy the conditions $\phi(\pm\infty) = v_\pm$. This problem is known to have a unique solution, which is also strictly monotonic (Van Duyn and Peletier 1977).

First it was shown that the solutions of system (5.4.62) and (5.4.63) as well as those of (5.4.64) and (5.4.65) with the same limiting values of the initial conditions at $x \to \pm\infty$ satisfy the asymptotic behaviour

$$\|v(x,t) - \bar{v}(x+x_0,t)\|_{L_2(x)} + \|v(x,t) - \bar{v}(x+x_0,t)\|_{L_\infty(x)} = O(1)t^{-1/2} \tag{5.4.69}$$

as $t \to \infty$, where the translation x_0 in the solution of the porous medium equation is uniquely given by

$$\int_{-\infty}^{\infty} [v(x,0) - \bar{v}(x+x_0,0)]dx = \frac{u_+ - u_-}{-\alpha}. \tag{5.4.70}$$

The corresponding solution u tends to $\bar{u}$ in the following sense. Defining any smooth function $m_0(x)$ with compact support and requiring that $\int_{-\infty}^{\infty} m_0(x)dx = 1$, one may write

$$m(x,t) = -\frac{u_+ - u_-}{\alpha} m_0(x)e^{-\alpha t}. \tag{5.4.71}$$

Then it was shown by Hsiao and Liu (1992) that $u(x,t)$ tends to $\bar{u}(x,t)$ in the following sense;

$$\|(u-\bar{u}-\hat{u})(x,t)\|_{L_2(x)} + \|(u-\bar{u}-\hat{u})(x,t)\|_{L_\infty(x)} = O(1)t^{-1/2} \qquad (5.4.72)$$

as $t \to \infty$, where

$$\hat{u}(x,t) = u_- e^{-\alpha t} + \int_{-\infty}^{x} m_t(\eta,t)d\eta. \qquad (5.4.73)$$

For the special case when the end conditions v_+ and v_- coincide, one may choose $\bar{v}$ to be a multiple of the heat kernel, namely,

$$\bar{v} = v_- + \frac{1}{\sqrt{4\pi t}} e^{-(x-p'(v_0)t)^2/4t} \int_{-\infty}^{\infty} v(y,0)dy. \qquad (5.4.74)$$

If, furthermore, $u_+ = u_-$, then (5.4.71) shows that $m \equiv 0$. Moreover, if $(u_+,v_+) = (u_-,v_-) = (0,0)$ and

$$\int_{-\infty}^{\infty} v(y,0)dy = 0, \qquad (5.4.75)$$

then $\bar{u} = \bar{v} = 0, m = 0$ and the results of Hsiao and Liu (1992) reduce to those of Matzumura (1978). The analysis of Hsiao and Liu (1992) involves energy methods and uses the decay estimates for the self-similar solution of the parabolic equation (5.4.64). The main contribution of the present study is again to show how the damping term in the basic hyperbolic system (5.4.62) and (5.4.63) produces diffusive effects as t becomes large.

In an interesting study, Gallay and Raugel (1998) studied the large time behaviour of small solutions of the damped nonlinear wave equation

$$\varepsilon u_{\tau\tau} + u_\tau = (a(\xi)u_\xi)_\xi + \mathcal{N}(u,u_\xi,u_\tau), \quad \xi \in \mathbb{R}, \ \tau \ge 0, \ \varepsilon > 0. \qquad (5.4.76)$$

Here $\varepsilon > 0$ need not be small. They assumed that (i) $a(\xi) \to a_\pm > 0$ as $\xi \to \pm\infty$ and (ii) $\mathcal{N} \to 0$ sufficiently fast as $u \to 0$. They showed that the large time asymptotic expansion in powers of $\tau^{-1/2}$ (up to second order) of the solution of (5.4.76) is given by a linear parabolic equation. This parabolic equation depends on a_+ and a_- only. We also refer to Nishihara (1996, 1997), Gallay and Raugel (2000), and Gallay and Wayne (2002) for some more related studies on large time aymptotics.

5.5 Large time behaviour of solutions of unsteady boundary layer equations for an incompressible fluid

We studied gas dynamic equations with damping for an isentropic flow in Section 5.4 and arrived at their asymptotic behaviour which was shown to be governed by a nonlinear diffusive equation. The damping effect in these hyperbolic equations over long time manifests itself as diffusion. In the present section and the next we take up the study of boundary layer equations in two steps. First we show, following closely the work of Oleinik (1966a), how under appropriate conditions the horizontal particle velocity for the unsteady flow in the boundary layer tends to that for a steady flow as time tends to infinity. Later, in Section 5.6, we study the asymptotic behaviour of steady boundary layer equations, subject to a class of initial and boundary conditions and demonstrate that their solution tends to a self-similar solution governed by a solution of the Falkner–Skan equation, a third-order nonlinear ODE subject to 'reduced' boundary conditions, as the horizontal distance becomes large. In the second asymptotic study (Section 5.6) it is the large distance behaviour (Serrin 1967) with which we are concerned. Thus, we study the asymptotic behaviour of unsteady boundary layer equations in two steps: first for large time and then for large horizontal distance. The latter is governed by a self-similar solution of the steady system. How such a solution actually evolves remains to be investigated numerically.

First we pose the initial boundary value problem and boundary value problem for the unsteady and steady boundary layer equations, respectively.

The unsteady two-dimensional viscous incompressible flow is governed by the system

$$u_t + u u_x + v u_y = -\frac{1}{\rho} p_x + v u_{yy}, \tag{5.5.1}$$

$$u_x + v_y = 0, \tag{5.5.2}$$

which holds in the domain $D := \{0 \leq t < \infty, \ 0 \leq x \leq x_0, \ 0 \leq y < \infty\}$. The initial and boundary conditions for this flow are taken to be

$$u|_{t=0} = u_0(x,y), \tag{5.5.3}$$

$$u|_{y=0} = 0, \quad v|_{y=0} = v_0(t,x), \quad u|_{x=0} = u_1(t,y), \quad \lim_{y \to \infty} u(t,x,y) = U(t,x). \tag{5.5.4}$$

The pressure $p(t,x)$ and the 'free stream velocity' $U(t,x)$ are related by Bernoulli's law

$$-\frac{1}{\rho} p_x = U_t + U U_x. \tag{5.5.5}$$

It is assumed that the functions $p(t,x)$, $U(t,x)$, and $v_0(t,x)$ tend, respectively, to $p^\infty(x)$, $U^\infty(x)$, and $v_0^\infty(x)$ uniformly in x as $t \to \infty$. Similarly, the horizontal velocity $u|_{x=0} = u_1(t,y)$ is equal to $u_1^\infty(y)$ for $t > t_1 \geq 0$. It is also assumed that the system (5.5.1)–(5.5.5), subject to the limiting behaviour of the boundary and initial conditions stated above, has a solution for which $u_y > 0$ for $0 \leq y < \infty$, and $u(t,x,y)$ and u_y have continuous and bounded first-order derivatives with respect to t, x, and y in D. Besides, u_{yyy} and v_y must exist and satisfy the condition

$$\left[u_{yyy} u_y - (u_{yy})^2 \right] (u_y)^{-3} < K \tag{5.5.6}$$

in D, where K is some constant. It was shown by Oleinik (1963a, 1966b) that these assumptions hold physically provided x_0 in the definition of the domain D is chosen to be sufficiently small. We refer to Oleinik (1966b) for proof of the existence and uniqueness of the solution of the system (5.5.1)–(5.5.5)).

Oleinik (1966a) then posed the boundary value problem for the steady form of (5.5.1)–(5.5.2), namely,

$$uu_x + vu_y = -\frac{1}{\rho}p_x^\infty + vu_{yy},$$ (5.5.7)

$$u_x + v_y = 0,$$ (5.5.8)

which hold in the region $D^\infty := \{0 \le x \le x_0, 0 \le y < \infty\}$. The superscript '$\infty$' denotes the limiting form of the relevant entity as $t \to \infty$. The initial and boundary conditions relevant to the system (5.5.7) and (5.5.8) are

$$u|_{y=0} = 0, \quad v|_{y=0} = v_0^\infty(x), \quad u|_{x=0} = u_1^\infty(y), \quad \lim_{y \to \infty} u(x,y) = U^\infty(x).$$ (5.5.9)

The conditions (5.5.9) are assumed to be compatible with those for the unsteady case in the limit $t \to \infty$. It is assumed that the system (5.5.7)–(5.5.9) has a solution $u^\infty(x,y)$ and $v^\infty(x,y)$ for which $u_y^\infty > 0, 0 \le y < \infty$ and the functions $u^\infty(x,y)$ and $u_y^\infty(x,y)$ have continuous and bounded derivatives of first-order with respect to x and y in D^∞. It is further assumed that the derivatives u_{yyy}^∞ and v_y^∞ exist. Indeed, existence of the solution of (5.5.7)–(5.5.9) was proved by Oleinik (1963a) in the region D^∞ for some $x_0 > 0$ subject to the conditions that the limiting functions $p^\infty(x), v_0^\infty(x), u_1^\infty(y)$, and $U^\infty(x)$ satisfy certain smoothness requirements. The functions involved must also assume the same value at $(0,0)$ consistently and satisfy the conditions $u_1^\infty(y) > 0$ for $y > 0$ and $U^\infty(x) > 0$ for $x \ge 0$. The solution $u^\infty(x,y)$ and $v^\infty(x,y)$ of the problem thus formulated was shown to satisfy the conditions $u_y^\infty > 0$ for $y \ge 0$ if $\partial u_1^\infty/\partial y > 0$, when $y \ge 0$.

The main result of Oleinik (1966a) is that the horizontal velocity $u^\infty(x,y)$ satisfying the system (5.5.7)–(5.5.9) is the large time limit of the unsteady counterpart $u(t,x,y)$ governed by (5.5.1)–(5.5.5):

$$\lim_{t \to \infty} u(t,x,y) = u^\infty(x,y)$$ (5.5.10)

for all x,y in D^∞. For this purpose, the initial/boundary conditions for the two systems must be assumed to be compatible as described in the following. The analysis for the above result was carried out in terms of new variables which effectively eliminate v from the system (5.5.1) and (5.5.2). Thus, introducing

$$\tau = t, \quad \xi = x, \quad \eta = u(t,x,y)$$ (5.5.11)

as the new independent variables and $w = u_y$ as the new dependent variable, the system (5.5.1)–(5.5.2) changes to a single second-order PDE for $w = w(\tau, \xi, \eta)$,

$$vw^2 w_{\eta\eta} - w_\tau - \eta w_\xi + \frac{1}{\rho}p_x w_\eta = 0,$$ (5.5.12)

which holds in $\Omega = \{0 \le \tau < \infty, \ 0 \le \xi \le x_0, \ 0 \le \eta \le U(\tau, \xi)\}$. The conditions (5.5.3) and (5.5.4) now become

$$w|_{\tau=0} = (u_0)_y \equiv w_0(\xi, \eta), \quad w|_{\xi=0} = (u_1)_y \equiv w_1(\tau, \eta), \quad w|_{\eta=U(\tau,\xi)} = 0 \quad (5.5.13)$$

and (5.5.5) is replaced by the compatibility condition

$$(vww_\eta - \frac{1}{\rho}p_x - v_0 w)|_{\eta=0} = 0. \tag{5.5.14}$$

For the steady system (5.5.7) and (5.5.8), we again introduce

$$\xi = x, \quad \eta = u(x,y) \tag{5.5.15}$$

as the independent variables and $w = \partial u/\partial y$ as the dependent variable. We then obtain a single second-order PDE for w,

$$vw^2 w_{\eta\eta} - \eta w_\xi + \frac{1}{\rho}(p^\infty)_x w_\eta = 0 \tag{5.5.16}$$

(cf. (5.5.12)) which holds in the region $\Omega^\infty := \{0 \leq \xi \leq x_0, \ 0 \leq \eta \leq U^\infty(\xi)\}$. The boundary conditions

$$w|_{\xi=0} = \frac{\partial u_1^\infty}{\partial y} \equiv w_1^\infty(\eta), \quad w|_{\eta=U^\infty(\xi)} = 0, \tag{5.5.17}$$

$$(vww_\eta - \frac{1}{\rho}(p^\infty)_x - v_0^\infty w)|_{\eta=0} = 0 \tag{5.5.18}$$

(cf. (5.5.14)) apply on the boundary of the region Ω^∞. The solution of (5.5.16) subject to the conditions (5.5.17) and (5.5.18) is referred to as $w^\infty(\xi, \eta)$ in consonance with the notation introduced earlier.

We define the difference function

$$V(\tau, \xi, \eta) = w(\tau, \xi, \eta) - w^\infty(\xi, \eta) \tag{5.5.19}$$

over the region Ω_1, which is the intersection of the region Ω with the cylinder $\{0 \leq \tau < \infty, \ 0 \leq \xi \leq x_0, \ 0 \leq \eta \leq U^\infty(\xi)\}$. The functions w and w^∞ are governed by the systems (5.5.12)–(5.5.14) and (5.5.16)–(5.5.18), respectively. Using (5.5.12) and (5.5.16), we get the following equation governing V,

$$v(w^\infty)^2 V_{\eta\eta} - V_\tau - \eta V_\xi + \frac{1}{\rho}(p^\infty)_x V_\eta + v(w+w^\infty)w_{\eta\eta}V = \Phi(\tau, \xi, \eta), \quad (5.5.20)$$

where

$$\Phi(\tau, \xi, \eta) = \frac{1}{\rho}(p_x^\infty - p_x)w_\eta. \tag{5.5.21}$$

Referring to (5.5.13), (5.5.14) and (5.5.17), (5.5.18), we infer that $V(\tau, \xi, \eta)$ must satisfy the following conditions.

$$V|_{\tau=0} = w_0(\xi, \eta) - w^\infty(\xi, \eta), \quad V|_{\xi=0} = w_1(\tau, \eta) - w_1^\infty(\eta), \quad (5.5.22)$$

$$(vw^{\infty}V_{\eta} + (vw_{\eta} - v_0^{\infty})V)|_{\eta=0} = \Psi(\tau, \xi),\qquad(5.5.23)$$

where

$$\Psi(\tau, \xi) = \left[\left(\frac{1}{\rho}p_x - \frac{1}{\rho}p_x^{\infty}\right) + (v_0 - v_0^{\infty})w\right]_{\eta=0}.\qquad(5.5.24)$$

Because we have assumed that

$$\frac{\partial p(t,x)}{\partial x} \to \frac{\partial p^{\infty}(x)}{\partial x} \quad \text{and} \quad v_0(t,x) \to v_0^{\infty}(x) \text{ as } t \to \infty$$

uniformly in x and w and w_{η} are bounded in Ω, it follows from (5.5.21) and (5.5.24) that $|\Phi(\tau, \xi, \eta)| < \varepsilon$ and $|\Psi(\tau, \xi)| < \varepsilon$ for $\tau > \tau_1$, a sufficiently large number; here ε is a (small) arbitrary positive number. Moreover,

$$V|_{\xi=0} = \frac{\partial u_1}{\partial y} - \frac{\partial u_1^{\infty}}{\partial y} = 0\qquad(5.5.25)$$

as $\tau > \tau_1$ becomes sufficiently large.

To show the convergence of the unsteady solution to the steady solution for large τ, Oleinik (1966a) introduced the function V_1 by writing

$$V = e^{\beta\xi}\phi(\alpha\eta)V_1(\tau, \xi, \eta),\qquad(5.5.26)$$

where $\alpha, \beta > 0$ are sufficiently large numbers chosen later; the function $\phi(s), s \geq 0$ is defined such that $\phi(s) = 3 - e^s, 0 \leq s \leq 1/2$ and $1 \leq \phi(s) \leq 3$ for all s. The function V_1 is shown to tend to zero as $\tau \to \infty$ uniformly in ξ and η, implying that $w(\tau, \xi, \eta) \to w^{\infty}(\xi, \eta)$ as $\tau \to \infty$. Substituting $V(\tau, \xi, \eta)$ from (5.5.26) into (5.5.20) we get the following equation for V_1,

$$L(V_1) \equiv v(w^{\infty})^2 V_{1\eta\eta} - V_{1\tau} - \eta V_{1\xi} + \left(\frac{1}{\rho}p_x^{\infty} + 2v\alpha(w^{\infty})^2\frac{\phi'}{\phi}\right)V_{1\eta} + cV_1$$

$$= \Phi\frac{e^{-\beta\xi}}{\phi},\qquad(5.5.27)$$

where

$$c = v(w + w^{\infty})w_{\eta\eta} - \eta\beta + \frac{\alpha}{\rho}p_x^{\infty}\frac{\phi'}{\phi} + v(w^{\infty})^2\alpha^2\frac{\phi''}{\phi}.\qquad(5.5.28)$$

We easily check from the definition of the function ϕ that, for $\alpha\eta < 1/2$, we have $-2 < \phi' \leq -1, \phi'' \leq -1$, and $1 \leq \phi \leq 3$. We also observe from the nature of the solution $u^{\infty}(x,y)$ of the problem (5.5.7)–(5.5.9) (see Oleinik (1963a) or Serrin (1967)) that $w^{\infty}(\xi, \eta) \geq a > 0$ in $0 \leq \eta \leq \delta_1$, where a is some constant and $\delta_1 > 0$ is sufficiently small; besides, $w_{\eta\eta}$ is bounded. Therefore, it follows from (5.5.28) that, if we choose $\alpha > 0$ sufficiently large, we have

$$v(w+w^\infty)w_{\eta\eta} - \eta\beta + 2\frac{\alpha}{\rho}|p_x^\infty| - \frac{1}{3}va^2\alpha^2 < -M, \qquad (5.5.29)$$

where $M > 0$ is arbitrary. We easily check from (5.5.28) and (5.5.29) that $c < -M$ provided that $\alpha\eta < 1/2$, and $\eta < \delta_1$. Furthermore, we may choose $\beta > 0$ so large that $c < -M$ when $\eta > \min\left((1/2)\alpha^{-1}, \delta_1\right)$. Now we write the conditions on V_1 in accordance with (5.5.22), (5.5.23), and (5.5.26):

$$V_1|_{\tau=0} = (w_0(\xi,\eta) - w^\infty(\xi,\eta))\frac{e^{-\beta\xi}}{\phi}, \quad V_1|_{\xi=0} = \frac{1}{\phi}(w_1(\tau,\eta) - w_1^\infty(\eta)),$$
$$(5.5.30)$$

$$l(V_1) \equiv (vw^\infty V_{1\eta} - c_1 V_1)_{\eta=0} = \frac{1}{2}\Psi e^{-\beta\xi}, \qquad (5.5.31)$$

where

$$c_1 \equiv \left(\frac{1}{2}v\alpha w^\infty - vw_\eta + v_0^\infty\right)_{\eta=0}. \qquad (5.5.32)$$

We now require that

$$\alpha > \frac{2}{va}\left(\max|v_0^\infty| + v|w_\eta| + 1\right); \qquad (5.5.33)$$

this ensures that c_1 defined by (5.5.32) is greater that 1. Because $w^\infty(\xi,\eta) \to 0$ as $U^\infty(\xi) - \eta \to 0$ and $w(\tau,\xi,\eta) \to 0$ as $U(\tau,\xi) - \eta \to 0$ uniformly with τ, there exists $\kappa > 0$ sufficiently small such that $|U(\tau,\xi) - U^\infty(\xi)| \le \kappa$ for τ sufficiently large; we may then write

$$|V| = |w - w^\infty| \le \varepsilon \text{ for } \eta > U^\infty(\xi) - \kappa, \ \tau > \tau_2, \qquad (5.5.34)$$

where τ_2^{-1} and ε are sufficiently small. In view of (5.5.26) and the definition of the function ϕ (see below (5.5.26)), we have

$$|V_1| = \left|Ve^{-\beta\xi}\frac{1}{\phi}\right| \le \varepsilon \text{ for } \tau > \tau_2, \ \eta > U^\infty(\xi) - \kappa. \qquad (5.5.35)$$

Now consider the part of the domain Ω_1 (see below (5.5.19)) for which $\tau \ge \sigma$. We refer to this as G_σ.

It is now shown that in the domain Ω_1 we have

$$|V_1(\tau,\xi,\eta)| \le \delta + M_1 e^{-\gamma\tau}, \qquad (5.5.36)$$

where $\delta > 0$ is an arbitrary given number, $\gamma > 0$ is a constant less than M, and $M_1 > 0$ is a constant which depends on δ and γ. We consider the functions $W_\pm$ in G_σ defined by

$$W_+ = \delta + M_1 e^{-\gamma\tau} + V_1, \quad W_- = \delta + M_1 e^{-\gamma\tau} - V_1, \qquad (5.5.37)$$

where δ and γ have been defined earlier; M_1 is chosen presently. Using (5.5.27) we check that

$$L(W_\pm) = cM_1 e^{-\gamma\tau} + \gamma M_1 e^{-\gamma\tau} + c\delta \pm \Phi e^{-\beta\xi} \frac{1}{\phi}. \tag{5.5.38}$$

Because we have ensured that c defined by (5.5.28) is negative and less than $-M$ and $\gamma < M$ and because $\left|\phi^{-1}\Phi e^{-\beta\xi}\right| \to 0$ as $\tau \to \infty$ uniformly in ξ and η, we have

$$\pm \Phi e^{-\beta\xi}\phi^{-1} + c\delta < 0 \quad \text{when } \delta > 0 \tag{5.5.39}$$

provided that τ is chosen to be sufficiently large. Equation (5.5.38) now shows that $L(W_+)$ and $L(W_-)$ are both negative in G_σ if σ is chosen sufficiently large. It follows that $W_\pm$ cannot have a negative minimum in the region G_σ or when $\xi = x_0$, and also on $\tau = \tau_3$, where $\tau_3 > \sigma$, if we consider $W_\pm$ over $\sigma < \tau < \tau_3$.

Now we show that $W_+ \geq 0$ and $W_- \geq 0$ in G_σ provided that we choose σ sufficiently large. On the part of the boundary of G_σ where $\eta = U^\infty(\xi)$ or $\eta = U(\tau,\xi)$, we have $W_\pm > 0$; this is because for $\xi \sim 0$ we have $|V_1| < \varepsilon$ when $\eta > U(\xi) - \kappa$ and $\tau > \tau_2$ is sufficiently large. If we choose $\varepsilon < \delta$, and κ and τ_2^{-1} sufficiently small, we find from (5.5.37) that $W_\pm \geq 0$.

Now we consider the boundary $\eta = 0$. Here, we have (see (5.5.31))

$$l(W_\pm) = -c_1\left(\delta + M_1 e^{-\gamma\tau}\right) \pm \frac{1}{2}\Psi e^{-\beta\xi} < 0, \tag{5.5.40}$$

if we choose $\tau > \tau_4$ where τ_4 is sufficiently large. The inequality in (5.5.40) follows from the fact that $c_1 > 1$ and $\left|\Psi e^{-\beta\xi}\right| \to 0$ as $\tau \to \infty$ uniformly in ξ. Thus, $W_\pm$ cannot have a negative minimum when $\eta = 0$ and $\tau > \tau_4$. We may choose M_1 large enough to ensure that both W_+ and W_- are positive when $\tau = \sigma$ where $\sigma > \max(\tau_2, \tau_4)$.

Thus we have shown that, provided σ is sufficiently large, $W_\pm = \pm V_1 + M_1 e^{-\gamma\tau} + \delta \geq 0$ over G_σ; that is,

$$|V_1| \leq \delta + M_1 e^{-\gamma\tau} \quad \text{in } G_\sigma. \tag{5.5.41}$$

Now we may choose M_1 larger, if necessary, to ensure that (5.5.41) holds for all of Ω_1; hence the equality (5.5.36) follows. We, therefore, conclude from (5.5.19), (5.5.26), and (5.5.41) that, because δ is arbitrary, $w(\tau,\xi,\eta) \to w^\infty(\xi,\eta)$ uniformly in ξ and η as $\tau \to \infty$.

Now, we prove that

$$\lim_{t\to\infty} u(t,x,y) = u^\infty(x,y) \tag{5.5.42}$$

for all x,y in D^∞. From (5.5.4)$_4$, (5.5.9)$_4$, and (5.5.13)$_3$, we have

$$|U^\infty(x) - u^\infty(x,y)| < \varepsilon, \tag{5.5.43}$$

$$|U(t,x) - u(t,x,y)| < \varepsilon, \tag{5.5.44}$$

for $y > y_1$ and $w(\tau, \xi, \eta) \to 0$ when $U(\tau, \xi) - \eta \to 0$ uniformly with respect to τ. Therefore,

$$|u^{\infty}(x,y) - u(t,x,y)| \leq |u(t,x,y) - U(t,x)| + |U(t,x) - u^{\infty}(x,y)| < 2\varepsilon, \tag{5.5.45}$$

for $y > y_1$ and t sufficiently large. Thus, we obtain (5.5.42) for $y > y_1$. To prove (5.5.42) for $y \leq y_1$, we write

$$y = \int_0^{u(t,x,y)} \frac{ds}{w(t,x,s)}, \quad y = \int_0^{u^{\infty}(x,y)} \frac{ds}{w^{\infty}(x,s)} \tag{5.5.46}$$

(see below (5.5.11)). This implies that

$$
\begin{aligned}
0 &= \int_0^{u(t,x,y)} \frac{ds}{w(t,x,s)} - \int_0^{u^{\infty}(x,y)} \frac{ds}{w^{\infty}(x,s)} \\
&= \int_0^{u^{\infty}(x,y)} \left(\frac{1}{w} - \frac{1}{w^{\infty}} \right) ds + \int_{u^{\infty}(x,y)}^{u(t,x,y)} \frac{ds}{w(t,x,s)}.
\end{aligned} \tag{5.5.47}
$$

For $y \leq y_1$,

$$U(t,x) - u(t,x,y) > \kappa_1, \quad U^{\infty}(x) - u^{\infty}(x,y) > \kappa_1, \tag{5.5.48}$$

and

$$w(t,x,s) \geq a_1 > 0, \quad w^{\infty}(x,s) \geq a_1 > 0 \tag{5.5.49}$$

for $s < U(t,x) - \kappa_1$ and $s < U^{\infty}(x) - \kappa_1$, respectively. Then, by (5.5.47),

$$u(t,x,y) - u^{\infty}(x,y) = w(t,x,s_1) \int_0^{u^{\infty}(x,y)} \frac{w - w^{\infty}}{ww^{\infty}} ds, \quad y \leq y_1; \tag{5.5.50}$$

here, s_1 lies between $u(t,x,y)$ and $u^{\infty}(x,y)$. This, in turn, implies that

$$
\begin{aligned}
|u(t,x,y) - u^{\infty}(x,y)| &\leq |w(t,x,s_1)| \left| \int_0^{u^{\infty}(x,y)} \frac{w - w^{\infty}}{ww^{\infty}} ds \right|, \\
&\leq |w(t,x,s_1)| \frac{1}{a^2} \int_0^{u^{\infty}(x,y)} |w - w^{\infty}| ds, \\
&\leq \delta_2 + M_2 e^{-\gamma t}, \quad y \leq y_1,
\end{aligned} \tag{5.5.51}
$$

where δ_2 and M_2 are positive constants. Furthermore, inasmuch as δ_2 is arbitrarily small, we have (5.5.42) for $y \leq y_1$. It follows that

$$\lim_{t \to \infty} u(t,x,y) = u^{\infty}(x,y) \tag{5.5.52}$$

for all x and y in D^{∞}.

5.6 Asymptotic behaviour of velocity profiles in Prandtl boundary layer theory

We have shown in Section 5.5, following the work of Oleinik (1966a), how the solutions of unsteady two-dimensional boundary layer equations tend to those of the corresponding steady system. Here, we consider the asymptotic behaviour of the latter as $x \to \infty$. A most lucid exposition of asymptotics in (steady) boundary layers governed by Prandtl's equations was given by Serrin (1967). He explained pointedly the important role played by the similarity solutions:

> Perhaps the most fruitful source of information in this regard is the body of exact solutions derived under the assumption of similarity, of which the famous Blasius solution is typical. Nevertheless, in spite of the success of these particular solutions in predicting actual motions, a well known and primary problem has been present, namely, in what way are similar solutions unique or special among the totality of solutions of Prandtl's equations? What theoretical justification can be offered for the pre-eminent role of similar solutions in boundary layer theory?

This is the question which Serrin (1967) attempted to answer in a rigourous manner. We follow his work closely in the following. Similar results for nonlinear parabolic equations have been presented in Chapter 4.

Consider a steady two-dimensional flow past a rigid wall governed by the Prandtl equations

$$u_x + v_y = 0, \tag{5.6.1}$$

$$u u_x + v u_y = U U_x + v u_{yy}, \tag{5.6.2}$$

where x and y are coordinates in the horizontal and vertical directions: x denotes the length along the wall whereas y, the perpendicular distance from the wall. u and v are velocity components in the horizontal and vertical directions, respectively. v is the kinematic viscosity. $U = U(x) \geq 0$ represents the external streaming speed which is preassigned. It is related to the pressure $p = p(x)$ in the boundary layer via the Bernoulli relation

$$\frac{dp}{dx} + U U_x = 0. \tag{5.6.3}$$

The boundary conditions on the wall are

$$u = v = 0 \quad \text{on } y = 0. \tag{5.6.4}$$

The flow must merge with the external streaming conditions, requiring that

$$u \to U(x) \quad \text{as } y \to \infty, \text{ uniformly in } x. \tag{5.6.5}$$

One must also impose appropriate initial conditions at the leading edge:

$$u(0, y) = \tilde{u}(y), \quad 0 < y < \infty. \tag{5.6.6}$$

The function $\tilde{u}(y)$ is assumed to be nonnegative and continuous; it tends to $U(0)$ as $y \to \infty$. It is also assumed that $u \geq 0$ in some neighbourhood of the initial line $x = 0$. It is known (see Nickel 1958, Oleinik 1963b) that the above initial boundary value problem for appropriate functions $\tilde{u}(y)$ and $U(x)$ possesses a unique solution. How does this solution behave as $x \to \infty$?

Serrin (1967) considered specifically the streaming flow

$$U(x) = C(x+d)^m, \quad 0 \leq x < \infty, \tag{5.6.7}$$

where $C > 0$ and $d \geq 0$ are constants.

The similarity solution for the system (5.6.1) and (5.6.2) with the streaming function (5.6.7) is given by

$$\bar{u}(x,y) = U(x)f'(\zeta), \tag{5.6.8}$$

where $\zeta = y/g(x), g(x) = \sqrt{v(x+d)/U(x)}$. The function $f(\zeta)$ depends on m and is governed by the Falkner–Skan equation

$$f''' + \frac{m+1}{2}ff'' + m(1 - f'^2) = 0. \tag{5.6.9}$$

It satisfies the boundary conditions

$$f'(0) = 0, \quad f'(\infty) = 1, \quad \text{and} \quad f(0) = 0; \tag{5.6.10}$$

furthermore, f' is a monotonically increasing, concave function of ζ for all values of m (see Coppel 1960).

The main result that was proved by Serrin (1967) demonstrates the asymptotic character of the similarity solutions governed by (5.6.9) and (5.6.10): with $\tilde{u}(y)$ an arbitrary initial profile at $x = 0$, he assumed that the solution $u(x, y)$ of the Prandtl equations (5.6.1) and (5.6.2) subject to the initial/boundary condition (5.6.4)–(5.6.6) and the streaming flow (5.6.7) has a continuous derivative u_y in $0 < x < \infty, 0 \leq y < \infty$. Then, Serrin (1967) showed that

$$\left| \frac{u}{U} - f' \right| = o\left(\frac{1 + m \ln x}{x^m} \right) \quad \text{as } x \to \infty, \text{ uniformly in } y; \tag{5.6.11}$$

that is, the normalised velocity component u/U tends uniformly to the (derivative of) normalised similarity solution, f', of the Falkner–Skan equation (5.6.9) subject to (5.6.10) as x tends to infinity downstream, bringing out clearly the central position of the similarity solution for this class of problems. This solution is asymptotically 'independent' of the motion at the leading edge $x = 0$.

Serrin (1967) also proved the asymptotic uniqueness of the solution; he showed that this solution is independent of the state of motion at the initial point $x = 0$ provided that the free stream velocity U is twice continuously differentiable and obeys the inequality

$$C_1(x+d)^{2m-1} \leq UU_x \leq C_2(x+d)^{2n-1}, \quad 0 \leq x < \infty, \tag{5.6.12}$$

where C_1 and C_2 are positive constants and m and n are exponents satisfying the condition $m \leq n < 5m/3$.

The main assumption in the proof is that the parameter m in (5.6.7) is greater than or equal to zero ensuring a favourable pressure gradient, $dp/dx \leq 0$. This follows from (5.6.3) and (5.6.7).

Because $u \geq 0$ in some neighbourhood of $x=0$ and is continuously differentiable on $y=0$ and because it is assumed that $UU_x \geq 0$, it follows from a theorem of Velte (1960) that $u > 0$ in $0 < x < \infty, 0 < y < \infty$ and $u_y(x,0) > 0$ for $0 < x < \infty$. Serrin (1967) also assumed the initial line to be $x = 1$ rather than $x = 0$. With this choice, he relabelled the coordinates such that the new initial position is again called $x = 0$. It is further assumed that u and u_y are continuous in $0 \leq x < \infty, 0 \leq y < \infty$ and the initial profile $\tilde{u}(y)$ satisfies the conditions

$$\tilde{u}(0) = 0, \quad \tilde{u}_y(0) > 0 \text{ and } \tilde{u}(y) > 0 \text{ for } y > 0. \tag{5.6.13}$$

Inasmuch as $u > 0$, one may introduce the von Mises variables

$$x = x, \quad \psi = \psi(x,y) = \int_0^y u(x,t)dt \tag{5.6.14}$$

into the Prandtl system (5.6.1) and (5.6.2) and obtain

$$(u^2)_x = vu(u^2)_{\psi\psi} + (U^2)_x \tag{5.6.15}$$

for which the boundary and initial conditions become

$$u = 0 \text{ on } \psi = 0 \tag{5.6.16}$$

and $u \to U$ as $\psi \to \infty$, uniformly in x on any finite interval $0 \leq x \leq A$. Moreover,

$$u(0, \psi) = \tilde{u}(\psi) \quad (0 \leq \psi < \infty), \tag{5.6.17}$$

where $\tilde{u}(\psi)$ is the transformed initial condition (see (5.6.4)–(5.6.6)) (see Schlichting 1960).

Serrin (1967) proved his main result via several lemmas. We discuss them here informally.

Let u and $\bar{u}$ be two solutions of the boundary layer equations as above with free stream speeds $U(x)$ and $\bar{U}(x)$ and initial conditions $\tilde{u}(\psi)$ and $\tilde{\bar{u}}(\psi)$, respectively, such that

$$(U^2)_x \leq (\bar{U}^2)_x \text{ and } \tilde{u}(\psi)^2 \leq \tilde{\bar{u}}(\psi)^2 + a^2, \tag{5.6.18}$$

where $a > 0$ is some constant. Further let either $u_{yy} < 0$ or $\bar{u}_{yy} < 0$. Then, $u(x,\psi) \leq \bar{u}(x,\psi) + a$.

We consider first the case $u_{yy} < 0$. Because $\bar{u}$ satisfies

$$(\bar{u}^2)_x = v\bar{u}(\bar{u}^2)_{\psi\psi} + (\bar{U}^2)_x \tag{5.6.19}$$

(see (5.6.15)), the difference function $\phi(x,\psi) = \bar{u}^2 - u^2$ satisfies the equation

$$
\begin{aligned}
\phi_x &= v\bar{u}(\bar{u}^2)_{\psi\psi} - vu(u^2)_{\psi\psi} + (\bar{U}^2)_x - (U^2)_x \\
&= v\bar{u}\phi_{\psi\psi} + \frac{v(u^2)_{\psi\psi}}{u+\bar{u}}\phi + (\bar{U}^2 - U^2)_x \\
&= v\bar{u}\phi_{\psi\psi} + \alpha\phi + (\bar{U}^2 - U^2)_x, \qquad (5.6.20)
\end{aligned}
$$

where we have used the relation $(u^2)_{\psi\psi} = 2u_{yy}/u$ (see (5.6.1) and (5.6.14)) and have set $\alpha = 2vu_{yy}/u(u+\bar{u})$. The function ϕ satisfies the boundary conditions

$$
\phi = 0 \text{ on } \psi = 0, \quad \phi \geq -a^2 \text{ on } x = 0 \qquad (5.6.21)
$$

(see (5.6.16) and (5.6.18)$_2$). Furthermore, $\phi \to \bar{U}^2 - U^2$ as $\psi \to \infty$, uniformly in x on any finite interval. Because it is given that $(\bar{U}^2 - U^2)_x \geq 0$ and $\bar{U}^2(0) - U(0)^2 \geq -a^2$ (see (5.6.18) and (5.6.21)), it follows that $\bar{U}^2 - U^2 \geq -a^2$ for all x. Now it may be shown that $\phi(x,\psi) \geq -a^2$ for $0 \leq x < \infty$, $0 \leq \psi < \infty$. To that end, we assume the contrary, namely, that $\phi < -a^2$ at some point (x_0, ψ_0). Consider $\phi(x,\psi)$ on $R : 0 < x \leq x_0, 0 < \psi < \infty$. It follows, from (5.6.21) and the asymptotic behaviour of ϕ as $\psi \to \infty$, that ϕ assumes its absolute minimum in R. Let this point of minimum be (x_1, ψ_1). At this point, we have

$$
\phi_x \leq 0, \quad \phi_{\psi\psi} \geq 0, \quad \phi < 0 \qquad (5.6.22)
$$

and $\alpha = 2vu_{yy}/u(u+\bar{u}) < 0$ (because $u_{yy} < 0$) and $(\bar{U}^2 - U^2)_x \geq 0$. This contradicts (5.6.20) because the signs on the two sides are different. Thus, we have proved that $\phi = \bar{u}^2 - u^2 \geq -a^2$. The case $\bar{u}_{yy} < 0$ may be treated similarly.

Next we show how the constants C^* and C^{**} in the free stream velocities

$$
U^*(x) = C^*(x+d)^m, \quad U^{**}(x) = C^{**}(x+d)^n \qquad (5.6.23)
$$

may be chosen so that the corresponding solutions u and $\bar{u}$ of the Prandtl equations are bounded by the similarity solutions: $u^* \leq u \leq u^{**}$ and $u^* \leq \bar{u} \leq u^{**}$, where

$$
u^*(x,y) = U^*(x)f'(\zeta), \qquad (5.6.24)
$$

$$
u^{**}(x,y) = U^{**}(x)\hat{f}'(\hat{\zeta}), \qquad (5.6.25)
$$

$$
\zeta = \frac{y}{g^*(x)}, \quad g^*(x) = \sqrt{\frac{v(x+d)}{U^*(x)}}, \qquad (5.6.26)
$$

$$
\hat{\zeta} = \frac{y}{g^{**}(x)}, \quad g^{**}(x) = \sqrt{\frac{v(x+d)}{U^{**}(x)}}. \qquad (5.6.27)
$$

$(U^{*2})_x = 2mC^{*2}(x+d)^{2m-1}$ and $(U^{**2})_x = 2nC^{**2}(x+d)^{2n-1}$, thus it follows from (5.6.12) that we may satisfy the inequalities

$$
(U^{*2})_x \leq (U^2)_x \leq (U^{**2})_x \qquad (5.6.28)
$$

by choosing C^* and C^{**} appropriately. Assuming that $C^* < C^{**}$, we prove that

$$u^*(0, \psi) \leq \tilde{u}(\psi), \quad \tilde{u}(\psi) \leq u^{**}(0, \psi); \qquad (5.6.29)$$

the required result then follows from the statement involving (5.6.18) if we choose $a = 0$ therein. We prove the result for the function $\tilde{u}(\psi)$. The other part follows in a similar manner. From the assumption (5.6.13) regarding the solution, we have for $x = 0$ and y small, $\tilde{u} \approx by$, $\psi \approx \frac{1}{2}by^2$ and, therefore, $\tilde{u}^2 \approx 2b\psi$ where $b > 0$. In addition $\tilde{u} > 0$ for $\psi > 0$ and $\tilde{u}$ tends to $U(0)$ as ψ tends to infinity.

From the similarity form of the solution we have

$$u_y^*(0,0) = \frac{U^*(0)f''(0)}{g^*(0)} = \text{constant}.C^{*3/2}. \qquad (5.6.30)$$

It follows that

$$u^{*2} \approx \text{constant } C^{*3/2}\psi \qquad (5.6.31)$$

for small ψ. Because, by assumption, $u^*(0, \psi)$ tends to C^*d^m as ψ tends to infinity and $(u^{*2})_{\psi\psi} < 0$, it follows that C^* can be chosen so small that $u^*(0, \psi) \leq \tilde{u}(\psi)$. The similarity solution square u^{**2} is known to be a concave, monotonically increasing function of ψ. Besides, for the flow u^{**}, we have from (5.6.14) and the definition of u^{**},

$$\psi = U^{**}g^{**}\hat{f}(\hat{\zeta}). \qquad (5.6.32)$$

Because $\hat{f}'$ is monotonic and $\hat{f}'(\infty) = 1$, one may use the mean value theorem to show that $\hat{f}(\hat{\zeta}) < \hat{\zeta}$. Therefore, we have at $x = 0$, $\psi = 1$ the inequality

$$\hat{\zeta} > \hat{f}(\hat{\zeta}) = \{U^{**}g^{**}\}^{-1} = \frac{\text{constant}}{C^{**1/2}}. \qquad (5.6.33)$$

It follows that

$$u^{**} = U^{**}\hat{f}'(\hat{\zeta}) \geq \text{constant}.C^{**1/2}, \qquad (5.6.34)$$

where we assume that $C^{**} > 1$. Therefore, u^{**} at $x = 0$, $\psi = 1$ can be made arbitrarily large. Thus, the second inequality in (5.6.29) follows.

Now if we assume that $\left|\tilde{u}(\psi)^2 - \tilde{u}(\psi)^2\right| \leq a^2$ and $u_{yy} < 0$, we may prove $|u(x, \psi) - \bar{u}(x, \psi)| \leq a$ (see (5.6.18) and below). a, here, is positive .

The main proof for the asymptotic result consists of three parts. The first part requires the result that if $u_{yy} < 0$, then we have for $0 < x < \infty$,

$$y - \bar{y} \leq \frac{ag^*}{bU^*}\left(1 + \frac{a}{bU^*}\right)\ln\left(1 + \frac{bU^*}{a}\right) + \frac{a}{bU^*}\bar{y}, \qquad (5.6.35)$$

where $b = f'(1)$ and f is the Falkner–Skan function with the exponent m.

We recall that

$$y = \int_0^\psi \frac{d\psi}{u(x, \psi)}, \quad \bar{y} = \int_0^\psi \frac{d\psi}{\bar{u}(x, \psi)} \qquad (5.6.36)$$

are, respectively, the y coordinates associated with the flows u and $\bar{u}$ for a given value of ψ.

Because $u_{yy} < 0$, we may use (5.6.36) and write

$$y - \bar{y} = \int \left(\frac{1}{u} - \frac{1}{\bar{u}} \right) d\psi \leq \int \left(\frac{1}{u} - \frac{1}{u+a} \right) d\psi = a \int \frac{d\psi}{u(u+a)}. \qquad (5.6.37)$$

Letting $\psi_1 = U^* g^* f(1)$ and using the result $u^* \leq u$ proved earlier (see above (5.6.24)), we write (5.6.37) as

$$y - \bar{y} \leq a \int_0^{\psi_1} \frac{d\psi}{u^*(u^* + a)} + a \int_{\psi_1}^{\psi} \frac{d\psi}{u(u^* + a)}, \qquad (5.6.38)$$

where the second integral is absent if $\psi \leq \psi_1$. Because $d\psi/u^* = g^* d\zeta$, we have

$$\int_0^{\psi_1} \frac{d\psi}{u^*(u^* + a)} = \int_0^1 \frac{g^* d\zeta}{u^* + a} \leq \int_0^1 \frac{g^* d\zeta}{b\zeta U^* + a}, \qquad (5.6.39)$$

where we have used the fact that $u^* = U^* f'(\zeta) \geq U^* f'(1)\zeta, 0 < \zeta < 1, f'$ is concave and $f'(1) = b$. For the second integral in (5.6.38) we have

$$\int_{\psi_1}^{\psi} \frac{d\psi}{u(u^* + a)} \leq \int_{\psi_1}^{\psi} \frac{d\psi}{u(bU^* + a)} \leq \int_0^y \frac{dy}{bU^* + a}, \qquad (5.6.40)$$

because $u^* \geq U^* f'(1)$ for $\zeta > 1$. Combining (5.6.38)–(5.6.40), we have

$$y - \bar{y} \leq \frac{ag^*}{bU^*} \ln \left(1 + \frac{bU^*}{a} \right) + \frac{a}{bU^* + a} y. \qquad (5.6.41)$$

Rewriting (5.6.41), we have

$$y \leq \frac{bU^* + a}{bU^*} \left\{ \frac{ag^*}{bU^*} \ln \left(1 + \frac{bU^*}{a} \right) + \bar{y} \right\}.$$

Using this inequality in (5.6.41) we get (5.6.35). In a similar manner, one may prove that if $u_{yy} < 0$, then for $0 < x < \infty$, we also have

$$\bar{y} - y \leq \frac{ag^*}{bU^*} \left(1 + \frac{a}{bU^*} \right) \ln \left(1 + \frac{bU^*}{a} \right) + \frac{a}{bU^*} y. \qquad (5.6.42)$$

Next we show that, if $u_{yy} < 0$, then for $0 < x < \infty$,

$$|u(x, y) - u(x, \bar{y})| \leq \frac{aU^{**}}{bU^*} \left\{ 1 + f''(0) \left(\frac{U^{**}}{U^*} \right)^{1/2} \left(1 + \frac{a}{bU^*} \right) \ln \left(1 + \frac{bU^*}{a} \right) \right\}, \qquad (5.6.43)$$

where f is the Falkner–Skan function associated with the exponent n. Suppressing the dependence of u on x for convenience, we have

$$u(y) - u(\bar{y}) = \hat{u}_y(y - \bar{y}).$$ (5.6.44)

The function u has been assumed to be a (positive) concave function in y with $u(0) = 0$ and $u(\infty) = U$. Therefore, u_y is a positive decreasing function of y. Thus, for $y \geq \bar{y}$, we have

$$\hat{u}_y(y - \bar{y}) \leq u_y(\bar{y})(y - \bar{y})$$ (5.6.45)

whereas for $\bar{y} \geq y$,

$$\hat{u}_y(y - \bar{y}) \leq 0.$$ (5.6.46)

In view of (5.6.35) we have for both the cases (5.6.45) and (5.6.46) the inequality

$$\hat{u}_y(y - \bar{y}) \leq \frac{ag^* u_y(\bar{y})}{bU^*} \left(1 + \frac{a}{bU^*}\right) \ln\left(1 + \frac{bU^*}{a}\right) + \frac{a}{bU^*}\bar{y}u_y(\bar{y}).$$ (5.6.47)

Because $u_{yy} < 0$, we easily check that

$$\bar{y}u_y(\bar{y}) \leq U \quad \text{and} \quad u_y(\bar{y}) \leq u_y(0).$$ (5.6.48)

Using the result that $u^* \leq u \leq u^{**}$ proved earlier, we have

$$u \leq u^{**}, \quad u_y(0) \leq u_y^{**}(0) = \frac{U^{**} f''(0)}{g^{**}}.$$ (5.6.49)

From (5.6.44) and (5.6.47)–(5.6.49) it follows that $u(y) - u(\bar{y})$ is bounded by the RHS of (5.6.43).

To obtain the reverse inequality we observe that

$$u(\bar{y}) - u(y) = \hat{u}_y(\bar{y} - y) \leq \max(0, u_y(y)(\bar{y} - y)).$$ (5.6.50)

Now using (5.6.42), we obtain (5.6.43) as for the previous case.

In pursuit of the final result, we further show that, if $u_{yy} < 0$, then

$$\left|\frac{u - \bar{u}}{U}\right| = O\left\{\frac{a(1 + m \ln x + \ln_+ 1/a)}{x^{(5m - 3n)/2}}\right\} \quad \text{as } x \to \infty, \text{ uniformly in } y.$$ (5.6.51)

We observe that, for any positive number $\bar{y}$,

$$|u(x, \bar{y}) - \bar{u}(x, \bar{y})| \leq |u(x, \bar{y}) - u(x, y)| + |u(x, y) - \bar{u}(x, \bar{y})|.$$ (5.6.52)

Here y is such that

$$\psi = \int_0^y u(x, t)dt = \int_0^{\bar{y}} \bar{u}(x, t)dt.$$

We have already shown (see above (5.6.35)) that the second term on the RHS of (5.6.52) is less than or equal to a. The first term therein can be estimated by using (5.6.43). Here, one employs the simple result that $(1 + r^{-1})\ln(1 + r) \leq 2\ln 2 + \ln_+ r$ for any nonnegative number r; r in (5.6.43) is bU^*/a. U^* and U^{**} are explicitly given by (5.6.23). The RHS of (5.6.43) is evaluated in the limit $x \to \infty$. The multiplication factor implied in (5.6.51) involves C^*, C^{**}, and the given constants $f'(1)$ and $f''(0)$.

The final step in proving the estimate (5.6.11) requires the inequality

$$\left| \bar{u}(x_0, \psi)^2 - u(x_0, \psi)^2 \right| \le \varepsilon, \tag{5.6.53}$$

where $\varepsilon > 0$ and $x = x_0$ is a point far downstream. This inequality requires considerable technical detail and we refer the reader to the original paper of Serrin (1967). We assume this result in the following. We let $U(x) = C(x+d)^m$; the corresponding similarity solution is denoted by $\bar{u}$. When $m > 0$, this function satisfies (5.6.12) with $n = m$ and $C_1 = C_2 = 2mC^2$. Now we use (5.6.51) (where the solutions $\bar{u}$ and u interchange but this is only a notational matter). The estimate (5.6.51) with $\bar{u} = U f'$ becomes

$$\left| \frac{u}{U} - f' \right| = O\left\{ \frac{a(1 + m \ln x + \ln_+ 1/a)}{x^m} \right\} \quad \text{as } x \to \infty, \text{ uniformly in } y. \tag{5.6.54}$$

We assume that suitably downstream at $x = x_0$, say, (5.6.53) holds and then take x_0 as the new initial position. C^* and C^{**} need not be changed in this process. Now we let a in (5.6.54) be equal to $\varepsilon^{1/2}$. In the limit $x \to \infty$, we may choose ε arbitrarily small by choosing x_0 sufficiently large; the estimate (5.6.54) then reduces to (5.6.11).

We may observe that, if $m = 0$, the free stream speeds U, U^*, U^{**} are all constants and all the previous results continue to hold. Therefore (5.6.54) and hence (5.6.11) apply in this case too.

Our main purpose here was to briefly discuss Serrin's (1967) work to demonstrate the central position of the similarity solution as an asymptotic as $x \to \infty$. The second asymptotic result is also important. Let u and $\bar{u}$ be two solutions of the Prandtl system corresponding to the same streaming speed $U(x)$ but with different initial profiles $\tilde{u}(y)$ and $\tilde{\tilde{u}}(y)$. Assume further that $U(x)$ is twice continuously differentiable and satisfies the inequality

$$C_1(x+d)^{2m-1} \le UU_x \le C_2(x+d)^{2n-1}, \quad 0 \le x < \infty, \tag{5.6.55}$$

where C_1 and C_2 are positive constants; the exponents m and n obey the inequality $m \le n < 5m/3$. Then

$$\left| \frac{\bar{u} - u}{U} \right| = o(1) \quad \text{as } x \to \infty, \text{ uniformly in } y. \tag{5.6.56}$$

This result proves the asymptotic uniqueness of the normalised velocity profile for arbitrary conditions at the initial point $x = 0$. We refer the reader to Serrin (1967) for the proof of (5.6.56).

Now we summarise two interesting and related investigations which followed Serrin (1967). Peletier (1972) essentially studied the same problem as Serrin (1967). However, there were some interesting departures. He assumed the flow to be governed by (5.6.1) and (5.6.2), where, however, the exterior streaming speed was chosen to be

$$U(x) = U_0(x+1)^m, \tag{5.6.57}$$

where $U_0 > 0, m \geq 0$ (cf. equation (5.6.7)). The velocity $U(x)$ and pressure $p(x)$ in the boundary layer are related by Bernoulli's equation

$$p + \frac{1}{2}\rho U^2 = \text{constant}$$

or

$$\frac{dp}{dx} = -\rho U U_x \leq 0 \tag{5.6.58}$$

for the choice (5.6.57) of the free stream velocity. Thus, the adverse pressure gradient is excluded. The boundary conditions relevant to the flow are the same as in Serrin (1967), namely, (5.6.4) and the Prandtl's stream condition

$$u \to U \text{ as } y \to \infty, \text{ uniformly in } x. \tag{5.6.59}$$

As in Serrin (1967), the initial station $x = 0$ is chosen to be located at some distance from the leading edge where

$$u(0, y) = u_0(y), \quad 0 < y < \infty. \tag{5.6.60}$$

u_0 is assumed to be a smooth function with continuous and uniformly bounded first and second derivatives. Moreover, it is required that $u_0(0) = 0, 0 < u_0'(0) < \infty, u_0 > 0$ if $y > 0$, and $u_0 \to U(0)$ as $y \to \infty$. The existence of the solution of this boundary value problem was proved by Nickel (1958) and Oleinik (1963b). Nickel (1958) also proved that if $U(x)$ and $u_0(y)$ satisfy the conditions laid down above, then $u > 0$ in the entire domain, and $u_y(x, 0) > 0$ for $0 < x < \infty$. Peletier (1972) found the similarity solution of the Prandtl equations, expressed in von Mises variables, namely,

$$u_x = \frac{1}{2}v(u^2)_{\psi\psi} + u^{-1}U U_x \tag{5.6.61}$$

(cf. (5.6.15)), where u is now a function of x and ψ. The boundary conditions in these variables become

$$u(x, 0) = 0 \text{ for } 0 < x < \infty \tag{5.6.62}$$

and

$$u \to U \text{ as } \psi \to \infty, \text{ uniformly in } x. \tag{5.6.63}$$

The initial condition at $x = 0$ was imposed in the form

$$u(0, \psi) = u_0(\psi), \quad 0 < \psi < \infty. \tag{5.6.64}$$

Peletier (1972) sought the similarity solution of (5.6.61) in the form

$$u(x, \psi) = (x+1)^m f(\eta), \quad \eta = \psi(x+1)^{-(m+1)/2} \tag{5.6.65}$$

which would satisfy the conditions (5.6.62) and (5.6.63). This, when substituted into (5.6.61), leads to the second-order ODE

$$\frac{1}{2}v(f^2)'' + \frac{m+1}{2}\eta f' + m\left(\frac{U_0^2}{f} - f\right) = 0, \tag{5.6.66}$$

where the prime denotes differentiation with respect to η. The solution of equation (5.6.66) must satisfy the conditions

$$f(0) = 0, \quad f \to U_0 \text{ as } \eta \to \infty, \tag{5.6.67}$$

(see (5.6.57), (5.6.62), and (5.6.63)). Peletier (1972) also obtained asymptotic behaviour of the solution of (5.6.66) and (5.6.67) in the form

$$U_0 - f(\eta) = O\left\{\eta^{-1-2\beta} \exp\left(-\frac{m+1}{4vU_0}\eta^2\right)\right\} \text{ as } \eta \to \infty, \tag{5.6.68}$$

where $\beta = 2m/(m+1)$.

Unlike Serrin (1967) and Peletier (1972) obtained an estimate for the convergence of the velocity profile in terms of the arc length along the plate and the stream function rather than the physical variables x and y themselves. His main theorem may be stated as follows. Let $u(x,y)$ be a solution of the boundary layer equation (5.6.1) and (5.6.2) and let $u(0,y) = u_0(y)$ be a smooth function such that $u_0(0) = 0, 0 < u_0'(0) < \infty, u_0 > 0$ if $y > 0$, which further satisfies the following conditions as $y \to \infty$.

(i) $\quad \delta = \int_0^\infty \left|1 - \frac{u_0}{U_0}\right| dy < \infty,$ \hfill (5.6.69)

(ii) $U_0 - u_0(y) = O\left\{(y+\delta)^{-1-2\beta} \exp\left\{-\frac{(m+1)U_0}{4v}(y+\delta)^2\right\}\right\}$ as $y \to \infty$. \hfill (5.6.70)

(In the above, if $u_0 \leq U_0$ for $0 \leq y < \infty$, δ is the displacement thickness at $x = 0$.) Let $\bar{u}(x,y)$ be the similarity profile corresponding to the stream speed $U(x)$ (see (5.6.57) and (5.6.65)). Then,

$$\int_0^\infty \eta \left|\frac{u - \bar{u}}{U}\right| d\eta \leq M(x+1)^{-(2+\mu)m-1}, \quad x \geq 0, \tag{5.6.71}$$

where the positive constants M and μ depend only on $v, U(x)$, and the initial velocity profile $u_0(y)$. Moreover, $\mu \leq 1$; when $u_0 \leq U_0$, we may set $\mu = 1$. It was also shown that the power of $(x+1)$ in (5.6.71) is best possible for $m = 0$.

In another related study, Khusnutdinova (1970) generalised the work of Serrin (1967) to compare solutions which correspond not only to different initial conditions at $x = 0$ but also to different external streaming flows. Specifically, he considered solutions $u^1(x,y)$ and $u^2(x,y)$ of the system (5.6.1) and (5.6.2) which arise from two different initial conditions $u_0^i(y)(i = 1,2)$ and external streaming conditions $U_i(x)(i = 1,2)$, respectively, where

$$\lim_{x \to \infty} U_i(x) = U_\infty = \text{constant } (i = 1,2). \tag{5.6.72}$$

It was shown that if $\lim_{x \to \infty} |U_1(x) - U_2(x)| = \lim_{y \to \infty} |u_0^1(y) - u_0^2(y)| = 0$, then for $x \to \infty$, the difference between the solutions $u^1(x,y)$ and $u^2(x,y)$ tends uniformly to zero with respect to y, where $y \in [0,\infty)$. One consequence of this theorem is that the solution $u(x,y)$ of (5.6.1)–(5.6.5) in the boundary layer converges for large x to the well known Blasius solution

$$u_1 = U_\infty f'(\eta), \quad \eta = \frac{y\sqrt{U_\infty}}{\sqrt{2v(x+1)}}, \tag{5.6.73}$$

which describes flow past a plate in the longitudinal direction at velocity $U(x) \equiv U_\infty$. In this case, $f(\eta)$ is governed by the boundary value problem

$$f''' + ff'' = 0, \quad f(0) = 0, \quad f'(0) = 0, \quad f'(\infty) = 1. \tag{5.6.74}$$

The function f together with its first derivative is monotonically increasing. In this regard, the following theorem was proved by Khusnutdinova (1970). Let the following inequalities hold.

$$0 \leq u_0(y) \leq U(0), \quad u_0'(0) > 0, \quad u_0(0) = 0,$$
$$0 \leq \frac{dU}{dx} \leq \frac{M_0}{(x+1)^{\gamma_0+1}}, \quad \gamma_0 > 0.$$

Then, as $x \to \infty$, $|u(x,y) - u_1(x,y)| \to 0$, uniformly in $y \in [0,\infty)$; here, $u_1(x,y) = U_\infty f'(\eta)$, where $f(\eta)$ is the solution of the boundary value problem (5.6.74).

If, in addition, the inequalities

$$U(0)f'(y-N) \leq u_0(y), \quad y \in [N,\infty)$$

and

$$|u_0(y) - U(0)| \leq M_1 \exp\left(-\gamma_1 y^2\right), \quad y \in [0,\infty)$$

hold for some constants N, M_1, and $\gamma_1 > 0$, then

$$|u(x,y) - u_1(x,y)| \leq \frac{M}{(x+1)^\gamma}, \tag{5.6.75}$$

where M and $0 < \gamma < \gamma_0$ are some constants which depend only on the initial data of the problem. Khusnutdinova (1970) also worked with basic equations in terms of von Mises variables.

We may observe that the external flow conditions $U(x)$ lead to different forms of ODEs, Falkner–Skan or Blasius, governing self-similar flows, and thus characterising different asymptotic behaviour as x tends to ∞. Before proving the asymptotic nature of these solutions, Serrin (1967) and Khusnutdinova (1970) ensured the existence and uniqueness of the systems of PDEs and ODEs that were involved subject, of course, to the relevant initial and boundary conditions.

5.7 Conclusions

In this chapter, we have discussed the asymptotic behaviour of solutions of some physical problems arising from fluid mechanics. Section 5.1 presented the introduction to the chapter. Section 5.2 was concerned with the flow due to a strong explosion at the centre of an ideal gas sphere. It was assumed that the preshock density is $\rho_0 = kr^{-\omega}$; here r is the distance from the origin and k and ω are positive constants. An interesting feature of this problem is that the asymptotic flow is described by the self-similar solutions of first kind for $\omega < 3$ (Sedov–Taylor solutions) and by the self-similar solutions of the second kind for $\omega_g(\gamma) < \omega < \omega_c(\gamma)$; here ω_g and ω_c depend on the adiabatic index γ of the gas. This section followed the work of Waxman and Shvarts (1993). Section 5.3 dealt with the self-similar solutions of the second kind which describe a collapsing spherical cavity. We showed, by following Hunter (1960), that the numerical solution of the governing system of nonlinear partial differential equations with appropriate initial/initial boundary conditions converges to the relevant self-similar solutions of the second kind, for different values of γ, as the radius of the cavity tends to zero. We have also summarised the work of Thomas et al. (1986). In Section 5.4, we have presented a study of solutions of the compressible Euler equations with damping. Following Liu (1996), we constructed a family of solutions for the compressible flow with damping. As $t \to \infty$, these solutions converge to the Barenblatt solutions of the porous medium equation. This study justifies, in a limited sense, Darcy's law for the compressible flow for large time. In Section 5.5, we have studied, following Oleinik (1966a), the boundary layer equations for an unsteady flow of incompressible fluid. Under a certain set of conditions, it was shown that the large time behaviour of the longitudinal velocity component of the unsteady flow is described by the longitudinal velocity component of the steady flow. Section 5.6 was concerned with the study of boundary layer equations for the steady two-dimensional laminar flow of an incompressible viscous fluid past a rigid wall. Following Serrin (1967) closely, we clearly brought out the importance of similarity solutions governed by the Falkner–Skan differential equation. We have shown that the asymptotic behaviour (for large x) of the downstream velocity profile is described by the Falkner–Skan similarity solution when the streaming speed is $U(x) = c(x+d)^m$, $m \geq 0$.

References

Barenblatt, G. I. (1953) On a class of exact solutions of plane one dimensional problem of unsteady filtration of a gas in a porous medium, *Prikl. Mat. Mekh.* 17, 739–742.

Coppel, W. (1960) On a differential equation of boundary layer theory, *Phil. Trans. A* 253, 101–136.

Courant, R., Friedrichs, K. O. (1948) *Supersonic Flow and Shock Waves*, Wiley Interscience, New York.

Gallay, T., Raugel, G. (1998) Scaling variables and asymptotic expansions in damped wave equations, *J. Differential Eq.* 150, 42–97.

Gallay, T., Raugel, G. (2000) Scaling variables and stability of hyperbolic fronts, *SIAM J. Math. Anal.* 32, 1–29.

Gallay, T., Wayne, C. E. (2002) Long-time asymptotics of the Navier–Stokes and vorticity equations on $\mathbf{R}^3$, *Phil. Trans. R. Soc. Lond. A* 360, 2155–2188.

Guderley, G. (1942) Starke kugelige und zylindrische Verdichtungsstösse in der Nähe des Kugelmittelpunktes bzw. der Zylinderachse, *Luftfahrtforschüng*, 19, 302–312.

Hsiao, L., Liu, T. P. (1992) Convergence to nonlinear diffusive waves for solutions of a system of hyperbolic conservation laws with damping, *Comm. Math. Phys.* 143, 599–605.

Hunter, C. (1960) On the collapse of an empty cavity in water, *J. Fluid. Mech.* 8, 241–263.

Khusnutdinova, N. V. (1970) Asymptotic stability of the solutions of boundary layer equations, *Prikl. Mat. Mekh.* 34, 526–531.

Korobeinikov, V. P., Riazanov, E. V. (1959) Solutions of singular cases of point explosions in a gas, *Prikl. Mat. Mekh.* 23, 384–387.

Lazarus, R. B. (1982) One-dimensional stability of self-similar converging flows, *Phys. Fluids* 25, 1146–1155.

Liu, T. P. (1996) Compressible flow with damping and vacuum, *Japan J. Indust. Appl. Math.* 13, 25–32.

Matzumura, A. (1978) Nonlinear hyperbolic equations and related topics in fluid dynamics, Nishida, T. (Ed.) *Pub. Math. D'Orsay*, 53–57.

Nickel, K. (1958) Einzige Eigenschaften von Lösungen der Prandtlschen Grenzschicht-Differentialgleichungen, *Arch. Rat. Mech. Anal.* 2, 1–31.

Nishihara, K. (1996) Convergence rates to nonlinear diffusion waves for solutions of system of hyperbolic conservation laws with damping, *J. Differential Eq.* 131, 171–188.

Nishihara, K. (1997) Asymptotic behaviour of solutions of quasilinear hyperbolic equations with linear damping, *J. Differential Eq.* 137, 384–395.

Oleinik, O. A. (1963a) O sisteme uravnenii teorii pogranichnogo sloia (On the system of equations in boundary layer theory), *Zh. vychisl. matem. i matem. fiz.*, 3, 489–507.

Oleinik, O. A. (1963b) The Prandtl system of equations in boundary layer theory, *Dokl. Acad. Nauk. SSSR* 150, 28–32 (English trans. *Sov. Math.* 4, 583–586).

Oleinik, O. A. (1966a) Stability of solutions of a system of boundary layer equations for a non-steady flow of incompressible fluid, *Prikl. Mat. Mekh.*, 30, 417–423.

Oleinik, O. A. (1966b) A system of boundary layer equations for unsteady flow of an incompressible fluid, *Soviet Math. Dokl.* 7, 727–730.

Peletier, L. A. (1972) On the asymptotic behaviour of velocity profiles in laminar boundary layers, *Arch. Rat. Mech. Anal.* 45, 110–119.

Rayleigh, L. (1917) On the pressure developed in a liquid during the collapse of a spherical cavity, *Phil. Mag.* 34, 94–98.

Sachdev, P. L. (2004) *Shock Waves and Explosions*, Chapman & Hall/CRC Press, New York.

Schlichting, H. (1960) *Boundary Layer Theory*, McGraw–Hill, New York.

Sedov, L. I. (1946) Propagation of strong blast waves, *Prikl. Mat. Mekh.* 10, 241–250.

Serrin, J. (1967) Asymptotic behaviour of velocity profiles in the Prandtl boundary layer theory, *Proc. Roy. Soc. London Ser. A* 299, 491–507.

Taylor, G. I. (1950) The formation of a blast wave by a very intense explosion, *Proc. Roy. Soc. London Ser. A* 201, 159–174.

Thomas, L. P., Pais, V., Gratton, R., Diez, J. (1986) A numerical study on the transition to self-similar flow in collapsing cavities, *Phys. Fluids* 29, 676–679.

Van Duyn, C. J., Peletier, L. A. (1977) A class of similarity solutions of the nonlinear diffusion equation, *Nonlinear Anal. Theor. Meth. Appl.* 1, 223–233.

Van Dyke, M., Guttmann, A. J. (1982) The converging shock wave from a spherical or cylindrical piston, *J. Fluid Mech.* 120, 451–462.

Velte, W. (1960) Eine Anwendungen des Nirenbergschen Maximumsprinzips für parabolische Differentialgleichungen in der Grenzschichttheorie, *Arch. Rat. Mech. Anal.* 5, 420–431.

von Neumann, J. (1947) *Blast Waves*, Los Alamos Sci. Lab. Tech. series (Los Alamos, NM (1947, Vol. 7).

Waxman, E., Shvarts, D. (1993) Second-type self-similar solutions to the strong explosion problem, *Phys. Fluids A* 5, 1035–1046.

Whitham, G. B. (1974) *Linear and Nonlinear Waves*, John Wiley & Sons, New York.

Zel'dovich, Ya. B., Raizer, Yu. P. (1967) *Physics of Shock Waves and High-Temperature Hydrodynamic Phenomena*, Vol. 2, Academic Press, New York.

Index

P.L. Sachdev, Ch. Srinivasa Rao, *Large Time Asymptotics for Solutions of Nonlinear
Partial Differential Equations*, Springer Monographs in Mathematics,
DOI 10.1007/978-0-387-87809-6, © Springer Science+Business Media, LLC 2010